Modelling and Control of Switched Reluctance Machines

Edited by Rui Esteves Araújo
and José Roberto Camacho

Published in London, United Kingdom

IntechOpen

Supporting open minds since 2005

Modelling and Control of Switched Reluctance Machines
http://dx.doi.org/10.5772/intechopen.82219
Edited by Rui Esteves Araújo and José Roberto Camacho

Contributors
Ana Camila F. Mamede, José Roberto Camacho, Rui Esteves Esteves Araújo, Pedro Melo, Manuel Pereira, Pedro Lobato, Jordi Garcia-Amoros, Pere Andrada, Baldui Blanque, Cheng Gong, Thomas Habetler, Mahmoud Hamouda, Aleksas Stuikys, Jan Sykulski, Sílvio José Pinto Simões Mariano, Maria Do Rosário Alves Calado, Rui Pedro Mendes, José Salvado, António Espírito Santo, Alexander Kashuba, Alexander Petrushin, Dmitry Petrushin, Tárcio Barros, Marcelo Vinicius De Paula, Chang-Ming Liaw, Min-Ze Lu, Ping-Hong Jhou, Kuan-Yu Chou, László Számel, Joaquim A. Dente, Armando J. Pires, Pere Andrada, Thiago de Almada Lopes, Paulo Sergio Nascimento Filho, Ernesto Ruppert Filho, Pedro José Dos Santos Neto

First published in London, United Kingdom, 2020 by IntechOpen
IntechOpen is the global imprint of INTECHOPEN LIMITED, registered in England and Wales, registration number: 11086078, 7th floor, 10 Lower Thames Street, London, EC3R 6AF, United Kingdom
Printed in Croatia

British Library Cataloguing-in-Publication Data
A catalogue record for this book is available from the British Library

Additional hard and PDF copies can be obtained from orders@intechopen.com

Modelling and Control of Switched Reluctance Machines
Edited by Rui Esteves Araújo and José Roberto Camacho
p. cm.
Print ISBN 978-1-78984-454-2
Online ISBN 978-1-78984-455-9
eBook (PDF) ISBN 978-1-83968-160-8

An electronic version of this book is freely available, thanks to the support of libraries working with Knowledge Unlatched. KU is a collaborative initiative designed to make high quality books Open Access for the public good. More information about the initiative and links to the Open Access version can be found at www.knowledgeunlatched.org

For EU product safety concerns:
IN TECH d.o.o., Prolaz Marije Krucifikse Kozulić 3, 51000 Rijeka, Croatia, info@intechopen.com or visit our website at intechopen.com.

Meet the editors

Rui Esteves Araújo received a diploma, MSc, and PhD in Electrical Engineering from the University of Porto, Porto, Portugal, in 1987, 1992, and 2001, respectively. From 1987 to 1988, he was an electrotechnical engineer with the Project Department, Adira Company, Porto, Portugal, and from 1988 to 1989, he was a researcher with the Institute for Systems and Computer Engineering (INESC), Porto, Portugal. Since 1989, he has been with the University of Porto, where he is currently an assistant professor with the Department of Electrical and Computer Engineering, Faculty of Engineering. He is a senior researcher with the INESC TEC, focusing on control theory and its industrial applications to motion control, electric vehicles, and renewable energies.

José Roberto Camacho was born in Taquaritinga, SP, Brazil, in 1954. He received a BSc and MSc in Electrical Engineering from Universidade Federal de Uberlândia, MG, Brazil in 1978 and Universidade Federal de Santa Catarina, Florianópolis, SC, Brazil in 1987, respectively. He also received a PhD in Electrical Engineering from Canterbury University, New Zealand, in 1993. From 1979 to 1994, he was a lecturer at the School of Electrical Engineering, Universidade Federal de Uberlândia, MG, Brasil. Since 1994, he has been a professor at the same school and university. His research interests include alternative energy and electricity for rural areas, electromagnetism and electromagnetic devices, engineering mathematics, small hydroelectric plants, and new devices and techniques for energy production.

Contents

Preface

The past decade has witnessed a marked growth in the use of switched reluctance machines (SRMs) in a wide variety of consumer products and industrial applications. Prominent examples of such use are kitchen robots, vacuum cleaners, washing machines, machine tools, mechanical presses, electric vehicles, and textile machinery.

Indeed, over the last 50 years SRMs have proven to be an attractive option due to their excellent performance at high speeds and intermittent operation. SRM design is based on simple construction with windings and a magnetic core. Its activation depends on an external electronic circuit, without the need for any mechanical switching system, making an SRM a robust machine in any environment, particularly one that requires operation at high speed or intermittently. The first prototypes SRMs appeared in the middle of the nineteenth century, and due to initial lack of knowledge, noise problems and torque oscillations attributed to non-linearities and problems in the smooth coupling of the electronic switching circuit with the magnetic circuit of the machine. Only in the mid-twentieth century, with the computational improvement of the analysis of magnetic circuits and the rapid evolution of electronic drive circuits by specialists in power electronics, these machines finally found their practical applications in electric drives.

Despite the visible successes of SRMs, there is still a substantial misunderstanding in terms of their intrinsic potential, how they compare to other electric machines, and their strengths and main limitations. In part, the misunderstanding stems from the circumstances that the optimum drive waveform is not a pure sine wave, due to the relatively non-linear torque of the rotor in its displacement, and the inductance is greatly dependent on the position of the stator phase windings. This book examines improvements to the design, modeling, and control of SRMs in the aspects of software, hardware, electrical and magnetic circuits, and of the machine and drive systems.

This volume contains fourteen chapters written by experts in the field from Asia, Europe, South America, and the United States. The book is organized into two parts. The first part focuses on modeling and explains the essence of the mathematical models for numerical simulation and the ideas underlying the machine design methodologies. This part ends with a chapter dedicated to design of ultra-high speed SRMs. The second part covers the control techniques of SRM where the potential of the controllers presented is demonstrated in numerical and experimental results. The last chapter addresses the challenging topic of control of a linear SRM.

The idea to edit this book stems from a very good collaboration between the editors. It began in 2018 when Ana Mamede, a PhD student supervised by José Camacho, was invited to work as a mobility student at the Faculty of Engineering of the University of Porto (FEUP) in Portugal. Since then the editors have participated in joint research. Some chapters in this book are the result of that collaboration.

We would like to express our special thanks to the chapter authors for their contributions and cooperation throughout the publication of this book. We would also like to express our sincerest thanks to our families for their support during the many months dedicated to this project.

Finally, the editors would also like to acknowledge the staff of the IntechOpen, especially Ms. Ana Pantar for the opportunity and Mr. Edi Lipovic for their invaluable support in preparing this book.

Rui Esteves Araújo
Faculdade de Engenharia da Universidade do Porto,
Portugal

José Roberto Camacho
Universidade Federal de Uberlândia,
Brazil

Section 1

Modeling and Design

Chapter 1

Modeling and Simulation of Switched Reluctance Machines

Mahmoud Hamouda and László Számel

Abstract

This chapter discusses the modeling and simulation approaches for switched reluctance machines (SRMs). First, it presents the modeling methods for SRMs including analytical models, Artificial intelligence based models, and lookup tables based models. Furthermore, it introduces the finite element method (FEM) and experimental measurement methods to obtain high fidelity magnetic characteristics for SRMs. Step-by-step procedure is explained for SRM modeling and analysis using FEM. The direct and indirect measurement methods of SRM magnetic characteristics are included, comparison between the measured and FEM-calculated characteristics is achieved, and good agreement is seen. In addition, this chapter gives the mathematical modeling of SRM, and explains its model development using MATLAB/Simulink environment. Simulation and experimental results are obtained, a very good agreement is observed.

Keywords: switched reluctance machines, magnetic characteristics, analytical models, artificial intelligent models, lookup tables, finite element analysis, experimental measurement, MATLAB simulation

1. Introduction

Accurate modeling of switched reluctance machines (SRMs) is the key stone for developing and optimizing different control strategies. Accurate prediction of machine performance under transient and steady-state conditions requires precise knowledge of its magnetic characteristics. However, the doubly salient structure, deep magnetic saturation, switching form of supply, and highly nonlinearity make it very complicated to accurately model the magnetic characteristics of SRMs [1–4]. Several approaches are used to model the magnetic characteristics of SRMs including analytical models, artificial intelligent models, and lookup tables based models [5, 6].

The analytical models can be derived directly from machine geometry, and magnetic theory [4, 7–10]. They can also be driven from the previously obtained data using finite element analysis (FEA) or experimental measurements [11–15].

Intelligent techniques such as fuzzy logic and artificial neural networks (ANNs) are inherently suitable to model the nonlinear characteristics of SRMs. They have been reported for SRM modeling in [16–19]. However, the training needs high skills and a large number of given data. It should be noted that although the accuracies of aforementioned intelligence methods are relatively high, they still demand substantial measured samples to train the network or generate the rules.

For the lookup tables' techniques, the models are commonly based on interpolation and extrapolation. The accuracy of the lookup table methods heavily depends on the number of stored samples. The data can be obtained by FEA or measurements with efficient resolution to achieve a highly trusted model [20, 21].

The analytical functions and the intelligent approaches introduce errors in the model and even the ones capable of a high grade approximation are usable only on certain machines. The output quantities have values different from the real ones measured on the test bench, making the model unusable for the optimization of the geometry and/or control. Thus the need of building models based directly on the magnetization curves obtained by FEA or by measurements on a test bench, capable of taking into account all nonlinearities and eliminating all inaccuracies arose. In the early days, the process of modeling electromagnetic field of SRMs with FEM based software was considered slow and demanding, but nowadays with the evolution of computers the FEM analysis has become imperative in describing the behavior of SRMs. Therefore, this chapter focuses on the modeling of SRM using data obtained from FEA or measurements in form of lookup tables.

2. Analytical modeling of SRM

Analytical models play an important role to easy the machine analysis as the integrations and differentiations are easier to be performed analytically. They can be of great help in the initial estimations of machine torque, efficiency that is required for the better selection of machine drive, where a trade-off between model accuracy and computation time can be made [5]. For high performance SRM drive, sometimes accurate analytical models become indispensable for machine simulation and real-time implementation as it is may be the simplest.

Several researches have been directed to analytically model SRM directly from its physical information. In order to express the idea of analytical models, an example is explained as follows. In [7], an analytical model is derived based on a piecewise analysis of machine fundamental geometry and turns per phase. The flux is represented by Eq. (1) as follows:

$$\lambda(i,\theta) = a_1(\theta)\left[1 - e^{[a_2(\theta)i]}\right] + a_3(\theta)i \tag{1}$$

where i is the motor current, θ is the rotor position, and $a_1(\theta)$, $a_2(\theta)$, and $a_3(\theta)$ are the unknown coefficient that needs to be calculated. The incremental inductance can be obtained from flux derivative as:

$$l(i,\theta) = \frac{\partial\lambda(i,\theta)}{\partial i} = -a_1(\theta)\cdot a_2(\theta)\cdot e^{[a_2(\theta)i]} + a_3(\theta) \tag{2}$$

Unsaturated phase inductance (L) can be represented as a function of rotor angle as:

$$L(\theta) = -a_1(\theta)\cdot a_2(\theta) + a_3(\theta) \tag{3}$$

Equation (3) is rearranged as,

$$a_2(\theta) = \frac{a_3(\theta) - L(\theta)}{a_1(\theta)} \tag{4}$$

The unknown coefficients $a_1(\theta)$, $a_3(\theta)$, and $L(\theta)$ are functions of rotor angle and needed to be determined. **Figure 1** shows the proposed piecewise linear models for these coefficients. The angle θ_a and θ_u refer to the aligned and unaligned rotor positions respectively. From θ_a to θ_1, the stator and rotor pole arcs are fully covered. After θ_1, the rotor pole arc starts to uncover stator pole arc. At θ_2, the pole arcs become fully uncovered. Eleven parameters are included to be determined in calculation process that are four Magnetization coefficients ($a_1(\theta_a)$, $a_3(\theta_a)$, $a_1(\theta_u)$, $a_3(\theta_u)$), three Inductive constant (L_{max}, L_{min}, L_{corner}), and four angular breakpoints (θ_1, θ_1', θ_2, θ_u). The angular breakpoints are found directly from motor design parameter. Inductive constant L_{max} is found from Eq. (3), while determination of L_{corner} and L_{min} require dimensional detail of the rotor and stator poles. The magnetization coefficient $a_1(\theta_a)$ and $a_3(\theta_a)$ are found iteratively. Step-by-step procedure of finding each parameter is covered in [7].

In [8], an analytical model for SRM is derived using the flux tube method. It divides the angle between the aligned and the unaligned positions into three regions. In [9, 10], the analytical model is derived from the equivalent magnetic circuits of SRM. In [4], a proposed method of determining the stator winding flux linkages and torque of a fully pitched mutual coupled SRM is presented.

A popular method for analytical model development of SRMs is to fit the previously obtained magnetic characteristics using analytical formulations. In [22], an exponential equation is used for SRM modeling. It was not enough to achieve an adequate model. Hence, an additional term depending on rotor position was introduced in [23]. In [5, 24], exponential functions are used for SRM modeling. It has a better accuracy, but requires intensive computation to find model parameters using least square method. In [25–27], Fourier series is used for SRM modeling. But the determination of Fourier series coefficients is complicated.

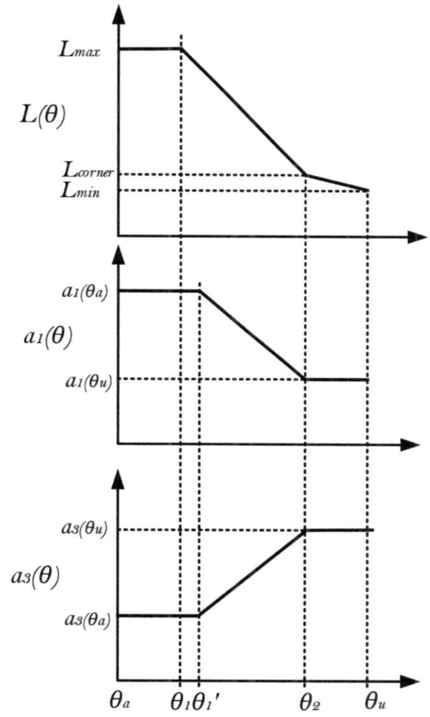

Figure 1.
Piecewise linear model assumed for unsaturated phase inductance, magnetization coefficient a_1 and a_3, as a function of rotor angle θ.

3. Artificial intelligence-based models

Among the artificial intelligence techniques, fuzzy logic and artificial neural networks (ANNs) are employed to model the nonlinear magnetic characteristics of SRMs. They have been reported in SRM modeling in [28–31]. In [28], a two-layer recurrent ANN is employed to identify the damper currents and resistance of phase winding from operating data. By this modeling method, the accurate nonlinear model can be obtained. Likewise, complex expressions and fitting algorithms are circumvented. In [29], a four-layer back-propagation (BP) ANN is applied to estimate the electromagnetic characteristics under the stator winding fault condition. Similarly, fuzzy logic systems also have strong nonlinear approximation ability. In [30], a fuzzy logic system is adopted to describe the electromagnetic characteristics, which shows high reliability and robustness. On this basis, an improved fuzzy logic system is implemented in [31] and it only requires 264 rules compared to nearly 1000 rules in [30] while maintaining a high accuracy. It should be noted that the intelligence methods requires a high number of measured samples to train the network or generate the rules.

4. Finite element analysis of SRM

It is well known that finite element method (FEM) is used to determine the magnetic vector potential over complex geometry with nonlinear magnetic characteristics such as SRMs. In the early days, the process of modeling electromagnetic field of SRMs with FEM based software was considered slow and demanding, but the recent programs for finite element analysis (FEA) make the calculation of SRM magnetization characteristics much easier and speed up computations by static magnetic field analysis. Lately, several software programs are available for FEA that can provide 2D or 3D analysis. The 3D software may require longer time but provides better accuracy. The 2D software can provide the required accuracy with proper settings, which can efficiently save time and effort. Hence, 2D FEA for SRMs is a good choice; it can provide accuracy similar to 3D FEA [13]. FEMM (Finite Element Method Magnetics) is a free 2D software for FEA, it has a basic advantages of being executed using MATLAB. Only an Octave is needed to link FEMM with MATLAB. It needs only 1/4 of the stator geometry to draw/represent the complete motor. The complete analysis and output data storage can be executed and plotted using MATLAB, which can provide an easy way for machine analysis and optimization. Therefore, FEMM is used in this work.

4.1 Equations used for FEA

A set of equations describing the problem is given below. The magnetic flux density B in a magnetic material can be given as [32],

$$B = \mu H = \frac{H}{\gamma} \tag{5}$$

where H is the magnetic field density, μ is permeability of the magnetic material and γ is the reluctivity of the magnetic material. From Ampere's law,

$$curl(B) = \mu J_o \tag{6}$$

where J_o is the current density.

Defining the magnetic vector potential A as,

$$B = curl(A) \tag{7}$$

$$curl(curl(A)) = \mu J_o \tag{8}$$

Setting $\nabla. A = 0$, and from the vector identity for the curl of the curl of the vector,

$$\nabla(\nabla.A) - \nabla^2 A = \mu J_o \tag{9}$$

This implies that,

$$\nabla^2 A = -\mu J_o \tag{10}$$

Using the assumption (c) below, the above expression can be written as,

$$\frac{\partial}{\partial x}\frac{\partial A}{\partial x} + \frac{\partial}{\partial y}\frac{\partial A}{\partial y} = -\mu J_o \tag{11}$$

The solution of Eq. (11) gives the magnetic vector potential A inside the motor. It is obtained using an interpolation technique to minimize the nonlinear energy functional,

$$F = \int_Q \left[\int_0^B H.dB - \int_0^A J_o.dA \right] dQ \tag{12}$$

where Q is the problem region of integration.

The entire problem region Q is subdivided into triangular finite elements. The elements are defined such that the sides of the triangles coincide with the boundary of each material. FEMM implements this by allowing the placement of nodes on each of the boundaries. The following assumptions are made to estimate the magnetic field inside SRM [14, 21, 32]:

The outer surface of the stator is treated as line of a zero magnetic vector potential. The nodes along this flux are defined using Dirichlet boundary condition.

a. The magnetic materials of the stator and rotor are isotropic

b. The magnetic vector potential and current density have z-directed components only.

c. The stator windings are identical and positioned symmetrically along the stator bore.

d. The stator and rotor are concentric and the air gap which separates them has constant width at aligned position.

e. The end effects, hysteresis effects, the skin effect are neglected.

4.2 Modeling of SRM using FEA

Modeling of SRM using FEMM requires three sequential stages. The first stage is to read the motor dimensions and plot its geometry. The second stage defines the materials properties over machine geometry, circuits, and boundary conditions. The final stage analyses the machine and calculates the desired magnetic characteristics. The code is written using FEM-Octave to link FEMM software with MATLAB software. The FEM-Octave file contains all the required functions to draw, define and analysis the studied geometry [33].

Before starting of geometry plotting, it is needed to open a new FEMM file with the desired settings. For SRM analysis, the new file is an electromagnetic problem with the following definitions:

```
%% Open a new magnetics problem
La =15.1; % The SRM stack length
addpath('C:\femm42\mfiles');
 savepath;
openfemm;
newdocument(0); % 0 means magnetics problem
mi_probdef(0,'centimeters','planar',1.e-8,la,30); % problem definitions
```

4.2.1 Geometry plotting

The FEM-based electromagnetic analysis of an SRM starts by creating the geometry of real machine. The geometrical dimensions of the studied 8/6 SRM are gathered in **Table 1**. **Figure 2** shows the cross-sectional area of SRM considering the symmetry along the length of the machine.

After the geometrical parameters are known, a lot of mathematical formulation should be deduced to define the Cartesian coordinates for each corner point in this geometry (in our case the SRM). The studied machine is an 8/6 SRM, it has 8 stator poles and 6 rotor poles as shown in **Figure 2**. The SRM has a symmetrical geometry. Only one rotor pole can be drawn and rotated five more times with 60° shift, the

Geometry parameter	Symbol	Value
Output power (kW)	—	4
Rated voltage (V)	—	600
Rated RMS current (A)	—	9
Rated speed (r/min)	—	1500
Turns per pole	N	88
Phase resistance (Ω)	R	0.64 Ω
Shaft diameters	D_{sh}	36
Bore diameters	D_{bore}	96.7
Stator outside diameter	D_{oy}	179.5
Height of rotor pole	h_{tr}	18.1
Height of stator pole	h_{ts}	29.3
Air-gap length	l_g	0.4
Rotor pole arc	β_r	21.5°
Stator pole arc	β_s	20.45°
Stack length	L_{st}	151

Table 1.
The design data of SRMs in mm.

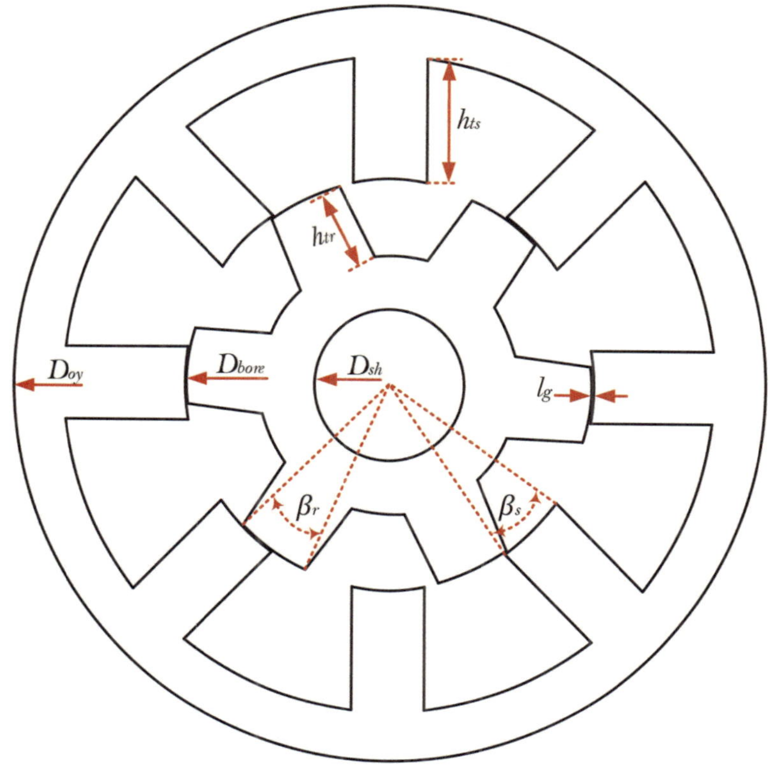

Figure 2.
The cross-sectional area of 8/6 SRM.

same can be done for the stator but the rotation will be seven more times with 45° shift. The procedure is well explained through the following **Figures 3–5**.

After the calculation of the required (x, y) coordinates for one rotor pole, it can be plotted as shown in **Figure 3(a)**. One rotor pole is drawn by the definition of 5 points (P1-p5). First, the five points need to be added to the FEMM file, after that the connection between points is achieved. Using the Octave functions (mi_addnode, mi_addsegment, and mi_addarc) to connect between the points as follows:

```
%% rotor points
mi_addnode(x_p1,y_p1);
mi_addnode(x_p2,y_p2);
mi_addnode(x_p3,y_p3);
mi_addnode(x_p4,y_p4);
mi_addnode(x_p5,y_p5);

%% connect between rotor points
 mi_addsegment(x_p1,y_p1,x_p2,y_p2); % segment P1 - p2
 mi_addsegment(x_p3,y_p3,x_p4,y_p4); % segment P3 - p4
 mi_addarc(x_p3,y_p3,x_p2,y_p2,theta32,5); % arc-segment P3 - p2
 mi_addarc(x_p5,y_p5,x_p4,y_p4,theta54,5); % arc-segment P5 - p4
```

The variable (theta32) is the angle of arc-segment that connects points P3 and p2. Noting that the arc is drawn counterclockwise direction with 5 segments. After drawing of one rotor pole, the other poles can be drawn by copying and rotation of this pole with 60° shift. The rotation function is (mi_copyrotate). It needs first to select the parts that will be rotated. For the rotor pole, the rotated parts are 2 segments (segment P1-p2, segment P4-p3,) and 2 arc-segments (arc-segment P2-p3, arc-segment P4-p5). The following MATLAB script explains the process:

```
%% Select the rotor parts that will be rotated
mi_clearselected; % clears all the selected parts
mi_selectsegment(x_p1,y_p1); % select segment P1 - p2
 mi_selectsegment(x_p4,y_p4); % select segment P4 - p3
 mi_selectarcsegment(x_p2,y_p2); % select arc-segment P2 - p3
 mi_selectarcsegment(x_p4,y_p4); % select arc-segment P4 - p5
 mi_copyrotate(0, 0, 60, 6) ; % rotate selected parts around (0,0)
mi_clearselected;
```

After the rotation, the rotor will be completely drawn as shown in **Figure 3(b)**. The same procedure can be done for the stator pole including the windings. One stator pole can be drawn by the definition of (x, y) coordinates of 13 points (P6-p18) as shown in **Figure 4(a)**. The final stator poles will be as shown in **Figure 4(b)**. The complete geometry of SRM is shown in **Figure 5**.

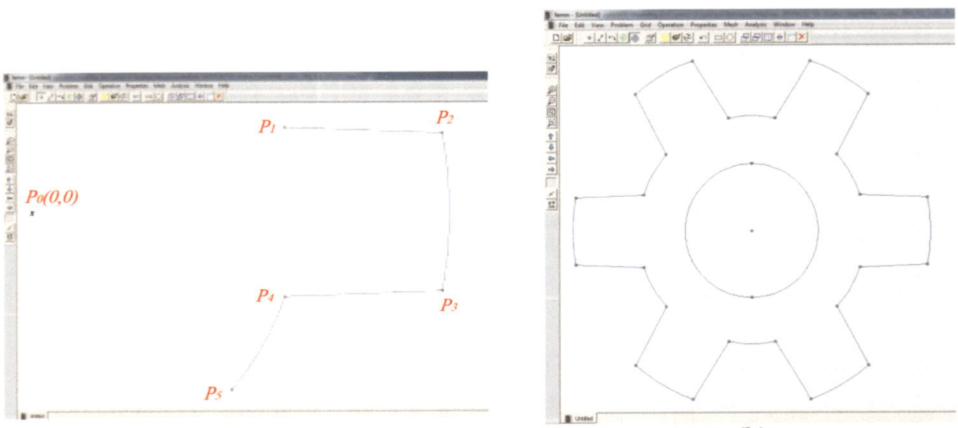

(a) (b)

Figure 3.
The rotor plotting for 8/6 SRM. (a) One rotor pole drawing (b) The complete rotor drawing.

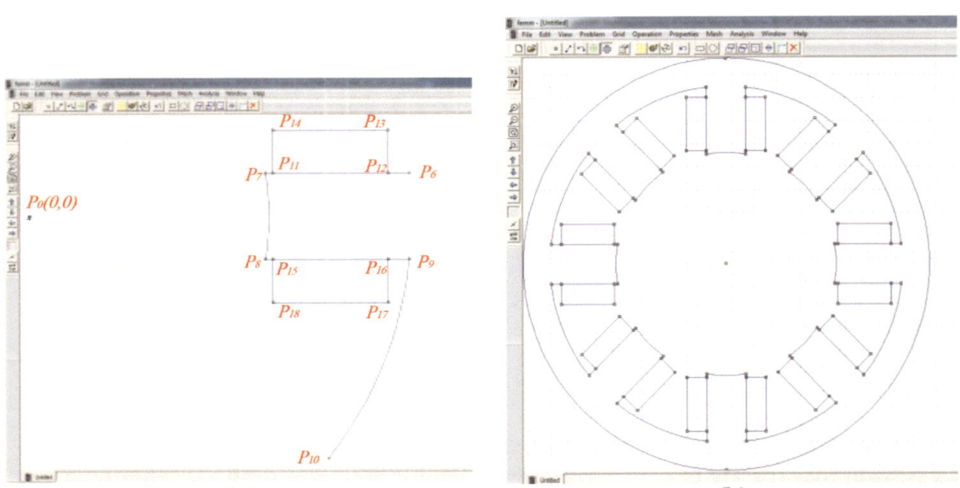

(a) (b)

Figure 4.
The stator plotting for 8/6 SRM. (a) One stator pole drawing (b) The complete stator drawing.

4.2.2 Definition of materials, circuits and boundary conditions

4.2.2.1 Definition of materials properties

After defining the geometry, it is necessary to attribute to each area its magnetic properties of the corresponding material. The areas of air and copper winding are set with a unitary relative magnetic permeability. It is important to note that the central circular area as shown in **Figure 3(b)** is not filled by air but corresponds to machine shaft that is made of a nonmagnetic material with a unitary relative magnetic permeability. Hence, the shaft can be considered as air. For the stator and rotor areas, the magnetization curve is very essential.

The FEMM has its materials library that contains the magnetic properties for quiet enough number of materials. For SRM, it is needed to define the magnetic properties for three materials. The air-gap can be defined as air, and the coils can be defined as copper (for example, 18 AWG). The type of steel core differs from machine to another; it should be defined carefully according to its steel type. Generally M-19, M-27, M-36, M-43, M-47, and M-50 are mostly used in rotary electric machines.

Before defining the material properties, it should be first added to the FEMM file using function (`mi_getmaterial`) as follows:

```
%% Add materials
mi_getmaterial('Air');
 mi_getmaterial('18 AWG'); % copper
 mi_getmaterial('Cold rolled low carbon strip steel'); % Steel core
```

The added materials are fully defined within the FEMM. So there is no need for more definitions. If it is necessary to change the material definitions, one can use function (`mi_modifymaterial`). If the material does not exist in FEMM materials library, one can use function (`mi_addmaterial`), but it is needed to define the material carefully.

Once all the materials are added and fully defined as desired, they should be included in the drawn geometry as shown in **Figure 6**. The shaft is defined as air, the rotor and stator iron are defined as cold rolled low carbon strip steel, the air-gap is defined as air, and the windings are defined as 18-AWG copper.

To achieve that, first you should add block label in each area, it is needed to set the block properties. This can be done by using (`mi_addblocklabel`, `mi_setblockprop`) functions.

4.2.2.2 Definition of current circuits

Now, it is time to define the current circuit for copper windings. In our case for 8/6 SRM, there are four independent phases (A, B, C, D). Each phase consists of two coils located at two opposite poles. The direction of current in motor windings must make magnetic field direction generated by both phase windings to be coherent. The flux will flow out from one pole to enter its opposite pole. According to the current direction in the coils, one pole is north and the other is south. It should be noted that, the current direction in FEMM is always positive, there is no negative currents, but it can be achieved if the number of turns is set to negative. The two coils on two opposite poles can be connected either in parallel or series.

Figure 7 shows the sign for the turns number for phase A. the pole on right is north and the other on left is south. If the turns are set wrong, for example both poles are north or both are south, the flux will diminish each other and the resultant flux will be almost zero.

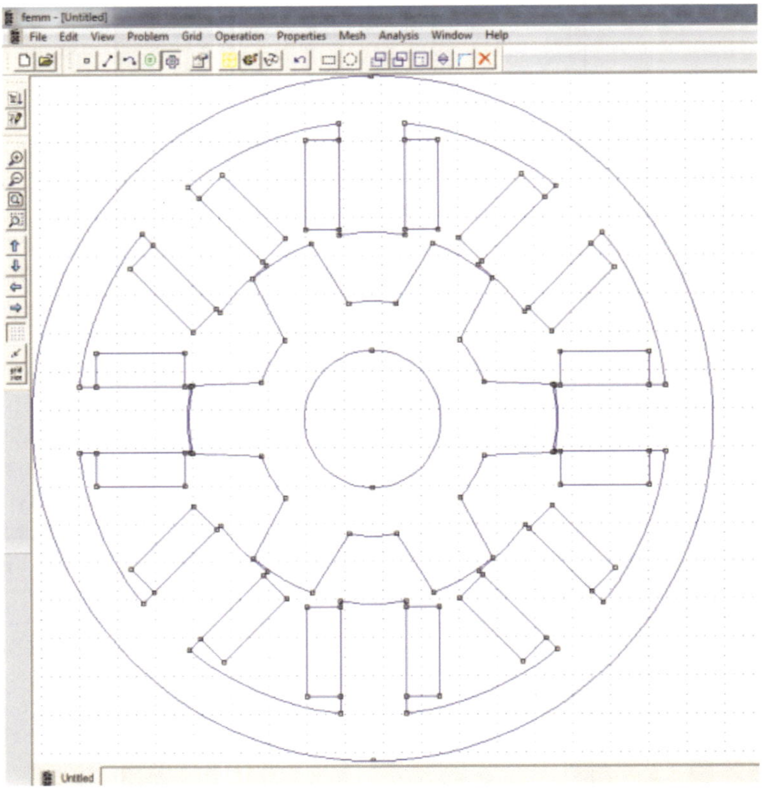

Figure 5.
The final drawn geometry of 8/6 SRM.

To add the circuit properties, the function (mi_addcircprop) can be used. Then the properties of the previously defined 18 AWG blocks can be set to include the circuits. A code example is given below:

```
 mi_addcircprop(A, i, 1); % add the circuit properties

%% To draw phase A (on the wright and above)
mi_selectlabel(x_a, y_a);
mi_setblockprop('18 AWG', auto_mesh, mesh_size, A, 30, group, -N);
 mi_clearselected ;

%% To draw phase A (on the wright and below)
mi_selectlabel(x_a, -y_a);
mi_setblockprop('18 AWG', auto_mesh, mesh_size, A, 30, group, N);
 mi_clearselected ;

%% To draw phase A (on the left and above)
mi_selectlabel(-x_a, y_a);
mi_setblockprop('18 AWG', auto_mesh, mesh_size, A, 30, group, -N);
 mi_clearselected ;

%% To draw phase A (on the wright and below)
mi_selectlabel(-x_a, -y_a);
mi_setblockprop('18 AWG', auto_mesh, mesh_size, A, 30, group, N);
 mi_clearselected ;
```

Each phase is defined by 4 copper blocks as shown in **Figure** 7. The four copper blocks should be included in the circuit with proper connection as described by the given code above. The (x, y) coordinates for each block are defined by (x_a, y_a). If

FEMM is set to automatically choose mesh size, then `auto_mesh = 1`. Otherwise `auto_mesh = 0` and `mesh_size` should be chosen as desired. `N` is the number of turns per pole.

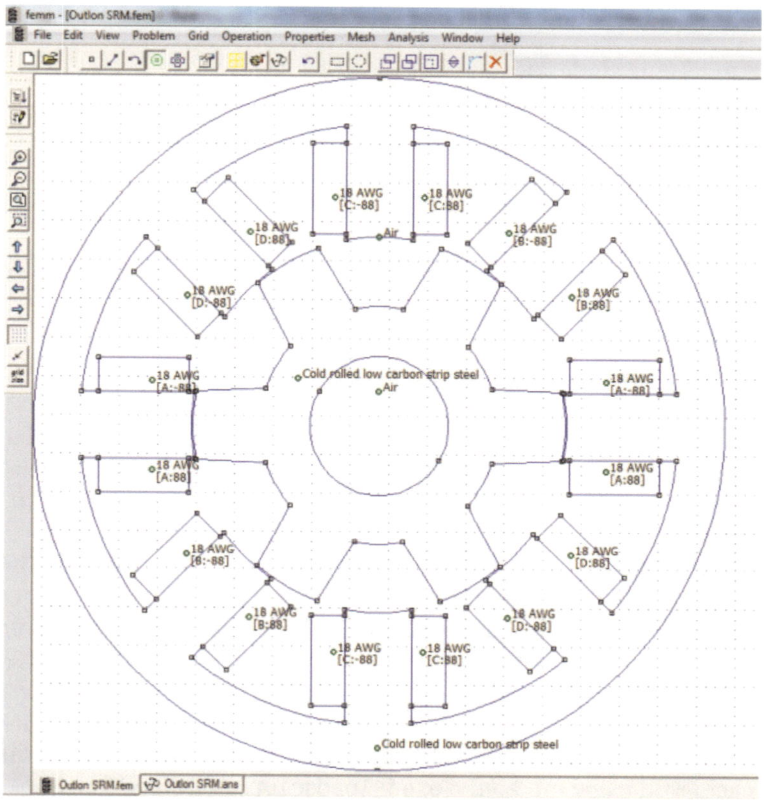

Figure 6.
The final drawn geometry with materials definitions.

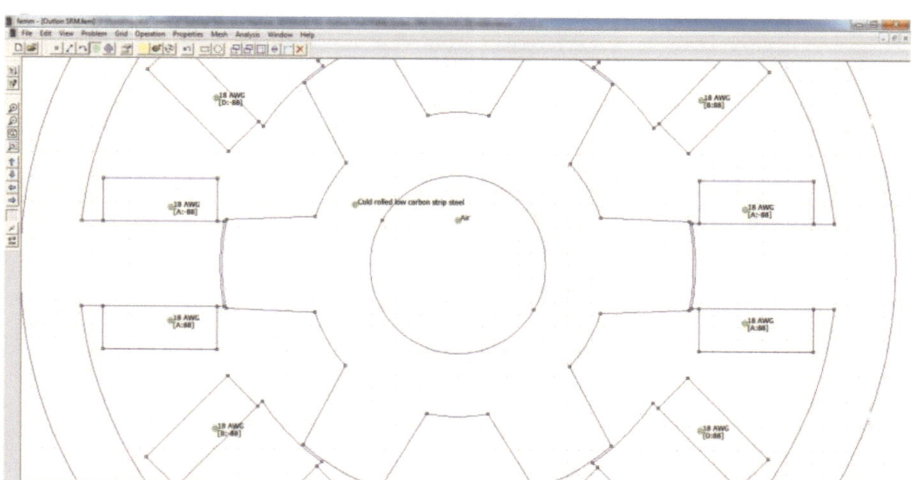

Figure 7.
Setting the number of turns for one phase.

4.2.2.3 Definition of boundary conditions

It is needed to specify the area where the magnetic field is confined within the motor. A negligible flux leakage can be observed outside the stator, so the yoke circumstance can be defined as a zero potential vector (A = 0). Dirichlet boundary conditions are the best choice for such problems.

```
%% Set boundary conditions
mi_addboundprop('A=0', 0, 0, 0, 0, 0, 0, 0, 0);
mi_selectarcsegment(Doy/2,0);
 mi_setarcsegmentprop(2.5,'A=0', 0, 2);
mi_clearselected;

 mi_selectarcsegment(-Doy/2,0);
 mi_setarcsegmentprop(2.5,'A=0', 0, 2);
mi_clearselected;
```

4.3 Results of FE analysis

In 2D FEA, the solution accuracy depends on field nature and mesh size. The mesh size for any component of the entire model can be controlled by FEMM. **Figure 8** shows two different mesh sizes at the unaligned rotor position ($\theta = 0°$) regarding phase A.

To obtain the solution, the value of phase current must be specified. This is done by applying a current driven source for a certain phase using `mi_setcurrent('A',i)` Octave function, and `i` is the value of the desired current. The phase current values used to obtain the model are between 0A and 30A with 1A increment. The rotor angles are changed from unaligned rotor position ($\theta = 0°$) to the aligned rotor position ($\theta = 30°$) with step of 0.5°. These current and angle resolutions are small enough to achieve high accuracy model for the studied 8/6 SRM. The FEMM analysis is achieved using `mi_analyze; mi_loadsolution` Octave functions.

Once a solution is obtained, the magnetic potential vector distribution is known, and hence the magnetic flux in each phase can be calculated. The flux density

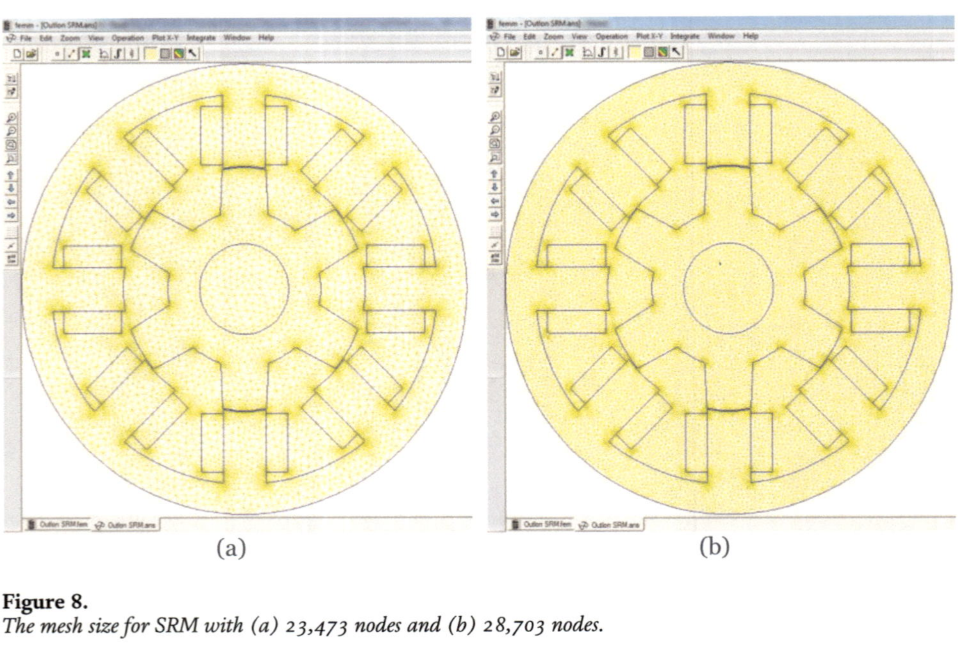

(a) (b)

Figure 8.
The mesh size for SRM with (a) 23,473 nodes and (b) 28,703 nodes.

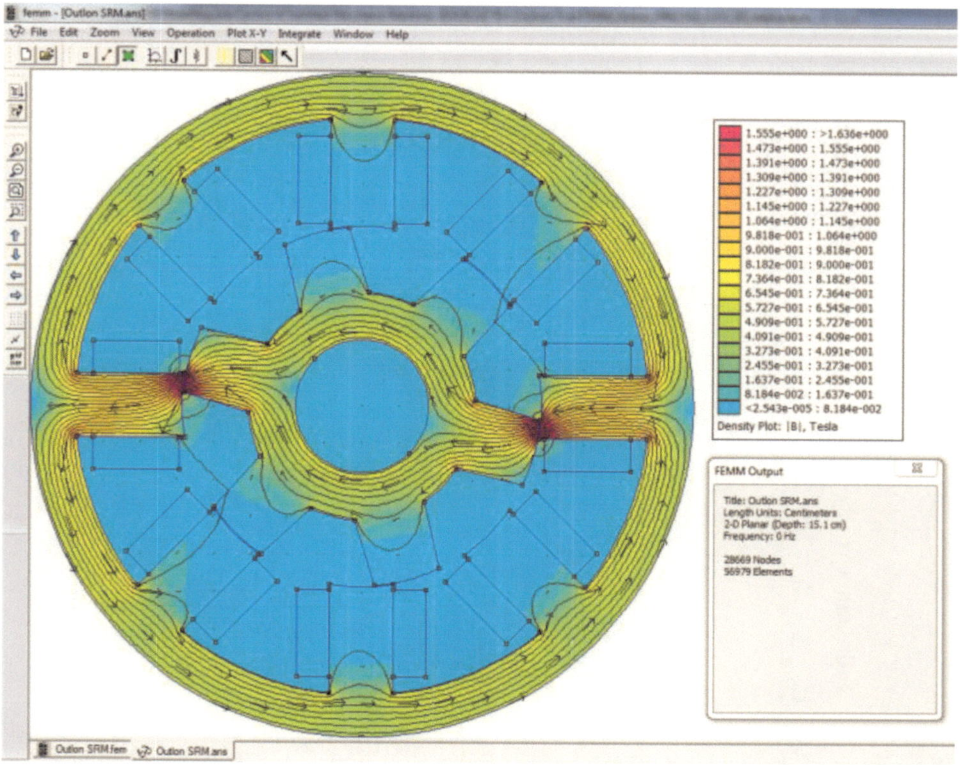

Figure 9.
The flux density distribution at 15° with excitation current 10 A.

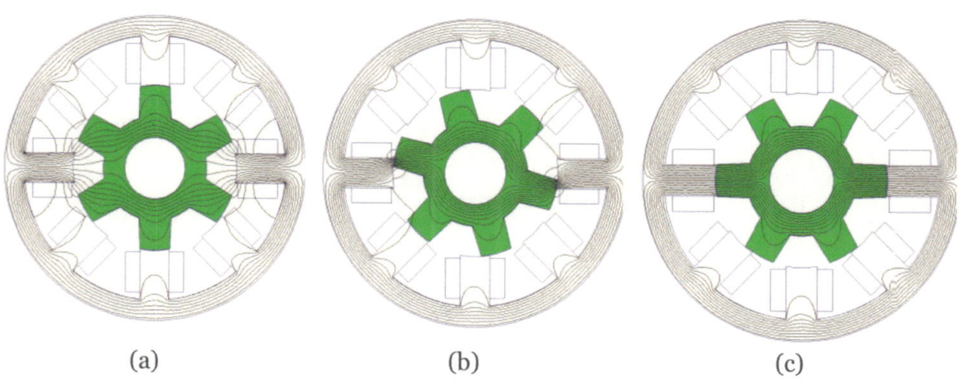

Figure 10.
The flux lines at different rotor positions with 10 A: (a) 0°, (b) 15°, (c) 30°.

distribution at θ = 15° with an excitation current of 10(A) is shown in **Figure 9**. The flux lines at different rotor positions are shown in **Figure 10**.

The Magnetic flux linkage in phase A as a function of rotor position and excitation current as obtained by the finite element model is shown in **Figure 11(a)**. Once the flux linkage is obtained, the inductance profile can be deduced according to Eq. (13) as shown in **Figure 11(b)**. The air-gap torque produced on phase A as a function of the rotor position and excitation current as calculated by the finite-element model is shown in **Figure 11(c)**. The doubly salient structure makes the flux, inductance and torque highly nonlinear functions of current magnitude (i) and rotor position (θ).

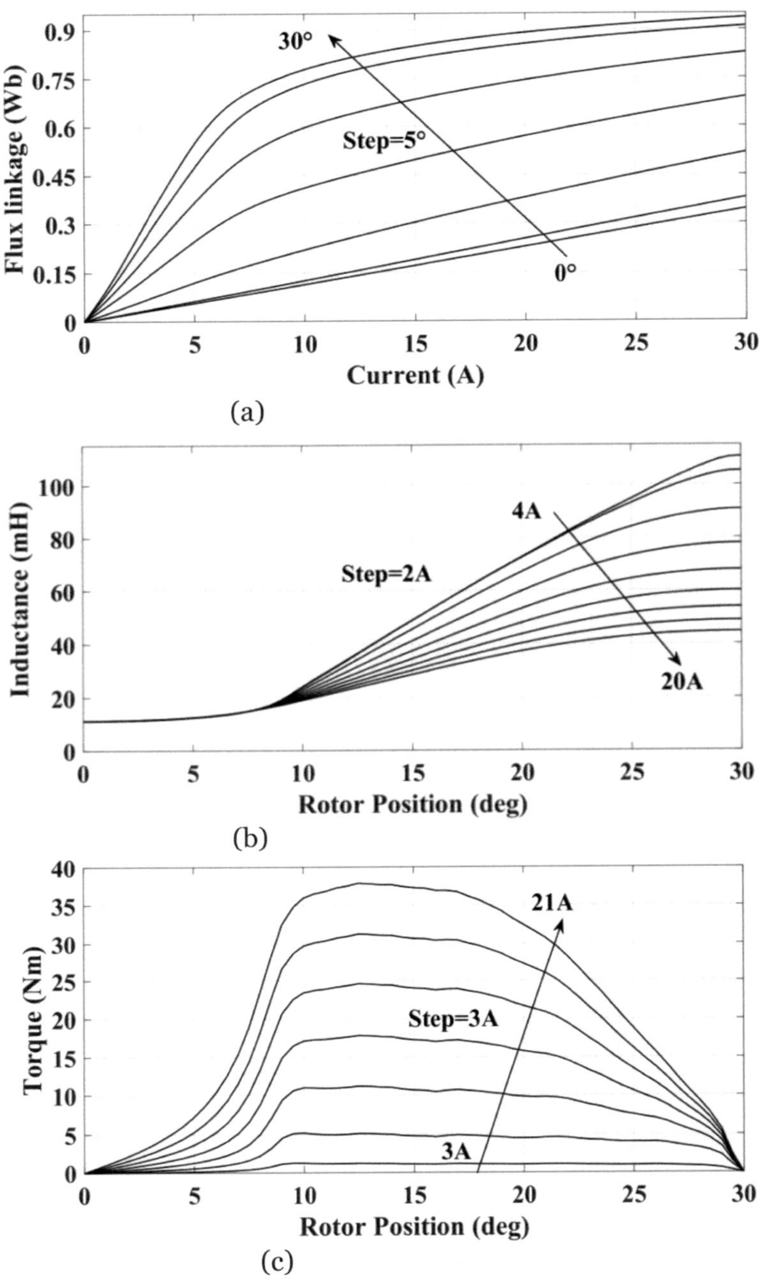

Figure 11.
The FEM-calculated characteristics for the studied 8/6 SRM. (a) Flux linkage for 8/6 SRM. (b) Inductance for 8/6 SRM. (c) Torque for 8/6 SRM.

$$L(i, \theta) = \lambda(i, \theta)/di \quad (13)$$

5. Experimental measurement methods

Generally, the experimental measurements can be categorized into direct and indirect methods [21, 24]. The direct methods utilize magnetic sensors to directly measure pole flux [24], or they may measure the pole flux directly after proper processing of induced voltage over search-coil that is mounted on stator pole [14].

The direct methods are rarely used because of the leakage flux that affects the accuracy [24]. On the other hand, indirect methods uses phase voltage and current to estimate flux [12]. They can provide simple structure, low cost, and better accuracy [24]. Hence, they are adopted in this work.

5.1 Measuring method and platform

The phase voltage (V) and current (i) are used to measure phase flux linkage indirectly. At a desired rotor position (θ), a pulsed dc voltage is applied to one phase winding, and the phase voltage and current are measured and recorded. The flux-linkage (λ) is calculated according to Eq. (10) or its discrete form in Eq. (11) [14].

$$\lambda(i,\theta) = \int (V - R\cdot i)\cdot dt + \lambda(0) \tag{14}$$

$$\lambda(n) = \sum_{k=1}^{n} [V(k) - R\cdot i(k)]\cdot T_s + \lambda(0) \tag{15}$$

where R, n, Ts are the phase resistance, adopted number of samples and sampling period, respectively. λ(0) is the initial pole flux. It equals to zero because SRM has no magnets.

The same principle is used for torque measurement, but the phase current and torque signals are measured and recorded directly while rotor is locked at a specific position. The measurement procedure of flux/torque should be repeated several times owing to the desired angle resolution to generate the complete flux/torque data. **Figure 12** shows the schematic diagram of measurement platform.

5.2 Measurement results

As SRM has an identical geometrical structure, only one phase can be used for measurement, and only half electric period of 30° mechanical degrees are adopted for measurement. The remaining part of characteristics can be estimated by proper

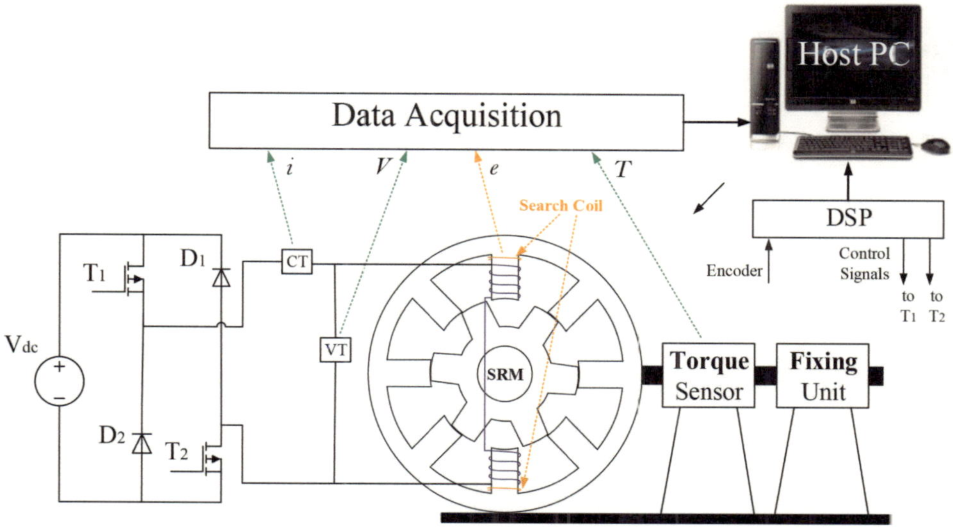

Figure 12.
The wiring diagram for the measurement platform.

mirroring. The measurement process starts at the unaligned position (θ = 0°) and ends at the aligned position (θ = 30°).

5.2.1 The measured results of flux-linkage

The tested 8/6 SRM is equipped with a search coil on stator poles. The search coil is used for verification purposes. The integration of voltage (e) induced on search coil gives directly the phase flux linkage. The measured phase voltage and current

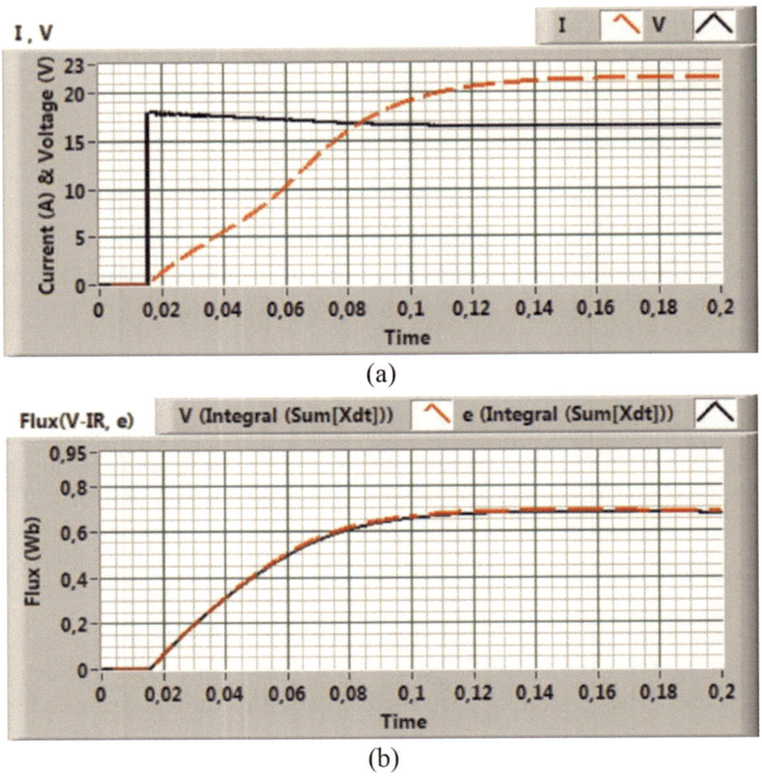

Figure 13.
The measured waveforms at 17° for (a) voltage and current, (b) flux.

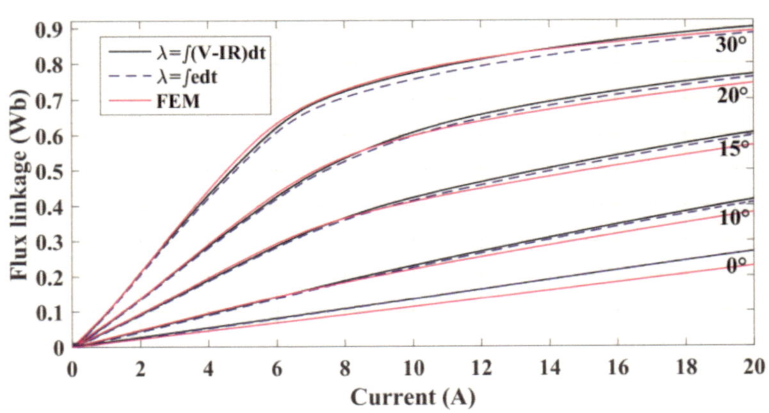

Figure 14.
The measured and FEM flux curves.

are given in **Figure 13(a)**. Their corresponding measured flux linkage is illustrated in **Figure 13(b)**. The adopted current for measurements is 20A with positioning step of 1°. **Figure 14** shows the obtained flux curve at five different rotor positions. It compares between the direct, indirect, and FEA methods. A very good agreement can be seen.

5.2.2 The measured results of static torque

The electromagnetic torque of the SRM is measured directly using a DRBK torque sensor. The DRBK has a limited sampling frequency of 1 kHz. This may affect the measurement accuracy. Better accuracy can be achieved if it is possible to measure reasonable number of samples. This can be achieved by increasing the measurement time. This time is the rising time of phase current. Adding an external inductance in series with phase winding can extend measurement time. The added inductance can increase time constant. Hence, current increase becomes slower allowing recording more torque samples. **Figure 15** shows the measured torque signal along with phase current at position of 16.5°. The measured torque curves are shown in **Figure 16**. **Figure 17** shows a comparison between measured and FEM-calculated torque characteristics, a very good agreement is observed.

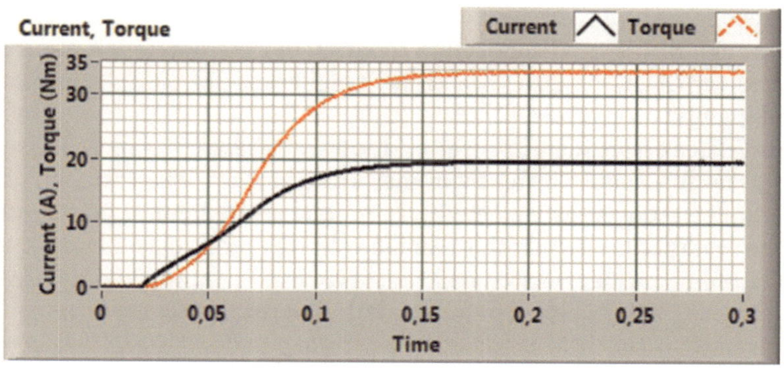

Figure 15.
The measured current and torque waveforms at 16.5°.

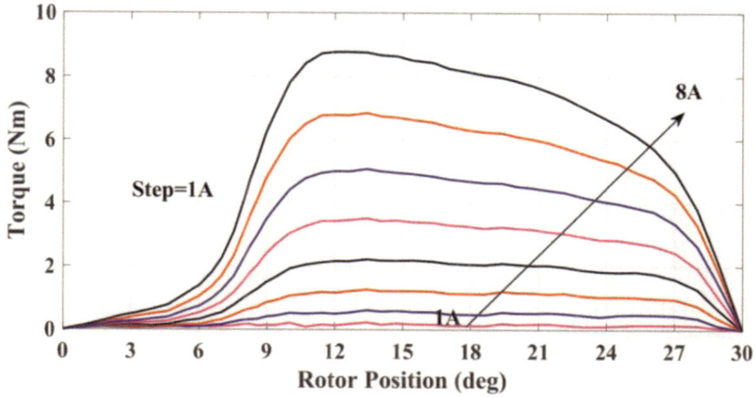

Figure 16.
The measured torque curves.

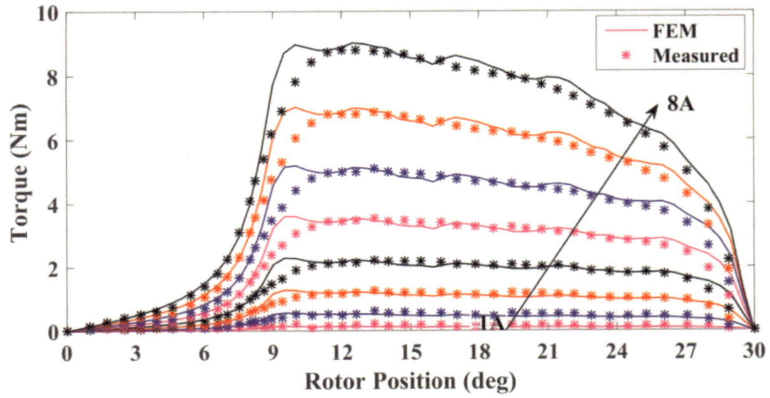

Figure 17.
The measured and FEM torque curves from 1A to 8A.

5.3 Error analysis and minimization

There are several introduced errors during measurement process such as signal errors, parameter errors, and calculation errors. These errors need to be well post-processed to reduce them.

The signal errors include sensor-offsets and nonlinearities, electronics noises, and the quantization errors. Therefore, the voltage and current transducers need precise inspection to compensate signal error. The quantization errors can be reduced through differential connection of DAQ. In this connection, the analog to digital converter (ADC) of DAQ can achieve higher resolution that can ensure minimized quantization errors.

The parameter errors include rotor position and phase resistance. These errors happen as measured values differ from their actual values. The variation of phase resistance can be reduced by two methods. A dc measurement of phase resistance can be done several times while the average value is reported as the real one. Besides, the temperature effect on coil resistance can be minimized by using very low frequency (< 1 Hz) voltage pulses [7, 10]. The rotor position error comes from the encoder itself as the SRM is locked in a certain position. Therefore, in order to reduce rotor position error, a small angular step encoder is advised.

The calculation errors occur mainly because of the numeric integration. These errors depend on integration method and the sampling frequency. If a too low sampling frequency is used, the voltage and current waveforms will be distorted. Hence, the truncation errors of the numerical calculations will increase. On the contrary, if a too high sampling frequency is adopted, the rounding error will increase because of the huge number of computations.

6. Model development using MATLAB/Simulink

The switched reluctance drive system simulation is more complex than DC and AC drives. The dynamic simulation is carried out based on its mathematical model that is given through Eq. (16) to Eq. (24). These equations are solved simultaneously using MATLAB/Simulink.

6.1 Mathematical model of SRM

Due to the double salient structure, the flux linkage $\lambda(i,\theta)$, inductance $L(i,\theta)$, and torque $T(i,\theta)$ are functions of both current magnitude (i) and rotor position (θ).

In ¾ phases SRMs, almost two phases conduct simultaneously. The incoming and outgoing phases are denoted as k^{th} and $(k\text{-}1)^{th}$ phases respectively. The phase voltage equations can be derived as follows [24, 34]:

$$v_k = Ri_k + \frac{d\lambda_k}{dt}, \quad v_{k-1} = Ri_{k-1} + \frac{d\lambda_{k-1}}{dt} \tag{16}$$

where v_k and v_{k-1} are the voltage for incoming and outgoing phases respectively. i_k and i_{k-1} *i* are the current for incoming and outgoing phases respectively. When mutual flux is considered, flux linkage for incoming (λ_k) and outgoing (λ_{k-1}) phases can be expressed as:

$$\lambda_k = \lambda_{k,k} + \lambda_{k,k-1}, \; \lambda_{k-1} = \lambda_{k-1,k-1} + \lambda_{k-1,k} \tag{17}$$

where $\lambda_{k,k}$ and $\lambda_{k-1,k-1}$ are the self-flux linkages of kth and $(k\text{-}1)^{th}$ phases respectively. $\lambda_{k,k-1}$ and $\lambda_{k-1,k}$ are the mutual flux linkages.

The flux linkage can be represented as Eq. (18) in terms of self and mutual inductances.

$$\begin{bmatrix} \lambda_k \\ \lambda_{k-1} \end{bmatrix} = \begin{bmatrix} L_{k,k} & M_{k,k-1} \\ M_{k-1,k} & L_{k-1,k-1} \end{bmatrix} \begin{bmatrix} i_k \\ i_{k-1} \end{bmatrix} \tag{18}$$

where $L_{k,k}$ and $L_{k-1,k-1}$ are the self-inductances of the kth and $(k\text{-}1)^{th}$ phases respectively. $M_{k,k-1}$ and $M_{k-1,k}$ are the mutual inductances.

Considering magnetic saturation, the phase voltage equations are derived as:

$$\begin{bmatrix} v_k \\ v_{k-1} \end{bmatrix} = R\begin{bmatrix} i_k \\ i_{k-1} \end{bmatrix} + \begin{bmatrix} L_{inc\_k} & M_{inc\_k,k-1} \\ M_{inc\_k-1,k} & L_{inc\_k-1} \end{bmatrix} \begin{bmatrix} \dfrac{di_k}{dt} \\ \dfrac{di_{k-1}}{dt} \end{bmatrix} + \omega_m \begin{bmatrix} \dfrac{\partial L_{k,k}}{\partial \theta} & \dfrac{\partial M_{k,k-1}}{\partial \theta} \\ \dfrac{\partial M_{k-1,k}}{\partial \theta} & \dfrac{\partial L_{k-1,k-1}}{\partial \theta} \end{bmatrix} \begin{bmatrix} i_k \\ i_{k-1} \end{bmatrix} \tag{19}$$

where ω_m is angular speed of SRM. L_{inc_k} and L_{inc_k-1} are the incremental inductances of the kth and $(k\text{-}1)^{th}$ phases respectively. $M_{inc\_k,k-1}$ and $M_{inc\_k-1,k}$ are the incremental mutual inductances. Incremental inductance and incremental mutual-inductance are given in Eq. (20) and Eq. (21). In linear magnetic region, the incremental inductance is equal to the self-inductance.

$$L_{inc\_k} = L_{k,k} + i_k \frac{\partial L_{k,k}}{\partial i_k}, L_{inc\_k-1} = L_{k-1,k-1} + i_{k-1} \frac{\partial L_{k-1,k-1}}{\partial i_{k-1}} \tag{20}$$

$$M_{inc\_k,k-1} = M_{k,k-1} + i_{k-1} \frac{\partial M_{k,k-1}}{\partial i_{k-1}}, M_{inc\_k-1,k} = M_{k-1,k} + i_k \frac{\partial M_{k-1,k}}{\partial i_k} \tag{21}$$

Electromagnetic torque of kth phase can be derived as:

$$T_{e(k)} = \frac{1}{2} \frac{\partial L_{k,k}}{\partial \theta} i_k^2 \tag{22}$$

where $T_{e(k)}$ is the torque produced by k^{th} phase. For SRM with m-phases, the total electromagnetic torque T_e can be represented as:

$$T_e = \sum_{k=1}^{m} T_{e(k)} \tag{23}$$

The equation for mechanical dynamics is expressed as:

$$T_e - T_L = B\omega_m + J\frac{d\omega_m}{dt} \tag{24}$$

where T_L is the load torque, B is the friction constant, J is the machine inertia.

6.2 Simulation of SRM drive

A simulation model for the experimentally tested 8/6 SRM is achieved using MATLAB Simulink environment. The model can use the data obtained from measurements or from FEA. As discussed earlier in Section 5, both are in a very good agreement, but the measured data have better accuracy. Therefore, the model uses the data obtained from measurements. Using these data, the unwanted approximations introduced by the analytic models are eliminated, the precision of the results is increased and the perspective on the machine is more accurate. To avoid the inaccuracies inherent to the analytical model, the characteristics of current versus rotor position and flux and of torque versus rotor position and current are stored in lookup tables and are introduced in the model of the SRM. In this manner an accurate model of the machine is obtained, capable of providing an objective perspective on the behavior of the SRM under different operating conditions. The model can be further used in the optimization of control and/or design. Although these models are very close to the real behavior of the machine, the mutual coupling between phases, reported by previous work as having a small influence on the machine's operation, is not taken into account [14, 21].

Modeling and simulation of one phase of SRM is shown in **Figure 18**. The inputs are the rotor position (θ) and the phase voltage (V). The outputs are phase current (*i*) and phase torque (T). The model utilizes the measurement data of flux and torque after proper rearrangement in form of lookup tables [20, 21]. The torque data can be rearranged easily as the torque is measured as a function of current and position T(i, θ). The hard task appears with the flux data λ(i, θ), as the model calculates current as a function of flux and position i(λ, θ). Hence, the measured flux data λ(i, θ) should be processed and rearranged to produce required current data i(λ, θ). This can be achieved by interpolation and extrapolation of flux data against current for different rotor positions. Due to the enormous measured

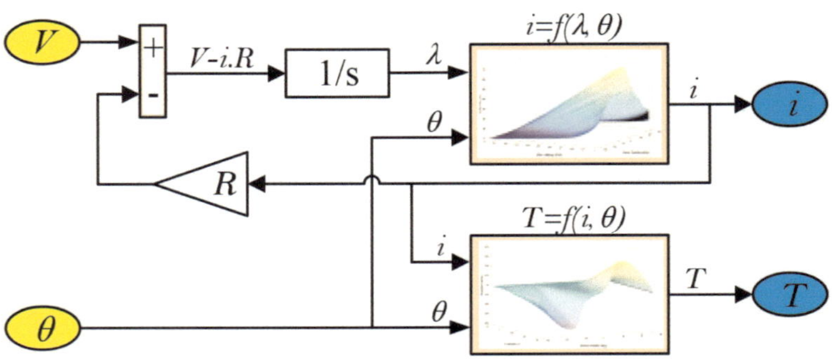

Figure 18.
Simulation of one phase of SRM.

samples of flux, the interpolation and extrapolation of current i(λ, θ) can ensure a sufficient accuracy.

The SRM cannot run directly from an AC or DC source; it requires continuous commutation from one phase to another. So, the motor itself and its converter are considered as one unit. For system simulation, the converter and the position sensing unit need also to be modeled. The most common converter for SRM is the asymmetric bridge converter as shown in **Figure 12** for one phase of SRM. The position sensing unit feeds the controller with the correct position for each phase. According to the phase position, the controller determines the firing angles for this phase. The control defines the switching state for each phase according to the switch-on (θ_{on}) and switch-off (θ_{off}) angles. The voltage $+V$ is applied when (θ_{on} = $<\theta < \theta_{off}$) and the voltage is $-V$ when (θ θ_{off}). The phase currents of the motor can

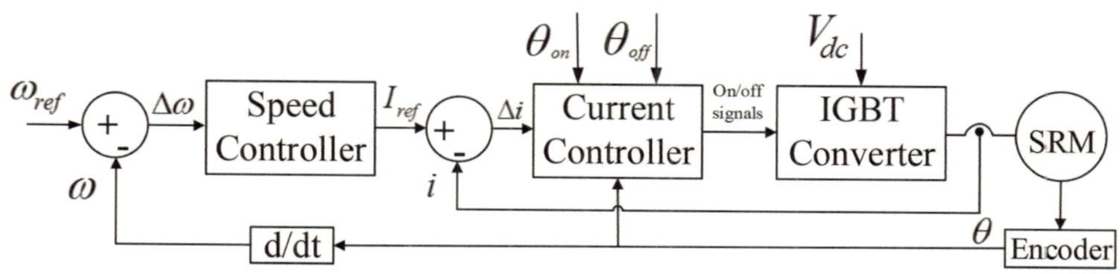

Figure 19.
Block diagram of SRM control system.

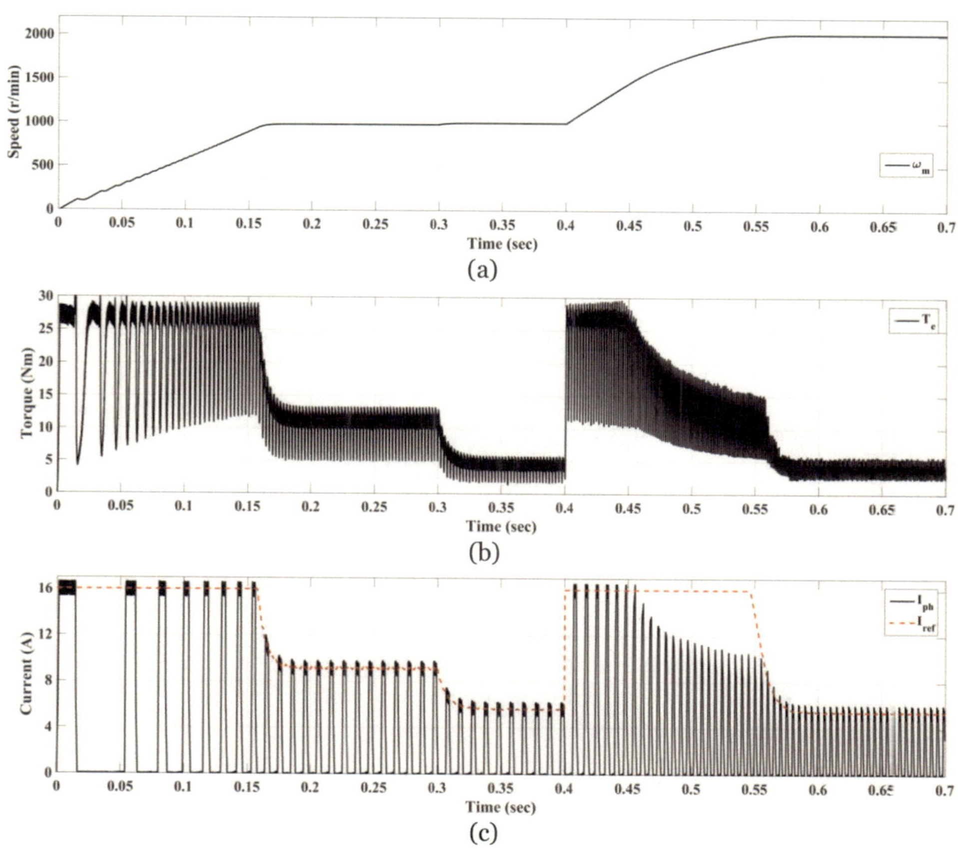

Figure 20.
Simulation results under transient conditions: (a) motor speed; (b) the total electromagnetic torque; and (c) the reference current and one phase current.

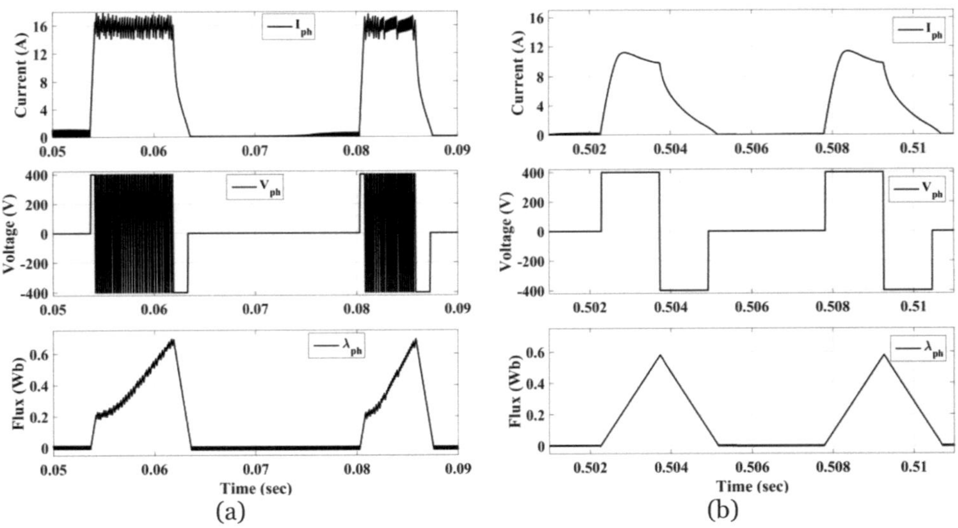

Figure 21.
Simulation results under (a) low speed and (b) high speed.

be limited using chopping process. The block diagram of SRM control system is shown in **Figure 19**. It has an outer loop speed controller that outputs the reference current signal (I_{ref}) according to speed error signal ($\Delta\omega$). An inner loop current controller is needed to regulate motor phase current (i) to track its reference current. A feedback of rotor position is essential for motor operation.

6.2.1 Simulation results

The SRM operates in two modes according to the motor speed. At low speeds the back-emf is very small compared to the applied voltage, the current increases rapidly after the instant of switching-on. The rapid increase of the phase current must be limited to an allowable value by Hysteresis Current Control (HCC). On the other hand, at high speeds the back-emf is considerable and the current is lower than the permissible value. In this case, the voltage is a single pulse.

Figure 20 shows the simulation results under transient conditions. The motor speed changes from 1000 r/min to 2000 r/min at 0.4 sec. The load torque changes from 10 Nm to 6 Nm at time 0.3 sec. The speed response is shown in **Figure 20(a)**. The total torque is given in **Figure 20(b)**, as noted it has a highly noticeable ripple that is the major drawback for SRMs. **Figure 20(c)** shows the performance of current controller as it regulates phase current to track its reference current.

The current profile under low and high speed is shown in **Figure 21**. For low speed, the current is regulated using HCC, the phase voltage is changed between +V and −V to achieve that control, the shape of phase flux is a little bit away to form a triangular shape as shown in **Figure 21(a)**. For high speeds, the phase voltage becomes a single pulse, the current limitation is no longer in use, the flux becomes a pure triangle as shown in **Figure 21(b)**.

7. Experimental verification

For the detailed experimental verification of the established model, a series of results are obtained. A comparison between the simulated and experimentally

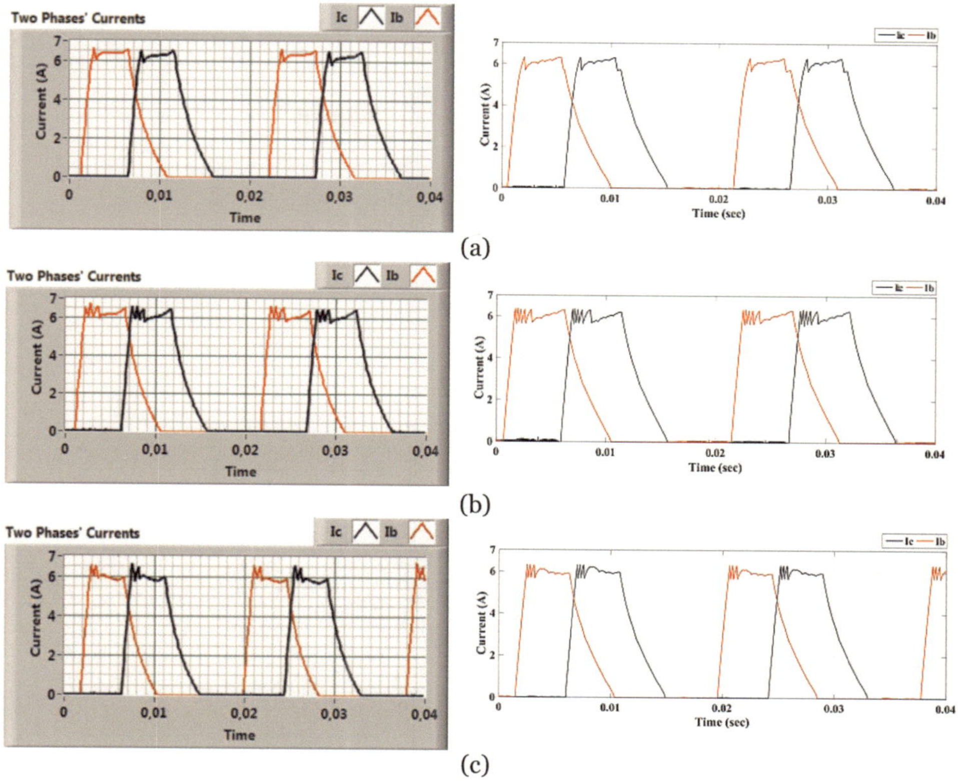

Figure 22.
Comparision between the measured and simulated current waveforms: (a) at 480 r/min, θ_{on} = 5°, θ_{off} = 20°; (b) at 481 r/min, θ_{on} = 3°, θ_{off} = 19°; and (c) at 551 r/min, θ_{on} = 3°, θ_{off} = 19°.

obtained current waveforms is achieved for different operating as given in **Figure 22**. As seen a very good agreement is achieved that reflects the modeling accuracy.

8. Conclusions

This chapter explains the modeling and simulation methods for SRMs. It discusses the analytical and intelligent methods for SRM modeling. Besides, it determines the static performance of an 8/6 switched reluctance machine as obtained from the two dimensional finite-element method analysis using FEMM and compares the results with experimental obtained ones. A very good agreement is observed. Then it gives the simulation of the SRM drive system in MATLAB/ Simulink environment.

Author details

Mahmoud Hamouda[1,2*] and László Számel[1]

1 Department of Electric Power Engineering, Budapest University of Technology and Economics, Budapest, Hungary

2 Electrical Engineering Department, Mansoura University, Mansoura, Egypt

*Address all correspondence to: m_hamouda26@mans.edu.eg

References

[1] Lin D, Zhou P, Stanton S, Cendes ZJ. An analytical circuit model of switched reluctance motors. IEEE Transactions on Magnetics. 2009; **45**(12):5368-5375

[2] Song S, Zhang M, Ge L. A new decoupled analytical modeling method for switched reluctance machine. IEEE Transactions on Magnetics. 2015;**51**(3). Article Number: 8100504

[3] Hamouda M, Számel L. Optimum control parameters of switched reluctance motor for torque production improvement over the entire speed range. Acta Polytechnica Hungarica. 2019;**16**(3):79-99

[4] Uddin W, Sozer Y. Analytical modeling of mutually coupled switched reluctance machines under saturation based on design geometry. IEEE Transactions on Industry Applications. 2017;**53**(5):4431-4440

[5] Chen H, Jiang D, Yang J, Shi L. A New analytical model for switched reluctance motors. IEEE Transactions on Magnetics. 2009;**45**(8)

[6] Yoopakdee C, Fuengwarodsakul NH. Analytic model of switched reluctance machine using combined Fourier-polynomial approximation technique. In: 2018 XIII International Conference on Electrical Machines (ICEM). Alexandroupoli, Greece; 2018

[7] Torrey DA, Niu XM, Unkauf EJ. Analytical modelling of variable-reluctance machine magnetisation characteristics. IEE Proceedings - Electric Power Applications. 1995;**142** (1):14-22

[8] Sheth NK, Rajagopal KR. Calculation of the flux-linkage characteristics of a switched reluctance motor by flux tube method. IEEE Transactions on Magnetics. 2005;**41**(10):4069-4072

[9] Vujičić V, Vukosavić SN. A simple nonlinear model of the switched reluctance motor. IEEE Transactions on Energy Conversion. 2000;**15**(4): 395-400

[10] Radimov N, Ben-Hail N, Rabinovici R. Simple model of switched-reluctance machine based only on aligned and unaligned position data. IEEE Transactions on Magnetics. 2004; **40**(3):1562-1572

[11] Zhong R, Xu Y, Cao Y, Guo X, Hua W, Xu S, et al. Accurate model of switched reluctance motor based on indirect measurement method and least square support vector machine. IET Electric Power Applications. 2016;**10**(9): 916-922

[12] Song S, Ge L, Ma S, Zhang M, Wang L. Accurate measurement and detailed evaluation of static electromagnetic characteristics of switched reluctance machines. IEEE Transactions on Instrumentation and Measurement. 2015;**64**(3):704-714

[13] Kiyota K, Kakishima T, Sugimoto H, Chiba A. Comparison of the test result and 3D-FEM analysis at the knee point of a 60 kW SRM for a HEV. IEEE Transactions on Magnetics. 2013;**49**(5): 2291-2294

[14] Hamouda M, Számel L. Accurate measurement and verification of static magnetization characteristics for switched reluctance motors. In: IEEE 19th International Middle East Power System Conference (MEPCON), Cairo, Egypt. 2017. pp. 993-998

[15] Zhang J, Radun AV. A new method to measure the switched reluctance motor's flux. IEEE Transactions on Industry Applications. 2006;**42**(5): 1171-1176

[16] Lachman T, Mohamad TR, Fong CH. Nonlinear modelling of

switched reluctance motors using artificial intelligence techniques. IEE Proceedings - Electric Power Applications. 2004;**151**(1):53-60

[17] Lin Z, Reay DS, Williams BW, He X. Online modeling for switched reluctance motors using B-spline neural networks. IEEE Transactions on Industrial Electronics. 2007;**54**(6): 3317-3322

[18] Ding W, Liang D. Modeling of a 6/4 switched reluctance motor using adaptive neural fuzzy inference system. IEEE Transactions on Magnetics. 2008; **44**(7):1796-1804

[19] Gouda E, Hamouda M, Amin ARA. Artificial intelligence based torque ripple minimization of switched reluctance motor drives. In: 2016 Eighteenth International Middle East Power Systems Conference (MEPCON), Cairo, Egypt. 2016. pp. 943-948

[20] Husain I, Hossain SA. Modeling, simulation, and control of switched reluctance motor drives. IEEE Transactions on Industrial Electronics. 2005;**52**(6):1625-1634

[21] Hamouda M, Számel L. Accurate magnetic characterization based model development for switched reluctance machine. Periodica Polytechnica Electrical Engineering and Computer Science. 2019;**63**(3):202-212

[22] Spong MI, Marino R, Peresada SM, Tayler DG. Feedback linearizing control of switched reluctance motors. IEEE Transactions on Automatic Control. 1987;**AC-32**(5):371-379

[23] Torrey DA, Lang JH. Modeling a nonlinear variable-reluctance motor drive. Proceedings of the Institution of Electrical Engineers. IEEE Proceedings B - Electric Power Applications. Sept. 1990;**137**(5):314-326

[24] Nasirian V, Kaboli S, Davoudi A, Moayedi S. High-fidelity magnetic characterization and analytical model development for switched reluctance machines. IEEE Transactions on Magnetics. 2013;**49**(4):1505-1515

[25] Xue XD, Cheng KWE, Ho SL, Kwok KF. Trigonometry based numerical method to compute nonlinear magnetic characteristics in switched reluctance motors. IEEE Transactions on Magnetics. 2007;**43**(4):1845-1848

[26] Khalil A, Husain I. A Fourier series generalized geometry-based analytical model of switched reluctance machines. IEEE Transactions on Industry Applications. 2007;**43**(3):673-684

[27] Song S, Zhang M, Ge L. A new decoupled analytical modeling method for switched reluctance machine. IEEE Transactions on Magnetics. 2015;**51**(3): 1-4

[28] Lu W, Keyhani A, Fardoun A. Neural network based modeling and parameter identification of switched reluctance motors. IEEE Transactions on Energy Conversion. 2003;**18**(2): 284-290

[29] Chen H, Han G, Yan W, Lu S, Chen Z. Modeling of a switched reluctance motor under stator winding fault condition. IEEE Transactions on Applied Superconductivity. 2016;**26**(4)

[30] Cheok AD, Ertugrul N. Use of fuzzy logic for modeling, estimation, and prediction in switched reluctance motor drives. IEEE Transactions on Industrial Electronics. 1999;**46**(6):1207-1224

[31] Lachman T, Mohamad TR, Fong CH. Nonlinear modelling of switching reluctance motors using artificial intelligence techniques. IET Electric Power Applications. 2004; **151**(1):23-60

[32] Hamouda M, Gouda E, Amin ARA. Performance Analysis of Switched Reluctance Machine. Mansoura

University, Mansoura city, Egypt; 2015. pp. 1-171

[33] Meeker D. Finite Element Method Magnetics: OctaveFEMM. Version 1.2, User's Manual; 2010. 59p. Available From: http://www.femm.info/Archives/doc/octavefemm.pdf

[34] Ye J. Advanced Control Methods for Torque Ripple Reduction and Performance Improvement in Switched Reluctance Motor Drives. Hamilton, Ontario: McMaster University; 2014

Chapter 2

Switched Reluctance Motor Modeling and Loss Estimation Review

Pedro Sousa Melo and Rui E. Araújo

Abstract

Switched reluctance machines (SRM) are an alternative to conventional and permanent magnet (PM)-based machines. They are simple, robust and fault tolerant, and able to reach very high speeds with high efficiency. However, they operate with high torque pulsation and are noisy. Also, a nonconventional power converter type and specific control schemes must be included. Furthermore, SRM operation is characterized by high nonlinear features, which makes it difficult to be modeled and controlled. SRM energy conversion principles are a keystone to understand its operation. SRM efficiency increases with speed, where core and mechanical losses are more significant. For this machine, core loss estimation is a complex task, due to the nonlinear behavior of the magnetic materials. In addition, flux waveforms are not sinusoidal and particular waveforms appear in different core sections. Empirical formulas are usually considered in core loss estimation, but this is insufficient for SRM.

Keywords: SRM geometric and magnetic features, modeling, energy balance, torque production, loss characterization

1. Introduction

Switched reluctance machines (SRM) have been recognized as an alternative to conventional AC and DC motor drives. They have a multi-domain potential, such as electric vehicle, household appliances, aircraft industry, and servo system, to name a few examples [1]. They are simple, rugged, and fault-tolerant machines, with low cost. Moreover, they can reach very high speeds with high efficiency, since there are no windings or permanent magnets (PM) in the rotor [2]. A considerable attention to SRM drives has been given by both industry and research community, mainly due to the absence of PM. Limitations on rare-earth element availability and their increasing cost are the main reasons, since those elements are fundamental in PM composition [1]. Nonetheless, there are some drawbacks that need to be tackled: they have high torque pulsation and noisy operation. Also, a nonconventional power converter type with specific control schemes must be included. Furthermore, its efficiency (low speed) and power density are lower than in PM machines [3, 4].

SRM modeling is fundamental for two basic reasons: design and control development. In the first one, reducing prototype costs, while optimizing its design (e.g., better efficiencies with reduced torque ripple and noise), is a crucial goal [5]. On the

IntechOpen

other hand, coupling the SRM model with power converter and load models is fundamental for developing better control algorithms [1].

SRM operate under strong nonlinear conditions, where modeling can be a very challenging task. Finite element analysis (FEA) is a very accurate but time-consuming method. Analytical models and lookup tables are much faster but have reduced accuracy. Therefore, SRM modeling must be addressed as a trade-off between simulation accuracy and computational speed [6].

For any electrical machine, loss estimation is fundamental for design optimization and control. Furthermore, an accurate loss evaluation is vital for achieving the desired operation conditions [7]. Like in other electric machines, SRM losses occur in stator windings, core materials (stator and rotor), and mechanical moving parts. In high-speed operation, where SRM has better performances, core losses have a significant weight, together with the mechanical ones. Iron loss estimation in SRM is a complex task, mainly due to nonlinearities of the magnetic core. In addition, flux density waveforms are not sinusoidal, and different ones must be considered in several magnetic circuit parts, depending on the machine geometry [8]. Moreover, those waveforms are conditioned by control features.

This chapter is structured as follows: Section 2 starts with an overview on reluctance machines. Next, fundamental geometric and magnetic features, for conventional SRM, are addressed. Operation principles are discussed, highlighting the correlation between machine features and control parameters. In Section 3, energy conversion is deeply analyzed, based on a lumped parameter model. The focus is given to torque production mechanism. Section 4 addresses SRM losses, where common methods for SRM loss estimation are discussed. The impact of sinusoidal and nonsinusoidal flux density waveforms is also discussed.

2. SRM characterization

This section intends to present a general SRM overview. Several subjects are addressed, starting with the main advantages and drawbacks. Geometric and magnetic features, as well the operation principles, are also discussed.

2.1 Reluctance machines (RM) and SRM basic features

For every RM, torque production is a consequence of rotor movement to a position where the self-inductance of the excited winding is maximized (i.e., minimum magnetic reluctance). The motion may be rotary or linear, and the rotor may be interior or exterior [9]. This is a different principle from conventional machines, such as induction, synchronous (cylindrical rotor), or DC motors, where torque development results from the combination of two magnetic fields (stator and rotor).

Stator and rotor magnetic circuits are made of laminated soft magnetic iron. Generally, windings are located only in stator, consisting in several electrically independent phase circuits. They may be excited separately (e.g., SRM and stepper motor) or together (e.g., synchronous reluctance motor, SRM, and stepper motor). The rotor has no windings or permanent magnets, and its shape has a significant role in maximizing the phase inductance variation with its position. In fact, all these machines have in common a magnetic circuit with deep anisotropic features. The rotor simplicity is the main advantage when compared to permanent magnet or rotor winding machines. The manufacturing cost can be lower than in other types of motors; also, the reliability and robustness are improved and rotor cooling is not so

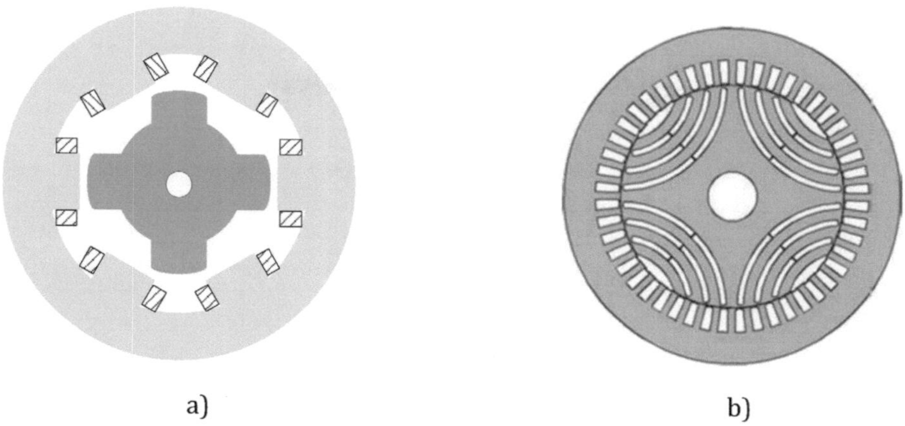

Figure 1.
(a) Three-phase SRM (6/4) and (b) Synch_RM.

critical [9]. **Figure 1** depicts the basic structure of SRM and synchronous reluctance motor (Synch_RM):

- SRM higher simplicity can be observed: stator winding coils are concentrated around its salient poles (N_s); usually, diametrically opposite windings are connected in series. Motor phases can be formed by one of these pairs or groups of pairs. The rotor geometry has also a set of salient poles (N_r)—in this case $N_s = 6$ (two poles/phase), while $N_r = 4$.

- Synch_RM has an AC conventional slotted stator, where the windings are sinusoidally distributed. **Figure 1(b)** represents a four-pole rotor with axially magnetic laminations.

Although SRM presents a similar structure when compared to variable reluctance (VR) stepper motors, there are important differences that should be highlighted: usually SRM have a small pole number, with a larger stepping angle. Most often, the rotor has a continuous movement, and its output power is much higher than in VR stepper motor [9].

In conventional SRM design, stator and rotor poles features are usually attached to the following conditions:

$$N_s > N_r \tag{1}$$

$$\beta_r \geq \beta_s \tag{2}$$

$$\varepsilon = \frac{1}{m} \cdot \frac{2\pi}{N_r} \tag{3}$$

$$\varepsilon < \text{Torque Zone} < \beta_s \tag{4}$$

$$m = \frac{N_s}{N_s - N_r} \tag{5}$$

where N_s and N_r are, respectively, the stator and rotor poles (even numbers); β_s and β_r are, respectively, the stator and rotor pole arcs; ε is the stroke angle, i.e., the rotor displacement due to a single pulse current; and m is the SRM phase number.

The denominator in Eq. (3) gives the total phase pulses/revolution, while Eq. (4) assures the motor can start, independently of rotor initial position. For an m-phase

m	N_s	N_r
3	6	4
3	12	8
4	8	6
5	10	8
6	12	10
7	14	12

Table 1.
SRM phase and poles configurations.

SRM, there are several (N_s; N_r) combinations, according to Eqs. (1) and (5). **Table 1** shows the most common configurations.

Usually, the average torque is higher for large N_s and N_r, with a smaller ripple. However, the number of power devices (switches and diodes) in the converter will increase. Therefore, the cost is higher, as well the switching losses. Conduction time is also higher, which increases copper losses. In addition, since phase current frequency (f_s) is given by $f_s = \omega_r{}^*N_r$ (ω_r: rotor angular speed), core losses will be higher. Nevertheless, torque ripple filtering becomes easier, due to its higher frequency [10].

2.2 SRM background and development

The SRM drives for industrial applications are relatively recent. Although the earliest recorded SRM is from 1838 [9], its actual development stage would not be possible without the semiconductor-based solid-state power switching technology in the early 1960s. This brought a huge interest in new machine development, as well in control methods, where the SRM is included.

In the last 40 years, different configurations of reluctance motor and switching methods have been investigated, both theoretically and experimentally. This is quite clear when the 67 worldwide registered patents on SRM before 1976 are compared to the 1755 ones, between 1976 and 1999 [11]. According to this reference, the total number of papers published in these periods increased from 11 to 1847, respectively. Presently, most significant SRM patents are the basis for its manufacture. The development of digital control and power electronics has made possible to exploit in an effective way the SRM characteristics [11].[1]

From the power conversion perspective, the SRM main disadvantage is related to the absence of permanent magnets and windings in the rotor: the machine power density is limited by the single stator excitation source [10]. Also, due to the magnetic circuit geometry, the machine is very sensitive to the effects of fringing fields and magnetic saturation. This introduces high nonlinear features that must be considered in SRM operation. A power converter and a controller must be integrated with the SRM, since phase currents are electronically commutated. The consequent pulsed magnetic fields have a major contribution in torque ripple and acoustic noise, which are the main drawbacks of SRM drives [12]. In order to tackle these drawbacks, research efforts have been focused on control strategies [13, 14], but machine design [15] and power converter structure [16] have also been addressed.

Another relevant SRM particularity is a small phase mutual inductance. In many applications it is considered null, i.e., each phase is magnetically independent from

[1] Chapter 2 of this reference includes an interesting and detailed overview on SRM historic development.

each other. This has a deep impact on the machine operation: since a phase fault has no influence in the other ones, SRM have a significant fault-tolerance capacity. As phase number increases, the smaller is the impact of a faulted phase. However, in high-speed operation, currents in adjacent phases can overlap for a significant portion of the conduction cycle. Therefore, the mutual flux linkages between phase windings can be most relevant, so their effects should be considered. This suggests that different model and control approaches must be addressed, whether SRM is operating in low or high speeds.

2.3 Geometric specifications

SRM phase inductance depends on both the current (due to saturation effect) and rotor position (θ). The unsaturated values of the aligned inductance (L_a) and the unaligned inductance (L_u) are fundamental in SRM basic theory, in particular the ratio L_a/L_u [9]. In addition, the machine geometric features have also a significant role, particularly the size and shape of stator and rotor poles, as well the air gaps. **Figure 2** depicts some of the most relevant geometric parameters, given by:

$$\tau_s = \frac{2\pi}{N_s} \tag{6}$$

$$\tau_r = \frac{2\pi}{N_r} \tag{7}$$

$$\theta_s = 2\pi\left(\frac{1}{N_r} - \frac{1}{N_s}\right), with\ N_s > N_r \tag{8}$$

The stator and rotor pole pitch are, respectively, τ_s and τ_r, where θ_s is the phase shift between successive stator phase inductances. It can easily be seen that θ_s = ε, by substituting Eq. (5) in Eq. (3).

2.4 Magnetic features

Figure 3 represents a set of magnetization curves, for one phase of an SRM, where each one is related to a different rotor position (θ). The aligned and

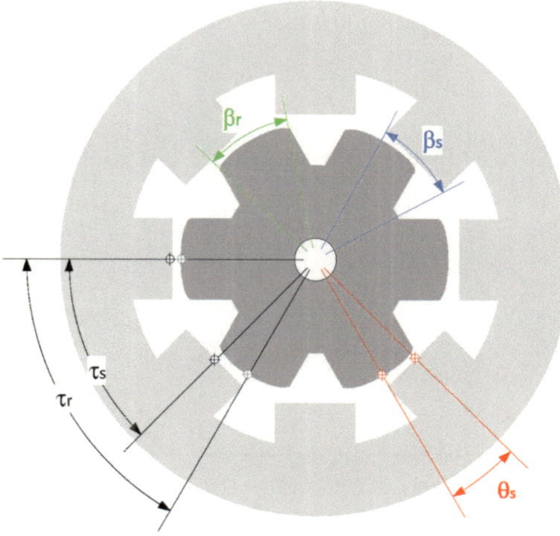

Figure 2.
SRM geometric features.

unaligned positions are the set limits. Also, the influence of magnetic saturation is visible.

Admitting uniform flux density, with no flux leakage and no saturation/hysteresis effect, the phase inductance can be approximated by:

$$L_s = N_s^2 \cdot A(\theta)\left[2 \cdot \left(\frac{l_p}{\mu_{fe}} \cdot \frac{A(\theta)}{A_p} + \frac{l_0}{\mu_0}\right) + \frac{l_y}{\mu_{fe}} \cdot \frac{A(\theta)}{A_y}\right]^{-1} \tag{9}$$

where: μ_{fe} and μ_0 are, respectively, the core material and the air magnetic permeability; A_p is the cross section of one stator and rotor magnetic pole (considering similar sections); A_y is the cross section of one yoke section (considering similar sections); l_p is the phase linkage flux longitudinal length in stator pole + rotor pole path; l_0 is the phase linkage flux longitudinal length in the air gap path; l_y, is the phase linkage flux longitudinal length in yoke path between two stator magnetic poles.

Under the previous conditions, for both the aligned and unaligned position, one has $A(\theta) = A_p$. The airgap term in Eq. (9) denominator is dominant, so l_0 has a profound impact in L_a and L_u. Moreover, since SRM anisotropic features play a fundamental role on its performance, maximizing L_a/L_u is crucial in SRM design. In order to increase L_a, the correspondent l_0 should be as small as possible. This highlights the fact that SRM aligned air gap must be tighter than for other machines. As a consequence, manufacture imperfection or rotor eccentricity will have a considerable negative impact in SRM operation [7].

From **Figure 3**, it can be seen that $L_s(\theta)$ gradient is more relevant in the overlapping region, meaning that this is the most relevant torque production zone. In the overlapping regions, L_s varies approximately linearly with θ. Thus, it is reasonable to admit $A(\theta)$ as:

$$A(\theta) = l \cdot R \cdot \theta, 0 < \theta < \beta_r \tag{10}$$

where $l \approx$ machine's axial length; R $\approx$ rotor pole radius; β_r rotor pole arc.

Nonetheless, it should be noted that $A(\theta)$ is in fact much dependent on the machine pole shape.

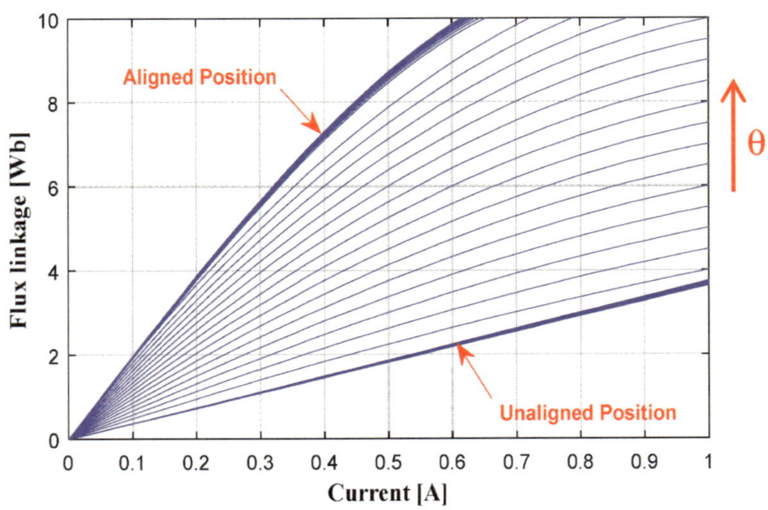

Figure 3.
SRM magnetic characteristics (single-phase) for different θ.

2.5 SRM operation fundamentals

Torque production conditions can be seen in **Figure 4**, where an elementary single-phase machine is depicted, for different rotor positions.

In (a), rotor poles are perfectly aligned with the stator phase poles, where its inductance is maximum. Thus, no torque is produced. The rotor unaligned position is depicted in (d): its interpolar axis is aligned with the stator poles. The torque has a null value, since the rotor poles are displaced in an equivalent position related to the stator poles (here the phase inductance is minimum, since magnetic reluctance is maximum). As the rotor moves from the unaligned position and comes closer to any stator pole, the torque is no longer zero and tries to align the rotor with that particular stator pole: in (b) and (c), the developed torque tends to bring the rotor back to (a) position—this is a stable equilibrium position, while (d) is an unstable equilibrium position [9]. Torque direction is contrary in (b) and (c), which highlights the fact that switching currents in stator windings must be implemented according to the rotor position—it becomes clear that stator phases must be energized for rotor position between maximum and minimum phase inductance. This elementary machine highlights three important SRM general features [9, 10]:

i. In order to avoid situations (a) and (d) in **Figure 4**, all SRM verify the condition: $N_s \neq N_r$. In those positions the motor is not able to start when stator phase is excited.

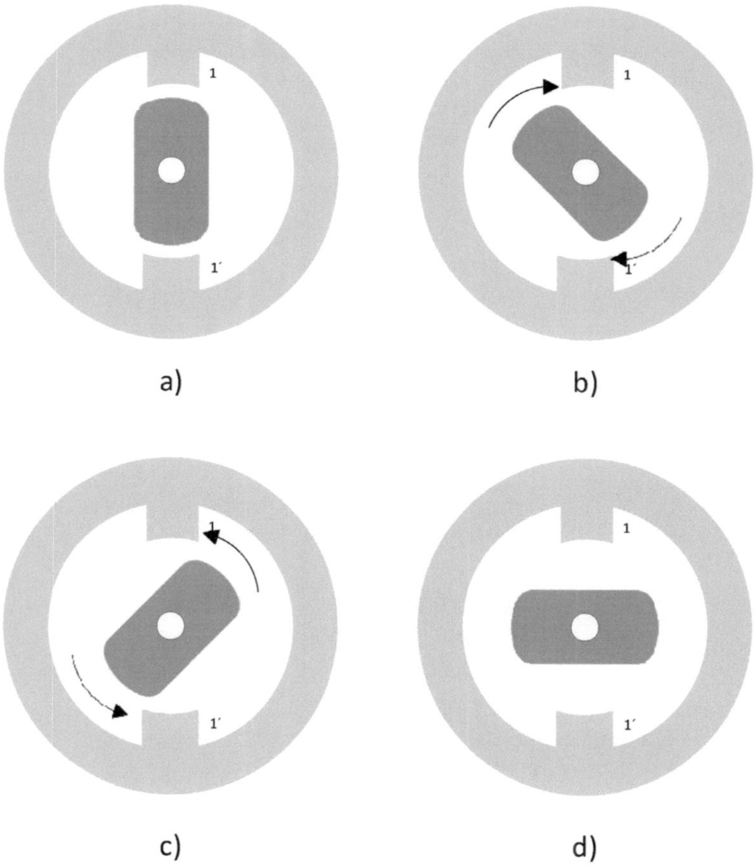

Figure 4.
Rotor positions of an elementary SRM.

ii. Torque is produced by sequentially switching the currents in the stator phase windings when there is a variation of reluctance, which depends on rotor position. This is the reason why a controlled power converter supply system is mandatory with rotor position feedback. Therefore, SRM operation is always implemented with a converter + controller (SRM drive).

iii. Winding current polarity has no influence in torque development, since this is a reluctance torque.

The fundamental concepts in correlating SRM rotor position to its control are depicted in **Figure 5**. Phase current (I(θ)) is represented for both motor and generator SRM modes, as well the torque production zones (T(θ)). Except for the red $L_s(\theta)$ curves which denote saturation effect, all the others assume ideal conditions (no saturation and hysteresis; ohmic winding resistance is zero).

It can be seen that L_s is a periodic function of θ, with period τ_r.[2] The following expressions can be considered:

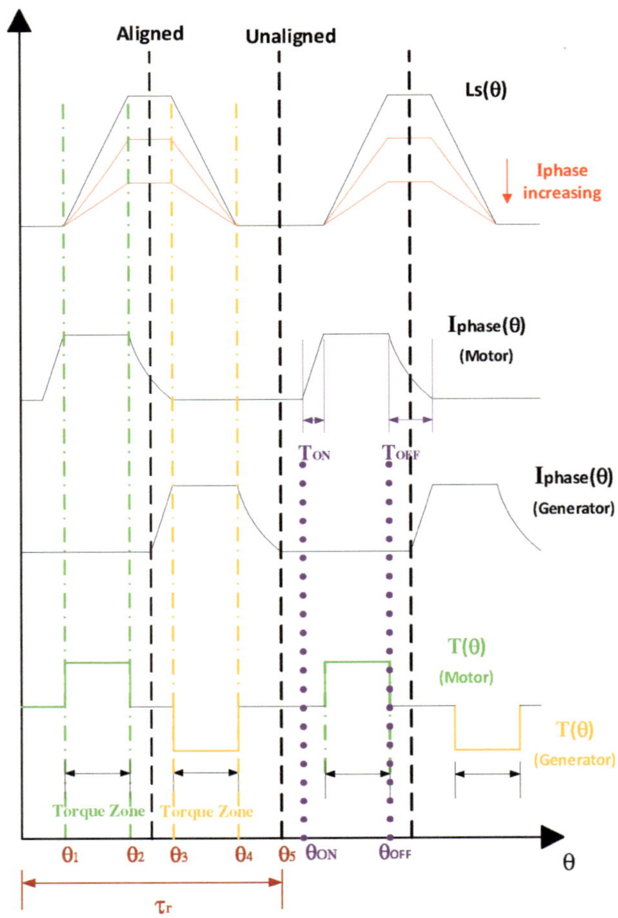

Figure 5.
L_s profile, current pulses (motor and generator), and torque zones.

[2] $L_s(\theta)$ origin is at the unaligned rotor position.

$$\theta_1 = \frac{\pi}{N_r} - \left(\frac{\beta_s + \beta_r}{2}\right) \tag{11}$$

$$\theta_2 - \theta_1 = \beta_s \tag{12}$$

$$\theta_3 - \theta_2 = \beta_r - \beta_s \tag{13}$$

$$\theta_4 - \theta_3 = \beta_s \tag{14}$$

$$\theta_5 - \theta_4 = \frac{\pi}{N_r} - \left(\frac{\beta_s + \beta_r}{2}\right) \tag{15}$$

The instantaneous torque is controlled by current amplitude regulation and the switching instants (θ_{ON} and θ_{OFF}). In addition, rotor speed has also a significant impact on control modes.

The $L_s(\theta)$ top flat zone is related to the difference between the stator and rotor pole arcs, respectively, β_s and β_r: as it tends to zero, the top flat becomes narrower. Although this is a "dead zone" since there is no energy conversion (torque is null), it might help to avoid braking torque in the demagnetizing period (T_{OFF}). It should be emphasized the unidirectional current polarity in both modes. Torque development is clearly associated to $dL_s/d\theta \neq 0$ (i.e., the overlapping poles regions); it is associated to motor (+) and generator (−) modes. This stresses the fact that current impulse rise (T_{ON}) and fall (T_{OFF}) periods have a fundamental role in SRM operation features: T_{ON} must be as small as possible, while current should reach its maximum when stator and rotor poles start to overlap. As for T_{OFF} (the defluxing period), it depends on how fast the energy stored in the linkage magnetic field can be turned to zero. Therefore, T_{OFF} limits must assure a zero current at the beginning of the $dL_s/d\theta < 0$, in order to avoid a braking torque. These are important issues that must be addressed by the drive controller.

The SRM core magnetic features, together with pole geometry, have a fundamental role in the $L_s(\theta)$ profile. This is crucial in estimating θ_{ON} and θ_{OFF} values. Moreover, magnetic characteristics are particularly influenced by local saturation in stator and rotor poles, which also depends on SRM operation conditions.

Phase current commutation, in addition to SRM geometry, is responsible for torque ripple components. This contributes for increasing the complexity demanded to the SRM drive controller.

In short, the phase current pulse waveform as a function of θ, in addition to θ_{ON} and θ_{OFF}, plays a keystone role in all SRM operation modes. Therefore, the machine must be integrated with the power converter and the controller, working as a single system.

3. SRM modeling and energy balance

SRM can be modeled in a similar way as for conventional electrical machines. Therefore, a model based on lumped parameters can be a first option. However, depending on the purpose (e.g., design or control), particular attention should be addressed when applying conventional methods for modeling this machine, due to its specific nonlinear operation features.

Nonetheless, this section intends to address the SRM energy conversion principles, where a lumped parameter approach is well suited. The mathematical model is based on the following equations:

- Voltage equation

$$V_{phase} = R_s \cdot i + \frac{d\Psi(i,\theta)}{dt} \tag{16}$$

where R_s, i, and $\Psi(i,\theta)$ are, respectively, the stator phase resistance and current and flux linkage.[3]

- Magnetic equation

$$\Psi(i,\theta) = L_s(i,\theta) \cdot i \tag{17}$$

where $L_s(i,\theta)$ is the stator phase self-inductance.

- Electromechanical equation (instantaneous torque)

$$T_e = \sum_{j=1}^{m} T_j(i,\theta) \tag{18}$$

where T_j is the torque developed by phase j and m is the SRM phase number.

- Mechanical equation

$$T_e - T_{load} = J\frac{d\omega r}{dt} + K_f \omega r \tag{19}$$

where T_{load} is the mechanical load, J is the system inertia, ωr is the instantaneous rotor mechanical angular speed, and K_f is the viscous friction coefficient, due to bearing lubricant fluid.

It should be noted that magnetic feature complexity is not entirely depicted in **Figure 3** (flux density waveforms in different SRM sections cannot be addressed). The high complexity of SRM modeling lies in Eq. (17), which has a significant impact on torque production, since it depends on the magnetic anisotropic features. In order to take a deeper insight on Eq. (18), the torque production mechanism is now analyzed with more detail.

3.1 Electromechanical energy conversion and torque production

Since SRM are electromechanical energy converters, a review on energy balance relationships is first addressed. An electromechanical system is based on an electrical system, a mechanical system, and a coupling field (i.e., electromagnetic and electrostatic fields), common to both systems. From this interaction, energy can be transferred from one system to the other. It should be noted that both coupling fields can exist simultaneously, while the electromechanical system may have several electrical and mechanical subsystems. Different losses by heat dissipation occur in the electrical and mechanical systems, as well in the coupling field. Several phenomena contribute to this [17]:

- Electrical losses, due to current-carrying conductor ohmic resistance
- Mechanical losses, due to friction of moving parts

[3] Phase mutual influences and leakage flux are neglected.

- Magnetic losses in the core ferromagnetic materials (magnetic fields) and dielectric losses (electric fields)

Figure 6 depicts the energy balance in an elementary electromechanical system—according to the energy conservation principle (first thermodynamic law)—where losses and stored energy components are represented.

The black dotted arrows represent the adopted convention for positive W's. The energy balances for the electric and mechanical systems are then expressed as:

$$W_E = W_{eL} + W_{eS} + W_{ef} \tag{20}$$

$$W_M = W_{mL} + W_{mS} + W_{mf} \tag{21}$$

where *WE* is the energy supplied by the electrical source, *WeL* is the electrical losses, *WeS* is the energy stored in the fields (electric or magnetic) not coupled with the mechanical system, and *Wef* is the energy transferred to the coupling field from the electrical system. In a similar way, *WM* is the energy supplied by the mechanical source, *WmL* is the mechanical losses, *WmS* is the energy stored in the moving parts of the mechanical system, and *Wmf* is the energy transferred to the coupling field from the mechanical system. The colored arrows represent the physical energy flux: except for the losses (irreversible phenomena), all the W's flux can be reversible, depending on the system operation. From **Figure 6** it can be seen that:

$$W_{ef} + W_{mf} = W_f + W_{fL} \tag{22}$$

where *Wf* is the energy stored in the coupling field and *WfL* is the loss energy within the coupling field. Combining Eq. (20) and Eq. (21) gives:

$$W_E + W_M = (W_{eL} + W_{mL}) + (W_{eS} + W_{mS}) + \left(W_{ef} + W_{mf}\right) \tag{23}$$

From this point, only electromagnetic coupling field are considered. Rearranging Eq. (23) gives:

$$W_E = (W_{eL} + W_{eS}) + \left(W_{ef} + W_{mf}\right) + (-W_M + W_{mL} + W_{mS}) \tag{24}$$

The magnetic coupling field will be addressed as a conservative field, in order to take advantage of a fundamental property: the energy stored in a conservative field depends only on state variable values, and not on the transition between states. The sum of potential and kinetic energy is constant; therefore, the losses are null (i.e., $W_{fL} = 0$). It should be noted that, for the purpose of this analysis, such approach is quite acceptable, since the most significant part of the stored energy in the coupling

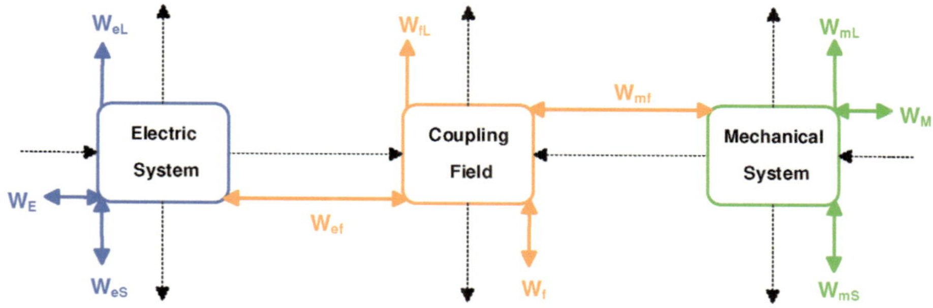

Figure 6.
Energy flux (colored arrows) in an electromechanical system.

fields is in the air gaps of the electromechanical device. Air is a conservative medium; thus the energy stored there can be returned to the electrical or mechanical systems. Finally, Eq. (24) can be written as:

$$W_E = W_{eL} + W_{eS} + W_f - W_{mf} \tag{25}$$

where:

$$W_f = W_{ef} + W_{mf} \tag{26}$$

$$-W_{mf} = -W_M + W_{mL} + W_{mS} \tag{27}$$

The voltage equation for a generic electric system (one phase) is given by Eq. (16). For convenience of analysis, it is written again as:

$$v = r \cdot i + \frac{d\Psi(i,\theta)}{dt} \tag{28}$$

Therefore, Eq. (25) can be expressed as:

$$\int v \cdot i \cdot dt = \int r \cdot i^2 \cdot dt + \int \frac{d\Psi(i,\theta)}{dt} \cdot i \cdot dt \tag{29}$$

Since $\Psi(i,\theta)$ is the flux linkage, the effect of leakage inductance is neglected. Therefore, $W_{eS} = 0$ and the following differential equation can be extracted:

$$dW_E = dW_{eL} + dW_f - dW_{mf} \tag{30}$$

$$dW_E = v \cdot i \cdot dt \tag{31}$$

$$dW_{eL} = r \cdot i^2 \cdot dt \tag{32}$$

$$dW_f - dW_{mf} = d\Psi(i,\theta) \cdot i \tag{33}$$

$$dW_{mf} = -T \cdot d\theta \tag{34}$$

It is important to stress that $\Psi(i,\theta)$ is a single-value function, since it represents a conservative field. Therefore, hysteresis is not included here, but saturation can be considered—this is the reason for having "i" as an independent variable. With θ as the other independent variable, one has:

$$d\Psi(i,\theta) = \frac{\partial\Psi(i,\Theta)}{\partial i} di + \frac{\partial\Psi(i,\Theta)}{\partial\theta} d\theta \tag{35}$$

Substituting this in Eq. (33) yields:

$$\underbrace{dW_{ef} + dW_{mf}}_{dW_f} - dW_{mf} = \frac{\partial\Psi(i,\Theta)}{\partial i} idi + \frac{\partial\Psi(i,\Theta)}{\partial\theta} id\theta =$$

$$= \underbrace{\frac{\partial\Psi(i,\Theta)}{\partial i} idi + [1-\alpha(i,\theta)] \cdot \frac{\partial\Psi(i,\Theta)}{\partial\theta} id\theta}_{dW_f} \cdot i + \underbrace{\alpha(i,\theta) \cdot \frac{\partial\Psi(i,\Theta)}{\partial\theta} id\theta}_{-dW_{mf}} \tag{36}$$

The function $\alpha(i, \theta)$ allows a general formulation for infinitesimal variations in the coupling field stored energy and the mechanical energy. However, analytical expressions for $\alpha(i, \theta)$ are difficult to get. At this point, it is convenient to define the coenergy function (W_c) as:

$$W_c(i,\theta) = \int \Psi(i,\theta) \cdot di \tag{37}$$

Under this condition, $W_f(i,\theta)$ is given by:

$$W_f(i,\theta) = \int i \cdot d\Psi \tag{38}$$

Both $W_c(i,\theta)$ and $W_f(i,\theta)$ are represented in **Figure 7** (i: $0 \rightarrow i_1$, θ = *constant*). From **Figure 7**, it can be seen that (for i = *constant*).

$$\Psi(i,\theta) \cdot i = W_f(i,\theta) + W_c(i,\theta) \tag{39}$$

Taking Eq. (39) θ derivative, one has:

$$\frac{\partial\Psi(i,\theta)}{\partial\theta} i = \frac{dW_f(i,\theta)}{d\theta} + \frac{dW_c(i,\theta)}{d\theta} \tag{40}$$

In order to take an insight over the torque production in an SRM, the analysis is now addressed taking di = 0. Under this condition, Eq. (36) is written as:

$$\frac{\partial\Psi(i,\theta)}{\partial\theta} i d\theta = dW_f - dW_{mf} \tag{41}$$

$$\frac{\partial\Psi(i,\theta)}{\partial\theta} i = \frac{dW_f(i,\theta)}{d\theta} - \frac{dW_{mf}(i,\theta)}{d\theta} \tag{42}$$

Finally, from Eq. (40), Eq. (41), and Eq. (34), the torque developed by a single phase (T) is given by:

$$T = \frac{dW_c(i,\theta)}{d\theta}, for\ i = constant \tag{43}$$

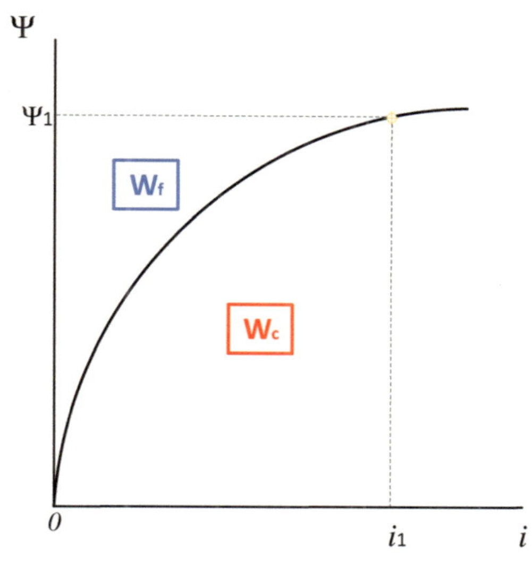

Figure 7.
Stored energy and coenergy in a magnetic field.

For a generic SRM with m phases, the total electromagnetic torque is:

$$T_e = \sum_{j=1}^{m} \frac{\partial W_c^j(i,\theta)}{\partial \theta}, \quad i = constant \tag{44}$$

where W^j_c is the coenergy related to the stored magnetic field of phase j and m is the SRM phase number. This highlights the fact that torque production is in close relation with the core magnetic properties, for each rotor position (θ). The average torque (T_{av}) can be calculated by integrating T_e:

$$T_{av} = \frac{1}{\varepsilon} \int_0^{\varepsilon} T_e \cdot d\theta \tag{45}$$

A geometric representation may help to clarify how energy is transferred between the electrical domain, the magnetic coupling field, and the mechanical domain. **Figure 8** includes two magnetic characteristics for two rotor positions, θ_a and θ_b.

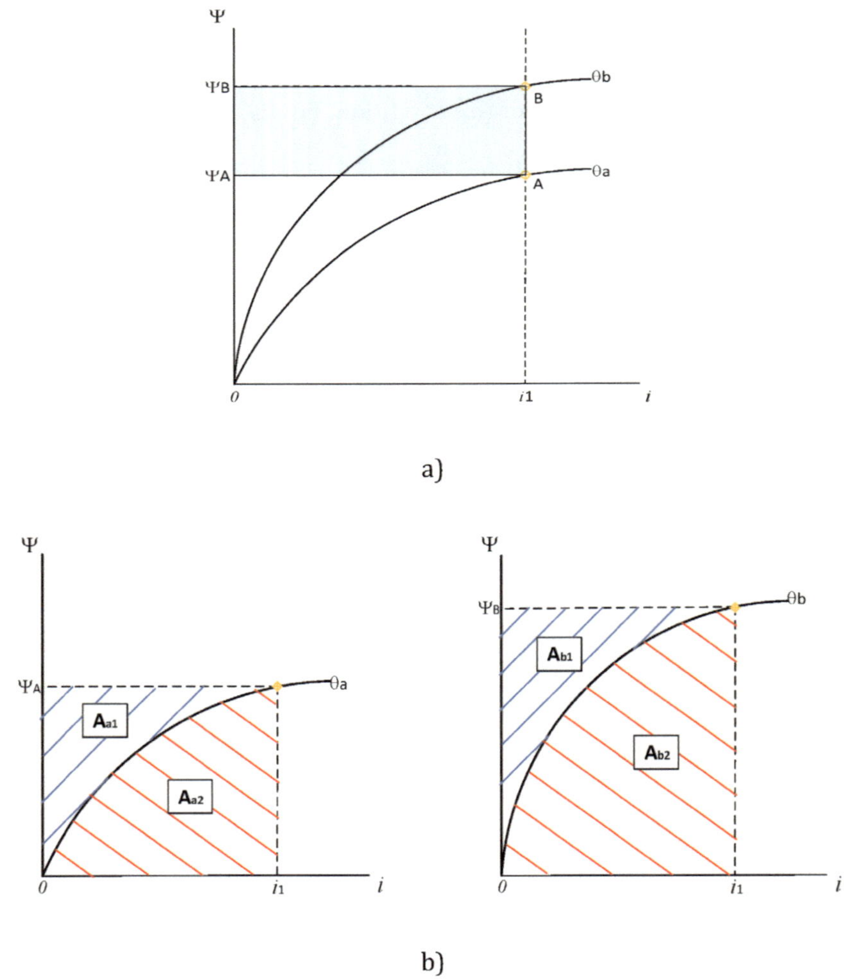

Figure 8.
(a) Magnetic characteristics for θ_a and θ_b (i = constant). (b) Stored magnetic and coenergy distribution for θ_a and θ_b (i = constant).

For $i = i_1$, $\Delta W_{source}(i_1, \theta) = i_1 \cdot (\Psi_B - \Psi_A)$ is the energy transferred to the magnetic coupling field from the electrical source, for a rotor position displacement from θ_a to θ_b (A → B in **Figure 8(a)**. This is related to the difference between rectangles $[0\text{-}i_1\text{-}B\text{-}\Psi_B]$ and $[0\text{-}i_1\text{-}A\text{-}\Psi_A]$ (i.e., the blue rectangle). From **Figure 8(b)**, it can be seen that this difference is given by:

$$\underbrace{(A_{b1} - A_{a1})}_{\Delta W_{stored\_coupling\_field}(i_1, \theta_A \to \theta_B)} + \underbrace{(A_{b2} - A_{a2})}_{\Delta W_c(i_1, \theta_A \to \theta_B)} \tag{46}$$

(A_{a2} and A_{b2} are the coenergy (W_c) related to, respectively, θ_a and θ_b.)
Therefore:

$$\Delta W_{source}(i_1, \theta_A \to \theta_B) = \Delta W_{stored\_coupling\_field}(i_1, \theta_A \to \theta_B) + \Delta W_c(i_1, \theta_A \to \theta_E) \tag{47}$$

For an infinitesimal energy change ($\theta_A \to \theta_A + d\theta$):

$$dW_{source}(i_1, \theta) = dW_{stored\_coupling\_field}(i_1, \theta) + dW_c(i_1, \theta) \tag{48}$$

which is similar to Eq. (41).

3.2 Linear magnetic circuit

If no magnetic saturation exists, the magnetization curves are straight lines (for a fixed θ), as represented in **Figure 9**.

Since $\Psi(i, \theta) = L_s(\theta) \cdot i$, from Eq. (35), one has:

$$d\Psi(i, \theta) = L_s(\theta) \cdot di + i \cdot \frac{dL_s(\theta)}{d\theta} \cdot d\theta \tag{49}$$

Therefore, from Eqs. (30) and (33), one has:

$$dW_{source}(i, \theta) = i \cdot L_s(\theta) \cdot di + i^2 \cdot \frac{dL_s(\theta)}{d\theta} \cdot d\theta \tag{50}$$

where $dW_{source}(i, \theta) = dW_E - dW_{el}$.

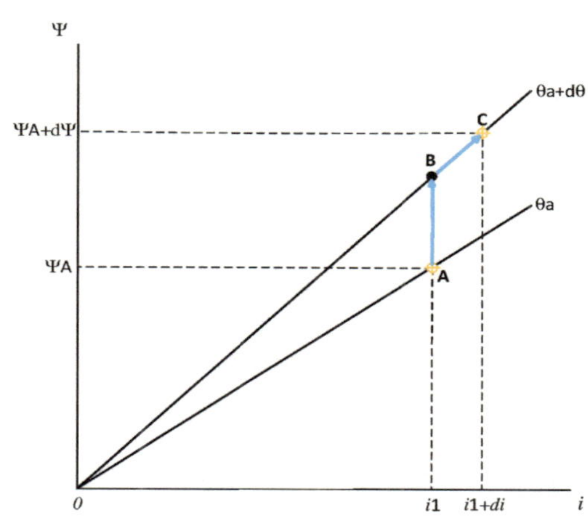

Figure 9.
Magnetic characteristics for θ_a and $\theta_a + d\theta$.

The first and second terms of the second member in Eq. (50) are attached to, respectively, the line segments BC and AB. From Eq. (36), one has:

$$dW_{source}(\mathrm{i},\theta) = i \cdot L_s(\Theta) \cdot di + (1-\alpha(\mathrm{i},\theta)) \cdot i^2 \cdot dL_s(\Theta) + \alpha(\mathrm{i},\theta) \cdot i^2 \cdot dL_s(\Theta) \quad (51)$$

From **Figure 9**, it can be seen that:

$$\alpha(\mathrm{i},\theta) = \frac{1}{2} \quad (52)$$

In other words, for *i =constant*, the energy transferred to the magnetic coupling field from the electrical source, as a result of a rotor displacement, is evenly distributed by the stored energy variation in the coupling field and by the coenergy variation (i.e., mechanical energy variation). Finally, Eq. (51) is written as:

$$dW_{source}(i,\theta) = \underbrace{i \cdot L_s(\theta) \cdot di + \frac{1}{2} i^2 \cdot dL_s(\theta)}_{dW_{stored\_coupling\_field}(i,\theta)} + \underbrace{\frac{1}{2} i^2 \cdot dL_s(\theta)}_{dW_c(i,\theta)=dW_{mechanical}(i,\theta)} \quad (53)$$

(53)

From Eq. (34), the instantaneous torque for a rotor position θ is given by:

$$\mathrm{T} = \frac{1}{2}\frac{dL_s(\Theta)}{d\Theta} \cdot i^2 \; (i = constant) \quad (54)$$

Some Section 2.5 observations are now more clear from this linear mathematic torque formulation:

- Phase current polarity has no influence in torque production. This is an important feature, since phase current pulses (and flux linkage) can be unipolar. In addition to a simpler controller (compared to other machines drives), iron losses are also smaller.

- Torque development is associated to the growth (+) or decline (−) of stator phase inductance curve ($L_s(\theta)$), respectively, for motor and generator modes.

4. SRM loss characterization

As for any electric machine, SRM models must include losses, since its performance is quite dependable on it. Moreover, all losses should be accurately estimated for different operation modes [18]. Compared to conventional AC machines, this is the most demanding task, particularly for the core losses, because flux density waveforms are not sinusoidal [19]. Similar to other machines, losses in an SRM are related to copper, iron core, and mechanical ones. The machine geometry and operation conditions make SRM loss characterization a nontrivial task.

4.1 Overview on developed SRM loss estimation

Different methods for core loss evaluation in SRM have been investigated by several authors. Using finite elements methods, flux density waveforms are obtained by simulation in [8]. The iron losses were then calculated based on the modified Steinmetz equation. In [18] an analytical method for calculating SRM core losses is proposed. The machine magnetization curves are extracted through finite element simulation, considering different control strategies. The flux density waveforms in the different parts of the SRM are derived from the flux density waveform of the stator pole, obtained by simulation. The specific core losses are separated into hysteresis, classical eddy current, and excess losses. Their calculation is based on the waveforms and time derivatives of the flux density in each zone. In [20], SRM efficiency is evaluated for continuous operation mode (i.e., current does not extinguish during each phase excitation), but only core losses are considered with more detail (copper losses are calculated through the current *rms* value and the DC phase resistance). Fourier flux density decomposition in the machine core sections is then applied, where core losses are calculated with the modified Steinmetz equation. In order to consider the minor loop influence, a correction factor is included. In [21] a SRM efficiency map is derived based on Fourier flux density decomposition. However, only copper and hysteresis losses are considered. In [22] core losses were calculated through simulation, based on analytic magnetic circuits, considering six different core materials. This approach was validated by comparing the simulation results to experimental ones. However, the paper does not provide information about the magnetic circuit approach.

In the following, an overview on iron loss estimation is addressed, from an engineering perspective. The evolution of the most relevant methods, as well as their merits and limitations, is discussed. Except for [22], the approaches mentioned in the previous paragraph are included here. In the last decades, considerable efforts have been addressing these issues. Nonetheless, further improvements are needed for a precise determination of the flux waveforms and correct calculation of iron loss at nonsinusoidal excitation and nonuniform flux distribution, for a wide range of operation scenarios. This is still an open research field.

As already stated, the focus will be given to iron loss. However, copper losses are usually more significant. Therefore, they are first addressed.

4.2 Copper losses

In the presence of nonuniform electric field (E) and current density (J) distributions, the power dissipated in an arbitrary volume (V) of a conductor is given by:

$$P = \iiint_V \mathbf{E} \cdot \mathbf{J} \, dV \tag{55}$$

The vector (Ohm's law) is expressed as[4] (vectors $\mathbf{E}, \mathbf{J}$):

$$\mathbf{E} = \rho \cdot \mathbf{J} \tag{56}$$

Thus, P may be formulated as follows:

$$P = \iiint_V \rho \cdot J^2 \, dV \tag{57}$$

where ρ is the conductor material resistivity (Ω·m). For uniform E and J distributions, Eq. (57) can be written as:

$$P = P_{DC} = R_{DC} \cdot I^2 \tag{58}$$

$I = J{\cdot}A$ and $R_{DC} = \rho{\cdot}l/A$ (R_{DC} is the ohmic resistance of the conductor volume (V), where l is the length of the current (I) path inside V and A is the cross-sectional area.

[4] Valid for homogeneous and isotropic materials.

An electrical machine winding is made of several coils, where different phenomena may have an impact on winding losses: ρ increases with temperature ($temp$), while the skin (k_{skin}) and proximity (k_{prox}) effects have a relevant impact on J distributions, particularly in high frequencies. All these effects should be included for an accurate winding loss characterization, meaning that Eq. (57) must be considered. For a lumped parameter model approach, as in Section 3, the previous effects are addressed by:

$$P_{AC} = R_{AC} \cdot I^2 \tag{59}$$

where R_{AC}=R_{DC} $(temp, k_{skin}, k_{prox})$.

Therefore, SRM total copper losses can be calculated based on Eq. (59), as the sum of all phase winding losses. Since phase current waveforms are periodic and not sinusoidal (with high frequency), Fourier series may be applied in order to get its harmonic components [23]. It follows that the total winding losses are calculated as:

$$P_w = m \sum_i R_{AC\_i} I_{i\_rms}^2 \tag{60}$$

where m is the phase number and R_{AC_i} and I_{i_rms} are, respectively, the phase AC resistance for the i-order harmonic and the current rms of the i-order harmonic.

In short, copper loss estimation is a relatively simple task, using the stator current. However, the skin and proximity effects require special attention, particularly for high-frequency components. Furthermore, for similar reasons discussed in the next section, Fourier analysis should be carefully applied. Particularly for high loads, temperature influence must also be included.

4.3 Core losses

Usually, lamination manufacturers provide core loss data under sinusoidal excitations in a limited frequency and flux density range. However, this kind of data is not adequate for predicting losses in electrical machines with nonsinusoidal flux waveforms. This requires loss information at high frequencies and high flux densities, particularly for high-speed operation, as in SRM [24].

The development of nonsinusoidal flux density machines and conventional machines fed by power converters (e.g., SR, brushless DC machines, and induction motors with PWM voltages) has motivated researcher's efforts to reach a deeper characterization of magnetic losses under such excitation waveforms. This can be done either by measurement or estimation [19, 24]. New challenges arise for both approaches, since sinusoidal-conventional methods are clearly insufficient. Moreover, since different flux density waveforms are related to particular electric machine configurations, this suggests that specific approaches for characterizing core losses must be addressed, according to the machine type [24].

SRM core loss modeling is a challenging task, since the nonsinusoidal flux density waveforms depend on motor design (geometry and number of stator and rotor poles, yoke geometry, number of phases), operating conditions (conduction angle and mechanical speed influence the magnetic saturation and lamination skin effect level), and the type of control used [18]. Core losses are more significant as speed increases, so for applications like hybrid and electric vehicles, they must be carefully addressed.

Different approaches have been proposed for calculating SRM core losses, from theoretical supported empirical models[5] to physical mathematical models [25]. The later are more accurate, since they take an insight on the core loss physics, aiming to

[5] In literature, they are referred as phenomenological models.

describe its nonlinear mechanism under distorted flux [20]. However, they are complex and computationally heavy. From an engineering perspective, empirical models are the preferred ones, due to its simplicity and faster processing. Parameter estimation is based on curve-fitting methods, validated by manufacturer iron sheet data and experimental results or through finite element modeling [20, 24]. Accuracy is much sensitive to parameter values, so their estimation must be attached to specific conditions (e.g., flux density and frequency ranges). Moreover, the manufacturing process of the machine has also a deep impact on core losses; this is very difficult to address in the parameter estimation process [25, 26]. Cutting and punching operations have a relevant influence in the material properties, since they can create inhomogeneous stresses inside the sheets. This depends on the alloy composite, whereas the grain size in the sheets seems to be the main influencing factor [25].

4.3.1 Core loss estimation

Electrical machine core losses can be addressed by three different approaches, in time or frequency domain: empirical equations, loss separation components, and hysteresis models. Only the first two are discussed here.

4.3.1.1 Empirical models based on Steinmetz equation

The Steinmetz coefficients depend on both frequency and flux density [27], so in a waveform with relevant harmonics, it might be difficult to find their values. Based on results of many tests, the classical Steinmetz equation was the first attempt to calculate core loss [28]:

$$P_{core} = C_m f^{\alpha} B_{max}^{\beta} \tag{61}$$

where B_{max} is the peak value of the flux density at the lamination, f = 1/T is the remagnetization frequency (T is the hysteretic cycle time interval), and coefficients Cm, α, and β are estimated by fitting the loss model to the lamination manufacturer or measured data. It must be pointed that Eq. (61) assumes sinusoidal flux densities, with uniform distribution across the lamination thickness. Over the years, several upgrades were performed, in order to account for nonsinusoidal waveforms.

The modified Steinmetz equation (MSE) aims to calculate core loss under arbitrary B waveforms. The macroscopic remagnetization rate dM/dt (which is proportional to dB/dt) is directly related to the core losses, as a consequence of wall domain motion [29]. Eq. (61) is then replaced by:

$$P_{core} = \left(C_m f_{eq}^{\alpha-1} B_{max}^{\beta}\right) f \tag{62}$$

with

$$f_{eq} = \frac{2}{\Delta B^2 \pi^2} \int_0^T \left(\frac{dB}{dt}\right)^2 dt \tag{63}$$

where $\Delta B = B_{max}\text{-}B_{min}$.

The MSE has the advantage to highlight the physical origin of the losses, with the same parameters as in Eq. (61). A disadvantage is that it underestimates losses for waveforms with a small fundamental frequency part. Another difficulty is the treatment of waveforms with multiple peaks, in which peak-to-peak amplitude is not an enough description [29]. In fact, this reflects the increasing MSE limitations, as the flux density moves away from the pure sinusoidal waveform.

The generalized Steinmetz equation (GSE) was also developed from the original Steinmetz equation [29], where losses are calculated in time domain. Here, the instantaneous iron loss is assumed to be a single-valued function of the rate dB/dt and B(t). The inclusion of B(t) allows to consider the DC bias influence in the loss calculation, without additional measurements or curve-fitting functions:

$$P_{core} = \frac{1}{T}\int_0^T C_m \left|\frac{dB}{dt}\right|^{\alpha} |B(t)|^{\beta-\alpha} dt \quad (64)$$

For different frequency ranges, different parameter values are necessary. Thus, in the presence of relevant harmonics, the accuracy decays. This is an important drawback, particularly in the presence of minor loops. To overcome this limitation, in [30] the flux density waveform is split into a major loop and minor loop(s), in order to consider the later one effect over loss calculation. Nonetheless, DC bias influence is no longer included. This approach was named improved GSE (iGSE). In [31] a test for loss calculation in nanocrystalline materials, with nonsinusoidal excitation voltage, is reported. Harmonic decomposition was considered, including the phase displacements. According to this reference, iGSE gives better results than GSE.

Both MSE and GSE are applied in time domain. It must be pointed out that the time evolution (history) of the flux density waveform is neglected. This has an impact on the physical phenomena insight, where loss evaluation may be affected.

4.3.1.2 Loss separation method

For a general scenario, the hysteresis loss density related to one cycle, in a particular core part, can be calculated by the following expression:

$$P_h = \frac{f}{m_v} \oint_{B_{min}}^{B_{max}} H dB \quad (65)$$

where *B* and *H* are, respectively, the magnetic flux density and the magnetic flux strength, *f* is the cycle frequency, and m_v is the density of the ferromagnetic material[6]. The eddy current loss density is derived from Maxwell equations, which is given by:

$$P_e = \frac{\sigma}{m_v} E^2 \quad (66)$$

where σ is the material electric conductivity[7]. However, in many situations, these equations are unpractical, even with finite element analysis. The complex nonlinear B(H) characteristic, which is also dependent on the lamination thickness, is the main reason [32]. So, empirical models for core loss evaluation are often considered. The most common is the Steinmetz equation [28]: the following expression for core loss density in a ferromagnetic material (formulated as the sum of hysteresis (P_h) and classic eddy current (P_e) losses) was first achieved:

---

6 For f [Hz] and m_v [kg/m^3], then P_h [W/kg].

7 For E [V/m] and σ [S/m], then P_e [W/kg].

$$P_{core} = P_h + P_e = k_h f B_{max}^{1,6} + k_e f^2 B_{max}^2 \tag{67}$$

which is valid in the range of 0.1 T < B_{max} < 1.5 T. Hysteresis and eddy current loss coefficients are, respectively, k_h and k_e, which can be extracted from measured data. They both depend on the core material; k_e also depends on the lamination thickness (d). In fact, k_e has an analytical formulation, which can be derived from Eq. (66):

$$P_e = \frac{\sigma \pi^2 d^2}{6 m_v} f^2 B_{max}^2 \tag{68}$$

An important remark is that Eq. (65) assumes uniform flux density distribution across the lamination thickness. It should be noted that for SRM this is a most relevant limitation. Over the years, Eq. (67) has been upgraded, which brought higher accuracy. Experimental data showed that the measured eddy current losses are higher than P_e. Based on statistical loss theory, Bertotti proposed an additional term to account for the excess losses (P_{ex}), which can be expressed as [33]:

$$P_{ex} = k_{ex} (f \cdot B_{max})^{\frac{3}{2}} \tag{69}$$

where k_{ex} is dependent on the material microstructure, the conductivity, and the cross-sectional area of the lamination. Different theories have been developed to explain excess losses, but this is still under discussion [32].

For the hysteresis term, the power of B_{max} was found later to be dependent on the material type, as well on B_{max}. Therefore, a more accurate expression was adopted for core loss estimation, with a modified hysteresis term (k_h, a, and b are its parameters) and including the excess losses:

$$P_{core} = P_h + P_e + P_{ex} = k_h f B_{max}^{a+b \cdot B_{max}} + k_e (f \cdot B_{max})^2 + k_{ex} (f \cdot B_{max})^{\frac{3}{2}} \tag{70}$$

Further experiments, together with finite element analysis, revealed that it predicts core losses with very good accuracy for $f \leq 1500$ Hz and B_{max} < 1 T. For higher flux densities, it gives good results for frequencies up to 400 Hz [32].

4.3.1.3 Core losses under nonsinusoidal and uniform flux densities

Most often, coefficients in Eq. (70) are obtained through curve-fitting techniques, based on manufacturer's core losses. Moreover, the hysteresis term in Eq. (70) is limited to symmetrically flux density variations about zero (i.e., $-B_{max} < B < B_{max}$) and, most important, without minor loops included in the main hysteresis loop [34]. In short, the previous expression is mostly suited for sinusoidal flux density waveforms, with uniform distributions in the core. For SRM, these conditions are not valid, so different approaches must be considered. Furthermore, parameter for empirical formulas related to nonsinusoidal flux losses must be based on measured core loss values. Manufacturer's data for sinusoidal flux may lead to large errors for nonsinusoidal core loss estimation. In order to consider nonsinusoidal waveforms, the product ($f \cdot B$) is substituted by (dB/dt) in these two terms:

$$P_{core} = k_h f B_{max}^{a+b \cdot B_{max}} + k_e \int_0^T \left(\frac{dB}{dt}\right)^2 dt + k_{ex} \int_0^T \left(\frac{dB}{dt}\right)^{\frac{3}{2}} dt \tag{71}$$

where $T = 1/f$.

However, since SRM usually operates at high frequencies, the influence of skin effect and saturation over eddy current and hysteresis losses cannot be disregarded: it leads to nonuniform field densities, which brings additional challenges for core loss modeling. In order to have accurate estimations, this must be addressed. In the following, a brief discussion about the impact on eddy current and hysteresis losses is addressed.

4.3.1.4 Eddy current losses

The skin effect in the magnetic core is due to the field created by eddy currents: for high frequencies, particularly for thick lamination core, the flux density at the lamination surface is higher than the one at its center; moreover, their waveforms are also displaced [24, 32]. Flux penetration in the lamination is measured by the skin depth penetration (δ): it is the distance from the steel surface, where the field has decreased by a factor of 1/e, and it is approximately given by:

$$\delta = \frac{1}{\sqrt{\pi f \mu \sigma}} \tag{72}$$

where f is the excitation frequency, μ the magnetic permeability, and σ is the material electric conductivity.

Therefore, the magnetic density field has a nonuniform distribution, and it is mainly concentrated on the lamination surface. Eddy current paths have now a higher resistance, meaning that its value and the correspondent losses are smaller, when compared to uniform field densities scenarios. Naturally, this is not foreseen by the previous expressions, which give excessive values. Several models to calculate eddy current loss in electrical machine laminations for nonuniform field density have been proposed [32]. A common approach for a periodic nonsinusoidal flux density waveform is to consider its Fourier series decomposition. The contribution of each harmonic component is calculated based on its frequency and magnitude, through Steinmetz expression. Frequency is particularly relevant, since it determines the skin effect magnitude of the individual harmonics. However, one must not forget that such an approach is based on the superposition principle. In this way, its effectiveness must be always confronted with experimental loss values.

4.3.1.5 Hysteresis losses

Due to SRM geometry and operating conditions (in particular, high frequencies, local saturation, and skin effect), the peak flux density may be very different in several parts of the lamination [24]. This causes local hysteresis loops, i.e., minor loops, in addition to the major loop (**Figure 10**).

As a consequence, several points inside the lamination have different local hysteresis power losses [24, 34], which may lead to core hot spots. It should be noted that a relevant harmonic content in the flux density waveform reflects a significant number of minor loops. Moreover, these additional losses may represent an important proportion of the total hysteresis loss, which stresses the importance of modeling them [35]. A most relevant conclusion in this reference is that minor loop positions inside the main cycle (i.e., the DC flux density value associated to the minor loop) have a significant influence in the hysteresis losses. It should be noted that none of this is considered in the first term of Eq. (71).

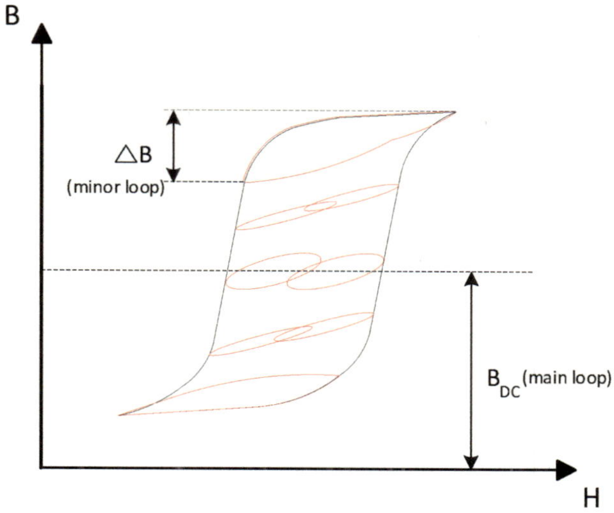

Figure 10.
Hysteresis loops (major in black, minors in red).

Based on experimental studies, minor loop hysteresis loss evaluation has been frequently addressed through empirical formulas [20, 24] (in [20] the Fourier series harmonic decomposition is considered). However, one must be aware that the effectiveness of this approach is attached to certain simplifications and/or to specific flux waveforms. This highlights the fact that for every kind of electrical machine, under specific operation conditions, a particular formula should be addressed. Therefore, a lot of work still need to be done, in order to get accurate methods to estimate hysteresis minor loop losses [24]. These are important issues to be addressed in SRM core loss characterization.

4.3.1.6 Rotational flux losses

Rotational flux densities (due to changes in the flux density vector direction, relatively to a given reference frame) may have important contributions for the total core losses in electrical machines.

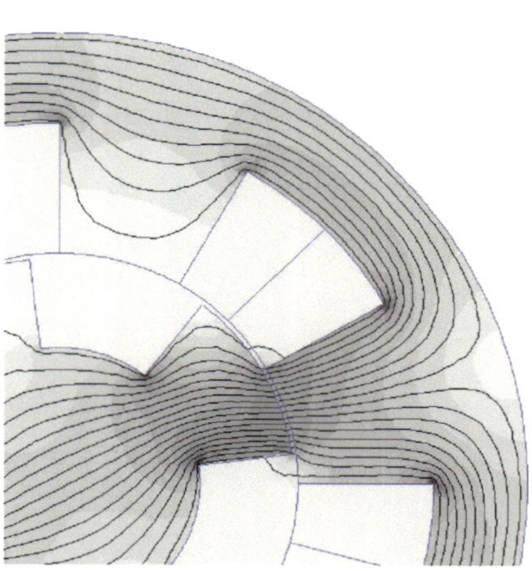

Figure 11.
Magnetic field distribution around stator and rotor tooth.

Particularly in SRM, such rotational flux density variations are well pronounced around the stator and rotor tooth, as represented in **Figure 11**, due to changes on their relative position.

Predicting rotational flux losses is much more complex than alternating flux, with a fixed magnetic axis. Moreover, measuring them is quite complex, since precisely controlled rotational flux density waveforms are difficult to induce in test samples [34].

Some models have been proposed for estimating iron loss under rotational conditions but are based on sinusoidal flux densities. Usually, the flux density vector is decoupled into two orthogonal components, for a given machine region. For each component, losses are independently calculated and then added [36]. Once again, this approach relies on the superposition principle. Due to hysteresis high nonlinearity, it seems reasonable to question the effectiveness of such approach.

5. Conclusions

SRM are simple, rugged, and fault-tolerant machines, with potential to compete with conventional motors. The major drawbacks are torque pulsation and acoustic noise, due to its high nonlinear operation features. In addition, the power converter requires a particular architecture, together with specific control methods. SRM modeling is a challenging task, due to its geometry and operation conditions. Nonetheless, this is a crucial task, since nowadays design and control developments are much dependent on simulation models. Independently of the modeling approach, a key issue is to understand the SRM electromechanical conversion process.

Accurate SRM loss characterization is crucial for designing and control algorithm development. Electrical machines core loss estimation has been supported by empirical models for sinusoidal waveforms. However, this is not suitable for SRM, since flux density waveforms are not sinusoidal and differ from one another, according to the core sections. Moreover, complex magnetic phenomena must be addressed, which are not considered in most empirical models.

This chapter has two main intentions. The first is to present a general view on conventional SRM, where a substantial effort is given on analytical description of electromechanical conversion. The second one is to give an overview on SRM losses—particularly, the iron loss—including the main challenges for characterizing and estimating them.

As a final remark, there are several open challenges in both subjects. A lot of research work still need to be done.

Author details

Pedro Sousa Melo[1]* and Rui E. Araújo[2]

1 School of Engineering, Polytechnic of Porto, Porto, Portugal

2 INESC TEC and Faculty of Engineering, University of Porto, Porto, Portugal

*Address all correspondence to: pma@isep.ipp.pt

References

[1] Gan C, Wu J, Sun Q, Kong W, Li H, Hu Y. A review on machine topologies and control techniques for low-noise switched reluctance motors in electric vehicle applications. IEEE Access. 2018; **6**:31430-31443

[2] Vrenken R et al. Switched reluctance motor drive for full electric vehicles—Part I: Analysis. In: 2013 8th International Conference and Exhibition on Ecological Vehicles and Renewable Energies (EVER), IEEE. 2013. pp. 1-7

[3] Haghbin S, Rabiei A, Grunditz E. Switched reluctance motor in electric or hybrid vehicle applications: A status review. In: 2013 8th IEEE Conference on Industrial Electronics and Applications (ICIEA); IEEE. 2013. pp. 1017-1022

[4] Rajashekara K. Present status and future trends in electric vehicle propulsion technologies. IEEE Journal of Emerging and Selected Topics in Power Electronics. 2013;**1**(1):3-10

[5] Mamede ACF, Camacho JR, Araújo RE. Influence of geometric dimensions on the performance of switched reluctance machine. Machines. 2019;7(4):71

[6] Burkhart B, Klein-Hessling A, Ralev I, Weiss CP, De Doncker RW. Technology, research and applications of switched reluctance drives. CPSS Transactions on Power Electronics and Applications. 2017;**2**(1):12-27

[7] Torkaman H, Moradi R, Hajihosseinlu A, Toulabi M. A comprehensive power loss evaluation for switched reluctance motor in presence of rotor asymmetry rotation: Theory, numerical analysis and experiments. Energy Conversion and Management. 2014;77:773-783

[8] Faiz J, Ganji B, Carstensen C, De Doncker R. Loss prediction in switched reluctance motors using finite element method. European Transactions on Electrical Power. 2009;**19**(5):731-748

[9] Miller TJE. Switched Reluctance Motors and their Control. Lebanon, OH: Magna Physics; 1993

[10] Krishnan R. Switched Reluctance Motor Drives: Modeling, Simulation, Analysis, Design, and Applications. CRC Press; 2001. ISBN: 978-084-930-838-3

[11] Miller TJE. Electronic control of switched reluctance machines. newnes. In: Power Engineering Series; 2001. ISBN: 978-075-065-073-1

[12] Ahn J-W. In: Lamchich MT, editor. Switched Reluctance Motor Torque Control. Croatia: IntechOpen; 2011, p. 53. ISBN: 978-953-307-428-3. Available from: http://www.intechopen.com/books/torque-control/switched-reluctance-motor

[13] Hannoun H, Hilairet M, Marchand C. Design of an SRM speed control strategy for a wide range of operating speeds. IEEE Transactions on Industrial Electronics. 2010;**57**(9): 2911-2921

[14] Vujicic VP. Minimization of torque ripple and copper losses in switched reluctance drive. IEEE Transactions on Power Electronics. 2012;**27**(1): 388-399

[15] Ishikawa T, Hashimoto Y, Kurita N. Optimum design of a switched reluctance motor fed by asymmetric bridge converter using experimental design method. IEEE Transactions on Magnetics. 2014;**50**(2):781-784

[16] Ye J, Emadi A. Power electronic converters for 12/8 switched reluctance motor drives: A comparative analysis. In: 2014 IEEE Transportation

Electrification Conference and Expo (ITEC); IEEE. 2014. pp. 1-6

[17] Krause PC, Wasynczuk O, Sudhoff SD, Pekarek S. Analysis of Electric Machinery and Drive Systems. Wiley Online Library; 2002

[18] Torrent M, Andrada P, Blanque B, Martinez E, Perat J, Sanchez J. Method for estimating core losses in switched reluctance motors. European Transactions on Electrical Power. 2011; **21**(1):757-771

[19] Bui MD, Schäfer U. Core losses measurement technique for high frequency and flux density of switched reluctance machines. In: 2012 XXth International Conference on Electrical Machines (ICEM); IEEE. 2012. pp. 1619-1624

[20] Parsapour A, Dehkordi BM, Moallem M. Predicting core losses and efficiency of SRM in continuous current mode of operation using improved analytical technique. Journal of Magnetism and Magnetic Materials. 2015;**378**:118-127

[21] Mokhtari H, Tara E. Efficiency map of a switched reluctance motor using finite element method in vehicular applications. In: 7th International Conference on Power Electronics, 2007. ICPE'07; IEEE. 2007. pp. 644-649

[22] Chindurza I, Dorrell DG, Cossar C. Assessing the core losses in switched reluctance machines. IEEE Transactions on Magnetics. 2005;**41**(10):3907-3909

[23] Rafajdus P, Hrabovcova V, Hudak P. Investigation of losses and efficiency in switched reluctance motor. In: 12th International Power Electronics and Motion Control Conference, 2006 "EPE-PEMC 2006"; IEEE. 2006. pp. 296-301

[24] Ibrahim M, Pillay P. Core loss prediction in electrical machine laminations considering skin effect and minor hysteresis loops. IEEE Transactions on Industry Applications. 2013;**49**(5):2061-2068

[25] Krings A, Nategh S, Stening A, Grop H, Wallmark O, Soulard J. Measurement and modeling of iron losses in electrical machines. In: 5th International Conference Magnetism and Metallurgy WMM'12, June 20th to 22nd, 2012, Ghent, Belgium; Gent University. 2012. pp. 101-119

[26] El-Kharahi E, El-Dessouki M, Lindh P, Pyrhönen J. Toward including the effect of manufacturing processes in the pre-estimated losses of the switched reluctance motor. Ain Shams Engineering Journal. 2015;**6**(1):121-131

[27] Ionel DM, Popescu M, Dellinger SJ, Miller T, Heideman RJ, McGilp MI. On the variation with flux and frequency of the core loss coefficients in electrical machines. IEEE Transactions on Industry Applications. 2006;**42**(3): 658-667

[28] Steinmetz C. On the law of hysteresis. Transactions of the American Institute of Electrical Engineers. 1892; **9**(1):1-64

[29] Li J, Abdallah T, Sullivan CR. Improved calculation of core loss with nonsinusoidal waveforms. In: Thirty-Sixth IAS Annual Meeting. Conference Record of the 2001 IEEE; Industry Applications Conference, 2001; IEEE, vol. 4. 2001. pp. 2203-2210

[30] Venkatachalam K, Sullivan CR, Abdallah T, Tacca H. Accurate prediction of ferrite core loss with nonsinusoidal waveforms using only Steinmetz parameters. In: IEEE Workshop on Computers in Power Electronics, vol. 2002. 2002. pp. 36-41

[31] Aguglia D, Neuhaus M. Laminated magnetic materials losses analysis under non-sinusoidal flux waveforms in power

electronics systems. In: 2013 15th European Conference on Power Electronics and Applications (EPE); IEEE. 2013. pp. 1-8

[32] Zhang Y, Pillay P, Ibrahim M, Cheng M-C. Magnetic characteristics and core losses in machine laminations: High-frequency loss prediction from low-frequency measurements. IEEE Transactions on Industry Applications. 2012;**48**(2):623-629

[33] Bertotti G. Hysteresis in Magnetism: For Physicists, Materials Scientists, and Engineers. Academic Press; 1998. ISBN: 978-012-093-270-2

[34] Calverley SD, Jewell GW, Saunders RJ. Prediction and measurement of core losses in a high-speed switched-reluctance machine. IEEE Transactions on Magnetics. 2005; **41**(11):4288-4298

[35] Simão C, Sadowski N, Batistela N, Kuo-Peng P. Simplified models for magnetic hysteresis losses evaluation in electromagnetic devices. In: Electric Machines and Drives Conference, 2009. IEMDC'09. IEEE International; IEEE. 2009. pp. 876-880

[36] Hernandez-Aramburo CA, Green TC, Smith AC. Estimating rotational iron losses in an induction machine. IEEE Transactions on Magnetics. 2003;**39**(6):3527-3533

Chapter 3

Review of Rotary Switched Reluctance Machine Design and Parameters Effect Analysis

Ana Camila Ferreira Mamede, José Roberto Camacho and Rui Esteves Araújo

Abstract

The switched reluctance machine (SRM) has gained much interest in industrial applications, wind power systems, and electric vehicles. This happened because its main disadvantages, such as the ripple in the torque, were overcome due to continuous research and its advantages, such as simple and robust construction, ability to operate at high speeds and variable speeds, insensitivity to high temperatures, and fault tolerance, have made the SRM the right machine for many applications. The SRM project is apparently similar to the traditional machine design, but diverges on several points due to the unique features of the SRM. Over the years, several authors have proposed different project methodologies for SRM, each with its own particularities and often contradicting each other. Thus, for a beginner designer, the SRM project is a challenge from choosing the right design methodology to choosing the values of some dimensions, which are often empirical. This chapter aims to offer the beginner designer a detailed review of the main SRM design methodologies. In addition, an effect analysis will provide useful insights on how each design variable affects machine performance. The designer will thus have important data on which to base his choices during the SRM design.

Keywords: switched reluctance machine, design, effect analysis, design of experiments, definitive screening design

1. Introduction

The switched reluctance machine (SRM) is a machine with double salience, that is, with stator and rotor poles, and this structure contributes to the production of a high output torque. Torque is produced by the alignment tendency of the poles; the rotor tends to move to a position where reluctance is minimized, and thus the inductance of the excited coil is maximized. Although the structure of the SRM is doubly salient, there is no winding or permanent magnets in the rotor [1], simplifying the structure of the machine and contributing to the low manufacturing cost.

SRMs differ from traditional electric machines both structurally and in their performance characteristics; however, there are some similarities. A crucial difference when comparing SRM to traditional machines is that the stator and rotor generally have different numbers of poles and cannot operate without an electronic converter [2]. Other features of the SRM are simplicity of construction, absence of coils in the rotor, tolerance to phase failures, robustness, and low production cost.

The performance of the switched reluctance machine depends heavily on its design and control, which allows reduced ripple in torque and improved torque-speed characteristics [3]. Few researchers have addressed the problem of the sensitivity of SRM performance in relation to its dimensions. In [4] the authors investigated how to mitigate the torque ripple through the variation of SRM geometric parameters based on finite element simulation results. In [5] the effect of the variation of the dimensions was analyzed in relation to copper losses, gains in the machine weight, torque ripple, and average torque value. The analyses were based on the method proposed in [6].

As the design of SRM for a particular application is a compromise between various performance criteria, improvement of a performance parameter may result in the degradation of other important features [7]. Consequently, the designer has to look for solutions that are feasible for all performance parameters. In order to deal with this trade-off and achieve an efficient design, the application of multi-objective optimization techniques to the SRM design seems to be the most appropriate approach. Optimization techniques are known for analyzing a large volume of machines, which demands a high computational cost. However, if the influence of each parameter on the performance of the machine is known, the number of parameters in the optimization can be reduced, making the process faster and more efficient.

Design of experiments (DOE) is a mathematical methodology used for planning and conducting experiments as well as analyzing and interpreting data obtained from the experiments. It is a branch of applied statistics that is used for conducting scientific studies of a system, process, or product in which input variables are manipulated to investigate its effects on the measured response variable [8]. Over past two decades, DOE has been a very useful tool traditionally used for improvement of product quality and reliability [9]. The usage of DOE has been expanded across many industries as part of the decision-making process either in new product development, manufacturing processes, or improvement. It is not used only in engineering field but has also been used in administration, marketing, hospitals, pharmaceutical [10], food industry [11], energy and architecture [12], and chromatography [13]. DOE is applicable to physical processes as well as computer simulation models [14].

There are several DOE methods, such as one-factor-at-a-time, factorial design, central composite design, Latin hypercube design, and others. Each method has its particularities, such as quantity of factors, levels of factors, and number of experiments required, and this is what determines the choice of a method for a specific experiment. Some methods, such as one-factor-at-a-time, for example, will become impractical when more design variables are considered, because the number of experiments increases exponentially with the number of design variables [15]. The design of electric machines generally involves a large number of design variables, and since the relationship between variables is often nonlinear, a three-level experiment is more appropriate because it allows a better evaluation of the curvature in the factor-response relationship [15, 16]. Therefore, with the intention of reducing the computational cost, a three-level class known as definitive screening design (DSD) will be used, which is capable of capturing nonlinearity and identifying the main active effects with a low number of experiments [16].

The main objective of this chapter is to present an effective and direct methodology for SRM design and to indicate the impact of each variable on the machine's performance, showing how and how much each design parameter affects the SRM response.

2. Structure and operation of SRM

The operational principle of SRM is based on the power effects of the magnetic circuit that tend to minimize the reluctance of the magnetic circuit [2]. The rotor poles tend to align to the diametrically opposite poles of the stator whenever they are excited by a current [6]. If a pair of stator poles is excited, the rotor tries to turn to align with the poles of the stator so that the reluctance of the magnetic circuit would reach the minimum and also the energy of the magnetic circuit would thus be minimized [2]. When the energy minimum is reached, the magnetic forces try to keep the rotor in a position where the energy minimum of the magnetic circuit is preserved. Now, to make the machine rotate again, the magnetic energy must be removed to another pair of poles [2].

Figure 1 shows a typical SRM with six stator poles and four rotor poles. Consider that the rotor poles r_1 and r_1' and stator poles c and c' are aligned. By applying a current in phase a, a flux is established through the stator poles a and a' and r_2 and r_2', which tends to pull the rotor poles r_2 and r_2' toward the stator poles a and a'. When they are aligned, the stator current of phase a is turned off, and phase b is excited [6].

Due to the operating principle presented above, the self-inductance of each phase of the SRM changes according to the rotor position and current value, as shown in **Figure 2**. The same SRM can operate as both a motor and generator, depending on the moment of current injection. **Figure 2** shows the current pulses for each one.

If the magnetic flux of the SRM varies according to the position of the rotor and current, a group of curves $\lambda(\theta, i)$, known as magnetization curves, can be obtained and are fundamental for the study of this machine. The effect of saturation is evident in this figure and plays a key role in the SRM. The SRM is designed to achieve a certain level of saturation to maximize energy density, torque, and machine performance. The area delimited by the flux linkage in the aligned and unaligned position determines the mechanical output power of the machine; the greater the difference between these curves, or the inductance values in those positions, the higher the average torque value developed by the machine [2, 6].

The calculation of the torque is a problem because of the dependence of the inductance of the magnetic circuit on both the rotor angle and the stator current. The dependence results from the excessive saturation of the magnetic circuit at certain rotor positions and with high stator currents (**Figure 3**) [2]. The torque developed is proportional to the area delimited by the curves in **Figure 3**, which

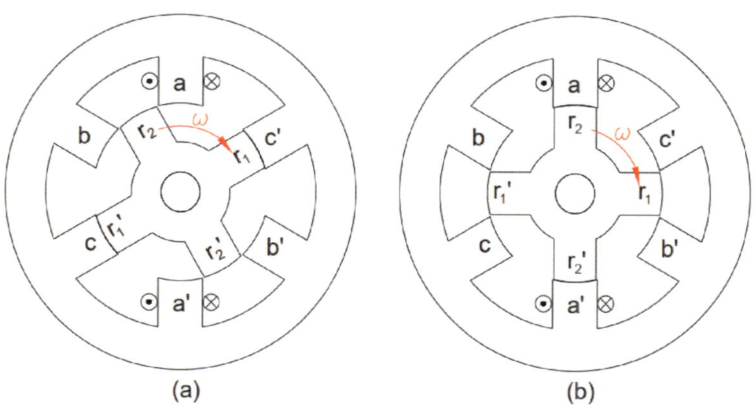

Figure 1.
Operation of an SRM. (a) Phase c aligned. (b) Phase a aligned. (Adapted from [6]).

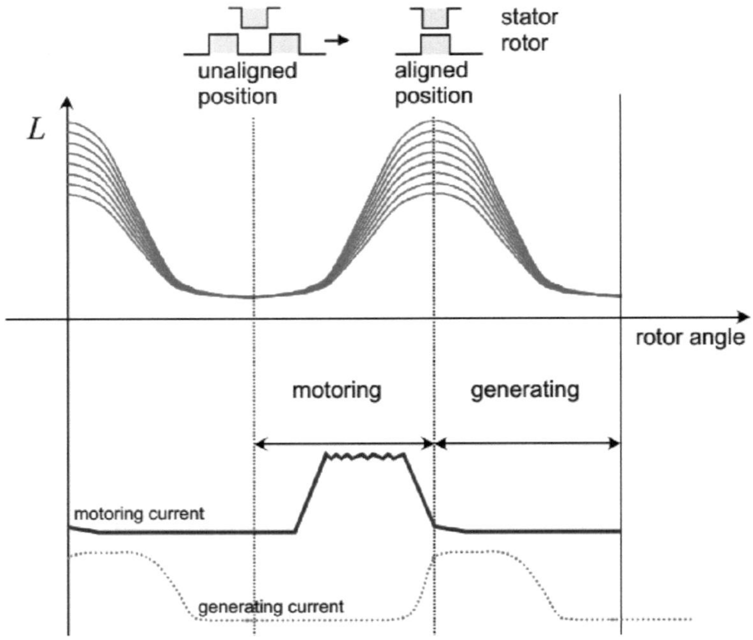

Figure 2.
Inductance of a saturating reluctance motor as a function of rotor angle with current as a parameter, and the current pulses of a motor drive and a generator drive, when the voltage of the intermediate circuit remains constant [2].

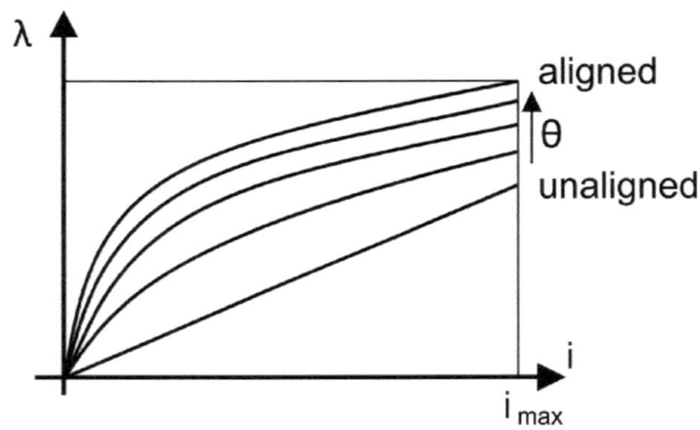

Figure 3.
Magnetization curves.

corresponds to the stored magnetic energy, and can be calculated in terms of co-energy, current, and rotor position by

$$T_e = \left. \frac{\delta W'_f(i,\theta)}{\delta\theta} \right|_{i=constant} \quad (1)$$

with the co-energy W'_f being defined by

$$W'_f = \int \lambda(\theta, i)\, di. \quad (2)$$

The average torque can be calculated by assuming that the current I_p is maintained constant between the aligned and unaligned positions of the rotor and that the magnetization curve is available [6]. The area between the curve in the aligned position and the unaligned position is denoted as δW_m and corresponds to the work done by the machine cycle, which can be calculated from the co-energies, as shown in Eq. (3):

$$\delta W_m = \delta W'_{f-aligned} - \delta W'_{f-unaligned} \tag{3}$$

The trapezoidal integration, as long as a reasonable number of points are considered in the curve, offers good results. The average torque can be obtained through the total work performed by a machine revolution by Eq. (4), where N_s and N_r are the stator and rotor number of poles, respectively:

$$T_{ave} = \frac{\delta W_m N_s N_r}{2\pi} \; N \cdot m \tag{4}$$

3. Design procedures

The general theory that governs the design of an electric machine includes studies in the areas of electromagnetism, winding arrangements, magnetic circuit behavior, inductance, and winding resistance [2]. The process of designing an electric machine takes into account all these studies and the empirical knowledge acquired over the years by designers and researchers.

The design of a rotating electrical machine starts by defining certain basic characteristics, such as the type of machine, type of construction, rated power, rated rotational speed, number of pole pairs, operating frequency, rated voltage and number of phases, intended duty cycle, enclosure class, and structure of the machine [2]. In machine design there are a considerable number of free parameters. The task of finding an ideal solution becomes extremely complex unless the number of these parameters is somehow limited [2].

The difficulties of the SRM design can be attributed to several factors, including the following: (1) high magnetic and control nonlinearities; (2) many flexible design parameters, including those that affect vibrations and acoustic noises; (3) interdependence on the design of the converter and control parameters; (4) variable-speed operations; (5) thermal management due to the concentrated coils; (6) difficulty of thermal rating prediction; and (7) possible high sensitivity of design to manufacturing tolerance [17]. Also, during sizing process it is important to establish proper flux density values in the magnetic core. Very low values have the outcome of an unsaturated machine and low energy density, whereas very high values will limit the power developed and cause the core to heat up [18].

Too low values have the outcome of an unsaturated machine, and due to this, poor power density and too high values will limit the developed power and cause core heating [18].

Over the years, research has been based mainly on the design methodologies proposed by [6, 19], proposing changes and optimizing the project according to their acquired knowledge and application requirements [20–23]. The main difference between these two methods is the output equation used for the SRM. In [19], the author proposes a simplified output equation, and the choice of some coefficients is made through a table of typical values based on the application. A more complex output equation is presented by [6]; however, some of its parameters are still chosen through ranges of typical values, which complicate the first steps in machine design.

The following design procedures are mainly based on [6, 24], presenting useful simplifications for manufacturing and some disagreements with other design methods. The process presented in this section corresponds to a simple and intuitive preliminary design of SRM for beginner designers. The calculation of the flux linkage at various machine positions will be done using the finite element method, performed in FEMM software.

3.1 Starting values

To design a machine for a specific application, one must know its characteristics, which will make up the SRM specifications. The design specifications for the SRM are the required output power P_{out} in W, the nominal speed n in rpm, the allowed peak current in A, and the available supply voltage for the system. From this data the mechanical torque of the machine is fixed and calculated through Eq. (5):

$$T_{mec} = \frac{P_{out}}{2\pi \frac{n}{60}} \tag{5}$$

Figure 4 shows all the dimensions that must be determined for the construction of an SRM, where β_s is the stator polar arc, β_r is the rotor polar arc, w_{sp} is the width of the stator pole, b_{sy} is the stator back iron thickness, b_{ry} is the rotor back iron thickness, h_s is the height of the stator pole, h_r is the height of the rotor pole, D_{sh} is the diameter of the shaft, D_i is the inner diameter, D_O is the outer diameter, and g is the length of the air gap.

3.2 Inner, outer, and shaft diameter and core length

In general, the starting point for designing a machine is to obtain the SRM power output equation; this process is presented by [6], and from the output equation, other dimensions are determined. However, the output equation presented by [6]

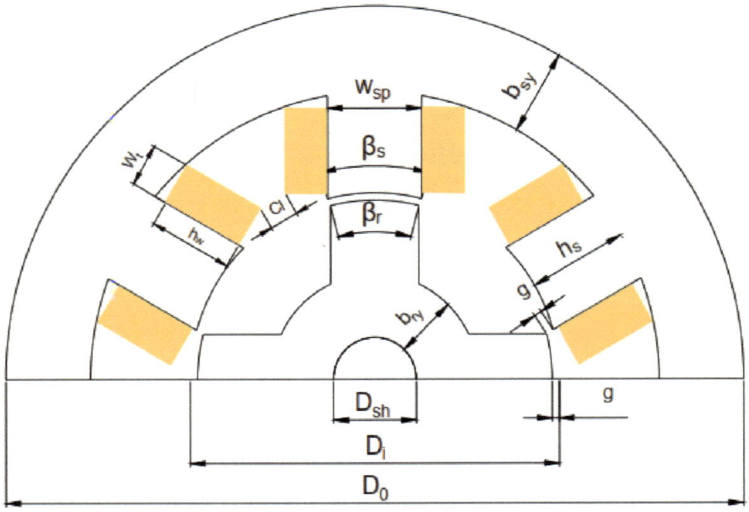

Figure 4.
SRM dimensions.

presents a series of parameters that must be determined arbitrarily so that the calculation of the inner diameter is possible (such as length, efficiency, duty cycle, and specific electric charge, among others), which makes the process a little bit ambiguous. This process can be simplified by using tables presented by international standards such as [25], which determine typical frame values for all rotating machines.

Thus, the initial values for the frame, inner diameter, external diameter, shaft diameter, and core length are selected from tables that relate the output power, nominal velocity, and frame size (*FS*) present in [25]. These values can be adjusted at the end of the sizing process. The outside diameter is calculated by

$$D_0 = (FS - x)2 \tag{6}$$

where x corresponds to the value of the foot of the machine, usually adopted as 3 mm. Normally, the internal diameter is 0.4–0.7 times the value of the external diameter [19].

The inner diameter is initially set equal to FS, but can be changed according to the need of the application and constructive limitations. In [6], the length of the core is determined as a multiple of the inner diameter (k_L). The value of k_L is decided by the nature of the motor application and space limitation. For non-servo applications, the interval for k_L is given by Eq. (8) and for servo applications by Eq. (9):

$$L = k_L D_i \tag{7}$$

$$(0.25 \le k_L \le 0.7) \tag{8}$$

$$(1.0 \le k_L \le 3.0) \tag{9}$$

Therefore, since the frame has already been chosen, the upper limit of the core length is already defined, and the lower limit can be determined by the above equations.

3.3 Air gap length

For switched reluctance machines, the air gap length determines the output torque and the volt-ampere requirement in the motor drive. However, the air gap cannot be made as small as possible due to manufacturing constraints, and it is roughly proportional to the motor size [23]. In [6] the air gap value should be chosen according to the size of the machine; for small machines, with power less than 1.0 hp, the air gap should vary between 0.18 and 0.25 mm. Integral horsepower machines, with power above 1.0 hp, may have air gaps from 0.3 to 0.5 mm [6]. For other authors, the air gap length should be selected to be about 0.5–1% of the rotor diameter [2, 19].

3.4 Selection of number of phases and number of poles

The number of phases of an SRM is usually determined through factors, such as directional capacity, reliability, cost, and high speed operation, as described below.

1. Starting capability: In a single-phase SRM, for example, the cost with the control is low; however, there are regions where the torque produced is null, and the consequence of this is the impossibility of the machine being activated in these positions if no artifice is used, such as a permanent magnet.

2. Directional capability: In some applications it is necessary to change the direction of operation of the SRM, and this determines the minimum number of phases for that application. For example, a 4/6 machine is capable of providing unidirectional rotation only, whereas a 6/4 is capable of two-direction rotation. The former case is a two-phase machine, and the latter case is a three-phase SRM [6].

3. Cost: The cost of an SRM relates the cost of the motor itself and the cost of its converter, so a high number of phases demand a high number of power devices in the converter, increasing the total cost of the application and restricting its use to some applications.

4. Reliability: A higher number of phases mean a higher reliability because a failure of one or more phases will not prevent the machine from continuing to operate. This factor may be highly relevant in critical applications where safety of human beings or successful mission completion is the predominant factor [6].

5. Power density: A higher number of phases tend to provide a higher power density in the machine.

6. Efficient high-speed operation: Efficiency is enhanced by reducing the core loss at high speed by decreasing the number of stator phases and lowering the number of phase switchings per revolution [6]. In high-speed operations it is necessary to keep the motor size smaller, which requires a great reduction in losses to maintain the thermal robustness [6].

The limiting factors in the number of poles selection are the number of converter power switches and their associated cost of gate drives and logic power supplies and the control requirement in terms of small rise and fall times of the phase currents [6]. The topologies 6/4 and 8/6, three and four phases, respectively, are common in industry because they are cost-effective; however, other topologies are possible. An increase in the pole number improves the operational accuracy of the motor and the quality of the torque, but simultaneously the structure and control of the converter switches get more complicated [2]. Many applications such as fans or pumps and even in off- and on-highway vehicle propulsion can stand higher commutation torque ripples, for example, than position servos.

The stator frequency for a phase is determined by the maximum machine speed and number of rotor poles. By increasing the number of rotor poles, the frequency in the stator increases in proportion, resulting in higher core losses and greater conduction time to provide the rise and fall of the current than that of an SRM drive with a smaller number of rotor poles [6]. Therefore, the selection of number of phases and number of poles is directly related to the application requirements and available budget.

3.5 Stator and rotor pole arc

The values of the stator pole arc and rotor pole arc are chosen to guarantee the proper starting of the machine and to shape the motor torque profile [6]. These requirements are inserted into the SRM project by offering a lower and upper limit for the values of the polar arcs. These restrictions are detailed in [1, 6, 19] and briefly described here.

In order to guarantee proper starting of the machine and to prevent the occurrence of parasitic currents due to the magnetic flux dispersion effect, the rotor polar arc must be larger than the polar arc of the stator [1, 6]:

$$\beta_r \geq \beta_s \tag{10}$$

Krishnan showed in [6] that the minimum value for polar arcs is set according to the number of poles of the machine by Eq. (11):

$$min\left(\beta_s, \beta_r\right) = \frac{4\pi}{N_s N_r} \tag{11}$$

The angle between the corners of the adjacent rotor poles must be greater than the polar arc of the stator or there will be an overlap of the stator and rotor poles in the non-aligned position [24]. This implies that the minimum inductance value will be greater, reducing the difference between the maximum and minimum values, which leads to a reduction in the torque value. This relation is presented in Eq. (12):

$$\frac{2\pi}{N_r} - \beta_r > \beta_s \tag{12}$$

The conditions presented in Eqs. (10)–(12) can be represented graphically in a triangle of possibilities. It is necessary for the values of the polar arcs of the machine to be within this triangle [24]. **Figure 5** shows the triangle of possibilities for an SRM 6/4 and an SRM 8/6. For example, for an SRM 8/6, if $\beta_s = 20°$, then $20° \leq \beta_r \leq 40°$. The feasible region restricts the possible combinations of stator and rotor pole angles but does not provide the best solution [26].

An optimum value can be solved for the pole size. This optimum value yields the maximum inductance ratio and simultaneously the maximum average torque. In addition, several other factors affecting the operation of the machine have to be taken into account, such as the torque ripple, starting torque, and the effects of saturation, and therefore no general solution can be attained [2]. Some authors have dedicated themselves to optimizing the values of the polar arcs [27–29].

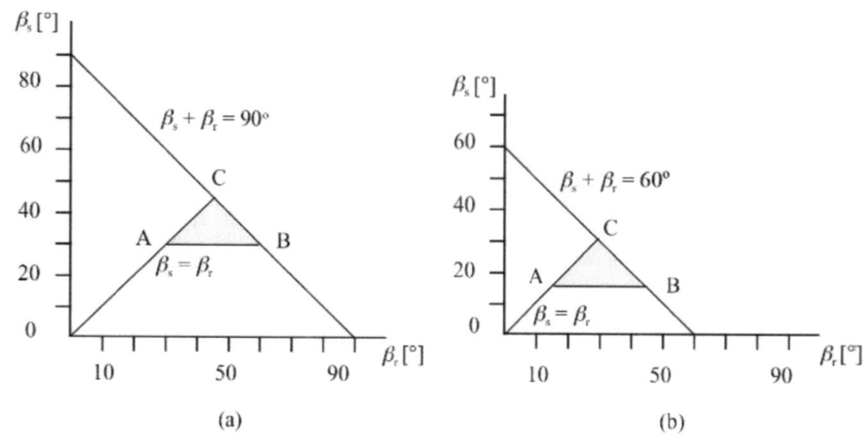

Figure 5.
Triangle of possibilities for β_s and β_r of (a) a three-phase 6/4 SRM and (b) a four-phase 8/6 SRM (Adapted from [19]).

3.6 Preliminary design

With the previously defined dimensions, the other dimensions of **Figure 1** can be calculated as shown below. To begin, it is necessary to have access to the B-H characteristics of the material used for the stator and rotor blades; an example is shown in **Figure 6**. From the curve the "knee" point is determined and used to limit the flux density inside the motor. SRM is designed to achieve saturation; a saturated machine has the potential to convert approximately twice as much energy as the unsaturated machine to the same dimensions and peak current [30]. Thus, the value of flux density is defined in order to maximize the energy density and machine performance. Assuming the flux density at the stator pole B_s is equal to B_{max}, the remainder of the machine can be projected by estimating the value of the flux density at the other points.

The stator pole width is determined by the stator polar arc and by the value of inner diameter as follows:

$$w_{sp} = D_i sin\left(\frac{\beta_s}{2}\right) \tag{13}$$

The stator back iron thickness has to be large enough to support half of the flux density passing through the stator pole. Therefore, the back iron thickness of the stator must be at least half the stator pole width. Due to considerations of mechanical robustness and minimization of vibration, it could have a value in the range of

$$w_{sp} > b_{sy} \geq 0.5w_{sp} \tag{14}$$

Next, the stator pole height can be calculated as

$$h_s = \frac{D_0 - 2b_{sy} - D_i}{2} \tag{15}$$

The rotor back iron thickness does not need be as much as the stator back iron thickness and neither has to be equal to the minimum value, which is equal to the minimum value of the stator back iron thickness. The rotor back iron thickness in terms of the stator pole width can be set in the range given below:

$$0.5w_{sp} < b_{ry} < 0.75w_{sp} \tag{16}$$

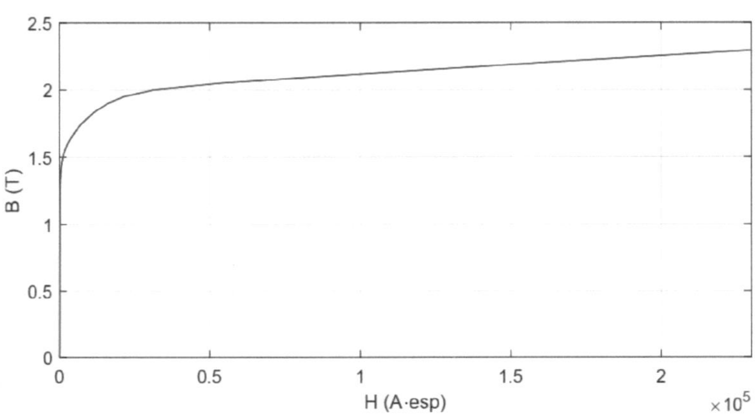

Figure 6.
B-H curve for M19 steel.

According to Vijayraghavan [24], securing the flux density at the stator back iron thickness B_y as approximately half of the B_{max} value and the flux density in the B_{rc} as approximately 80% of the maximum value is a good practice for SRM projects, because it reduces acoustic noise in the machine. This means choosing the value of b_{sy} equal to the value of w_{sp}, upper limit of Eq. (14), and b_{ry} equal to $0.625\,w_{sp}$. These values can be adopted in a preliminary design, but can be optimized due to their influence on the volume and cost of the machine.

The height of the rotor pole can be determined as follows:

$$h_r = \frac{D_i - 2g - D_{sh} - 2b_{ry}}{2} \tag{17}$$

With these dimensions, the next step is the design of the coils, and for this it is necessary to analyze the magnetic circuit of the SRM, shown in **Figure 7**. The reluctances of the stator pole, yoke, rotor pole, rotor core, and air gap are represented by $\mathfrak{R}_s$, $\mathfrak{R}_y$, $\mathfrak{R}_r$, $\mathfrak{R}_{rc}$, and $\mathfrak{R}_g$, respectively.

The equivalent reluctance of the equivalent magnetic circuit $\mathfrak{R}_{eq}$ in **Figure 7** is obtained as

$$\mathfrak{R}_{eq} = 2(\mathfrak{R}_s + \mathfrak{R}_r + \mathfrak{R}_g) + \frac{\mathfrak{R}_{rc} + \mathfrak{R}_y}{2} \tag{18}$$

The reluctance $\mathfrak{R}$ in a particular section can be calculated by Eq. (19), where H is the magnetic field intensity, l is the mean path length in the section, B is the flux density in the section, A is the area of the section, and φ is the flux in the section:

$$\mathfrak{R} = \frac{H\,l}{B\,A} = \frac{H\,l}{\varphi} \tag{19}$$

The magnetic circuit equation of aligned inductance can be written as

$$\mathfrak{I}_{alig} = NT\,i = \mathfrak{R}_{eq}\varphi \tag{20}$$

where NT is the number of turns per phase. The equations for calculating the length of sections, cross-sectional area, and respective value of B are shown in **Table 1**. Note that some values of flux density used follow the indications of [6, 24]. The value of H for each section is obtained through the B-H curve of the magnetic material when estimating the value of B for each section. We suggest the use of an interpolation algorithm.

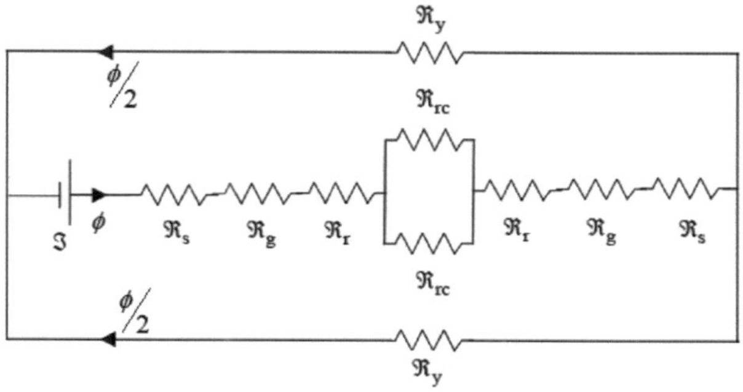

Figure 7.
Magnetic circuit of the SRM.

Part	Length	Cross section	Flux density
Stator pole	$l_s = h_s + \frac{b_{sy}}{2}$	$A_s = \frac{D_i}{2} L\beta_s$	$B_s = B_{max}$
Stator back iron	$l_y = \pi\frac{(D_0 - b_{sy})}{2}$	$A_y = b_{sy}L$	$B_y = \frac{B_s}{2}$
Rotor pole	$l_r = \frac{(D_i - D_{sh})}{4} + \frac{(h_r - g)}{2}$	$A_r = (\frac{D_i}{2} - g)L\beta_r$	$B_r = \frac{B_s n A_s}{A_r}$
Rotor back iron	$l_{rc} = \pi\left(\frac{(D_i + D_{sh})}{4} - \frac{(h_r + g)}{2}\right)$	$A_{rc} = \left(\frac{(D_i - D_{sh})}{2} - h_r - g\right)L$	$B_{rc} = 0.8B_s$
Air gap	$l_g = g$	$A_g = \frac{A_s + A_r}{2}$	$B_g = \frac{B_s A_s}{A_g}$

Table 1.
Magnetic path lengths, cross sections, and flux densities.

Now, the total ampere-turns $\mathfrak{I}$ required for the machine operation at full load can be calculated. The magnetic circuit equation is written as

$$\mathfrak{I} = NTi = 2(H_s l_s + H_r l_r + H_g l_g) + \frac{H_{rc} l_{rc}}{2} + \frac{H_y l_y}{2} \tag{21}$$

As the value of the peak current allowed in the machine was determined at the beginning of the design, the number of turns NT can be calculated through Eq. (21). The number of turns must be an integer and even. The value of the corrected current is then corrected through Eq. (22):

$$i_p{}^* = \frac{\mathfrak{I}}{NT} \tag{22}$$

The conductor area a_c can be calculated from the number of phases q, peak current value i_p, and current density J. The maximum permissible value of J is determined by the type of machine and cooling method employed [2, 6]. Usually, the value adopted for SRM during the preliminary project is $J = 6\ \text{A/mm}^2$:

$$a_c = \frac{i_p{}^*}{J\sqrt{q}} \tag{23}$$

The section of the normalized conductors is obtained by selecting the section closer but greater than the value obtained from Eq. (23). The wire diameter d_w including insulation is given by Eq. (24):

$$d_w = \sqrt{\frac{4a_c}{\pi}} + 0.1\text{mm} \tag{24}$$

The coil must be placed between two consecutive stator poles, and to ensure that the winding fits, it is necessary to perform the following calculations. If necessary, the stator dimensions or conductor size can be changed. Assuming a wedge of h_{wedge} is required to hold the windings in place; the stator pole arc length t_s at the closest point of the winding to the center of the shaft is given by Eq. (25):

$$t_s = \left(\frac{D_i}{2} + h_{wedge}\right)\beta_s \tag{25}$$

Accounting for the wedges that hold the windings in place leads to the calculation of a modified stator pole pitch λ_s as in Eq. (26):

$$\lambda_s = \frac{\pi\left(D_i + 2h_{wedge}\right)}{N_s} \tag{26}$$

The maximum height of the winding h_w which can be accommodated inclusive of the space required to place wedges that hold the windings in place is given by

$$h_w = h_s - h_{wedge}. \tag{27}$$

The number of layers that can be accommodated in this available winding height is given by Eq. (28), where f_f represents the field factor and is approximately equal to 0.95. The value of N_v is rounded off to the nearest lower integer:

$$N_v = \frac{h_w f_f}{d_w} \tag{28}$$

Now the number of horizontal layers required for winding is given by Eq. (29), and this value is rounded off to the nearest higher integer:

$$N_h = \frac{NT}{2N_v} + 1 \tag{29}$$

The width of the winding W_t is given by Eq. (30):

$$W_t = \frac{d_w N_h}{f_f} \tag{30}$$

The space between two stator pole tips at the bore is given by

$$Z = \lambda_s - t_s. \tag{31}$$

The clearance between the windings at the bore is given by

$$Cl = Z - 2W_t. \tag{32}$$

This value has to be positive and preferably greater than 3 mm. Naturally, the actual clearance between the windings will be slightly higher than the clearance calculated at the bore. If the clearance value is acceptable, the designer can proceed with the analysis, or else a different size conductor can be chosen, and the clearance can be checked again.

3.7 Design verification

The next step is to estimate the average torque value and check whether it matches the initial specifications, if the value obtained is not acceptable, it is necessary to modify the design. The average torque can be calculated from the values of the co-energies, according to Eq. (4), for which it is necessary to know the value of the flux linkage in the aligned and unaligned position for the peak current. These values can be determined analytically through different methods such as those presented by [6, 31] or can be obtained through a finite element simulation. Since there are several free and fast finite element simulation software packages available, we strongly suggest using them for the calculation of SRM characteristics.

The analysis of other characteristics of the SRM, such as calculation of losses in the copper and in the iron, by parasitic currents and hysteresis, is detailed by

[6, 24, 32–34]; however, they can also be obtained through finite element simulation. Thus, enough data are available to estimate the efficiency and other parameters of SRM. The control can be designed and the rest of the SRM characteristics determined.

3.8 Calculation of losses

The losses in the SRM are composed of copper losses and iron losses. The copper losses only appear in the stator, since there are no windings in the rotor, and can be calculated by

$$P_{cu} = qI^2R_s, \tag{33}$$

where R_s is the per-phase resistance of the stator winding and I is the rms value of the current given by

$$I = \frac{i_p}{\sqrt{q}} \tag{34}$$

Hence, the copper losses are

$$P_{cu} = i_p^2R_s. \tag{35}$$

The resistance of a single phase can be computed in terms of the specific resistivity ρ, mean length of the winding l_m, and area of cross section of the conductor a_c by Eq. (36):

$$R_s = \frac{\rho l_m}{a_c}NT \tag{36}$$

The mean length of a winding turn is given by Eq. (37):

$$l_m = 2L + 4W_t + 2D_i sin\left(\frac{\beta_s}{2}\right) \tag{37}$$

The iron losses can be split into two major portions, hysteresis and eddy current losses. Core losses are difficult to predict in the SRM due to the presence of flux densities with various frequencies in stator segments. Further, these flux densities are neither pure sinusoids nor constants [1, 33]. There are two papers that lead the calculation of core losses, the one developed in [33] and the one developed in [34].

The method proposed in [33] is simpler and provides a good estimate of the losses in the core, but does not clearly separate the hysteresis and eddy current losses. Manufacturers of lamination core steel provide data showing the variation of core loss (in watts per pound or kilogram) as a function of flux density and frequency [33]. Once the flux density and iron weight for each part of the machine are known, these data are used to calculate the losses in the core.

4. Dimensional effect analysis

The previous section presented the procedures for the preliminary design of the SRM, and it can be seen that many dimensions can be chosen within a range which, to a certain extent, influences the rest of the design and machine performance. Therefore, this section examines how and how much each dimension influences

Design variable	Type	Unit	Reference value	Range
D_i	Variable	mm	100	90–110
D_o	Variable	mm	190	185–195
L	Variable	mm	180	150–200
g	Variable	mm	0.4	0.3–0.5
β_s	Variable	deg	22.5	15–25
β_r	Variable	deg	22.5	15–30
b_{sy}	Variable	mm	19.509	9.7545–19.509
b_{ry}	Variable	mm	12.193	9.7545–14.632
B_{max}	Fixed	T	1.7	—
D_{sh}	Fixed	mm	28	—
Turns per phase	Fixed	units	152	—
Conductor area of cross section	17 AWG	—	—	—
Peak current	Fixed	A	12	—
Lamination material	M19	—	—	—

Table 2.
Details of 8/6 SRM analyzed.

SRM performance through a new DOE class, a small three-level design that is presented by [16]. This study is an important and useful tool for machine designers, since it allows a basis for decisions for the optimization of the SRM, reducing the search space and contributing to the design of more efficient machines.

Jones and Nachtsheim [16] proposed a new class of three-level designs that provide estimates of main effects that are not biased by any second-order effect, require only one more than twice as many runs as there are factors, and avoid confounding of any pair of second-order effects. The number of experiments required is very low compared to other methods, even if all the SRM design variables are considered, and this is a decisive factor in the choice of this DSD class. For m factors only $2m + 1$ experiments are required, which is much lower than the amount of experiments required for other classes, such as 3^m by factorial experiment and $2^m + 2m + 1$ required by central composite design. Thus, the computational cost of the following analysis can be greatly reduced.

The SRM studied is a four-phase motor with eight stator poles and six rotor poles. The effects on the variables D_i, D_O, L, g, β_s, β_r, b_{sy}, and b_{ry} will be analyzed, thus totaling 17 experiments performed. The responses chosen to analyze SRM performance are average torque and machine losses. The experiments performed are finite element simulation performed in FEMM software. The parameters of the SRM studied are shown in **Table 2**, which also presents machine details and the range adopted for each design variable according to the design guidelines shown in Section 3.

To perform these experiments, three software will be used. Minitab is a statistical software that will help design the experiment and get the results. MATLAB is used to perform the calculations and obtain the results of finite element simulations computed through FEMM.

4.1 Effects on average torque

The effects of changes in dimension values can be determined in several ways; in this work the statistical results obtained through Minitab will be presented. The Pareto graph is an important quality tool and allows visualization of the effects ranked by the magnitude of their contribution [35]. **Figure 8** shows the Pareto chart of the standardized effects for the average torque. The Pareto chart shows the absolute values of the standardized effects from the greatest effect to the smallest effect. The standardized effects are t-statistics that test the null hypothesis that the effect is zero. The greater the magnitude of the standardized effect, the greater the evidence against the null hypothesis, that is, the greater the probability that there is a significant difference.

From **Figure 8** we can determine which factors cause a greater impact on the average torque value; however, it is not possible to know how each dimension affects the response. The normal plot of standardized effects, shown in **Figure 9**, shows which effects are positive and negative. Positive effects, right of the line, increase the response when the factor changes from low to high value. Negative effects, left of the line, decrease the response when factor definitions change from low to high. Effects further from 0 are more statistically significant.

Through the results presented in **Figures 8** and **9**, it is possible to draw important conclusions about the influence of dimensions on the average torque value. The three dimensions that most influence the average torque value are the rotor polar

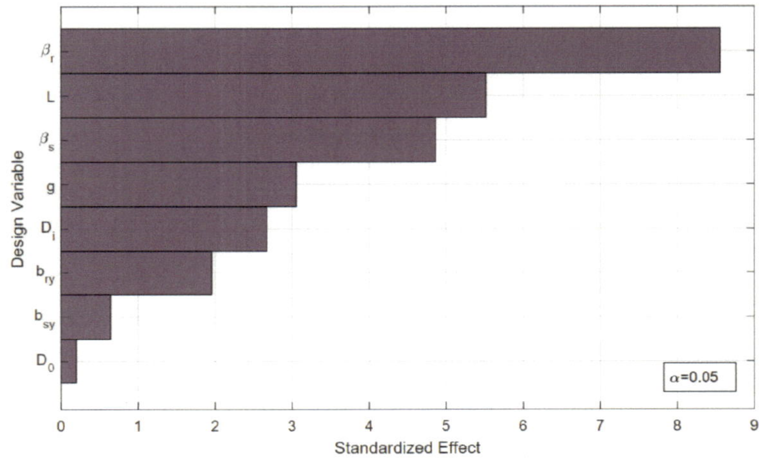

Figure 8.
Pareto chart of the standardized effects for average torque (T_{ave}).

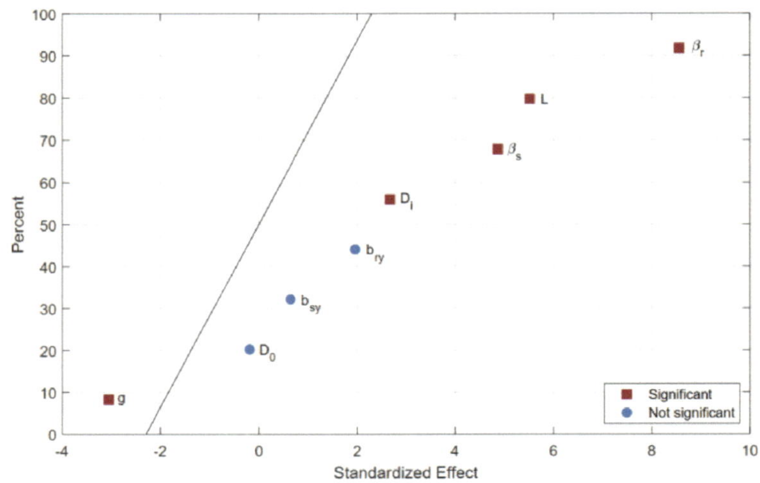

Figure 9.
Normal plot of the standardized effects for average torque (T_{ave}).

arc (β_r), the core length (L), and the stator polar arc (β_s). The polar arcs determine the energy conduction angle and thus the torque production interval and the average torque value (see **Figure 2**). In turn, the core length is a fundamental parameter and directly proportional to the torque value because it allows more energy to be converted into the machine. However, increasing the value of L corresponds to increasing the material volume and cost of the machine.

Of less importance, but still significant, the air gap length and inner diameter also play important roles. Of the five significant factors presented in **Figure 9**, only the air gap has a negative effect on the average torque value, i.e., as the air gap increase, the torque value decreases. The stator and rotor polar arcs, core length, and inner diameter have a positive effect.

4.2 Effects on total losses

When designing a machine, the designer should not be concerned only with the developed power and torque; the amount of losses and efficiency is often a decisive factor for the choice of the machine. Therefore, it is important to know how each dimension influences the value of the total losses, which is the sum of the losses in the windings and in the core. Losses were calculated as presented in Section 3, and in [33], mechanical losses are not considered. **Figure 10** shows the Pareto chart of standardized effects for losses, and **Figure 11** shows the normal plot of standardized effects.

According to the graphs presented, the stator polar arc and core length are the two dimensions that have the greatest impact on the losses value. This is because higher stator polar arc values cause increased losses in both copper and core as it increases the mean length of the winding, iron volume, and SRM saturation. The factor that causes the second major impact is core length as it has the greatest influence on copper losses, since the longer the core length, the longer the mean length of the winding.

The other dimensions have minor impact on the value of the losses, mainly because they cause change in only one part of the losses, copper losses, or core losses, so the total effect is smaller. Thus, the dimensions that affect the winding are more significant because copper losses represent the vast majority of losses of an electric machine.

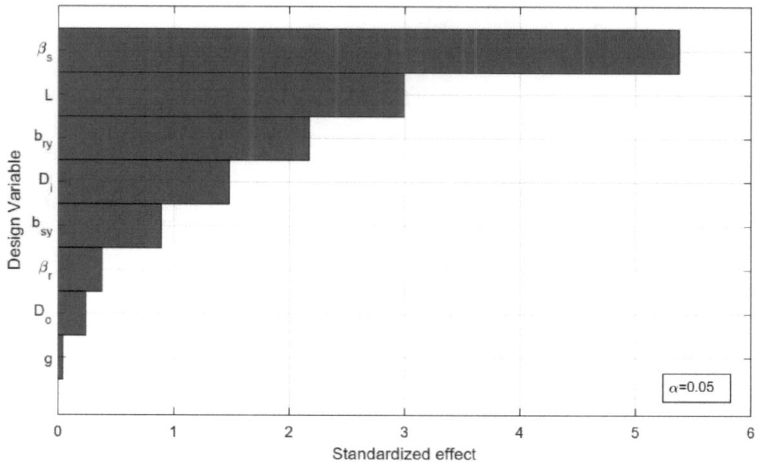

Figure 10.
Pareto chart of the standardized effects for total losses (P_{tl}).

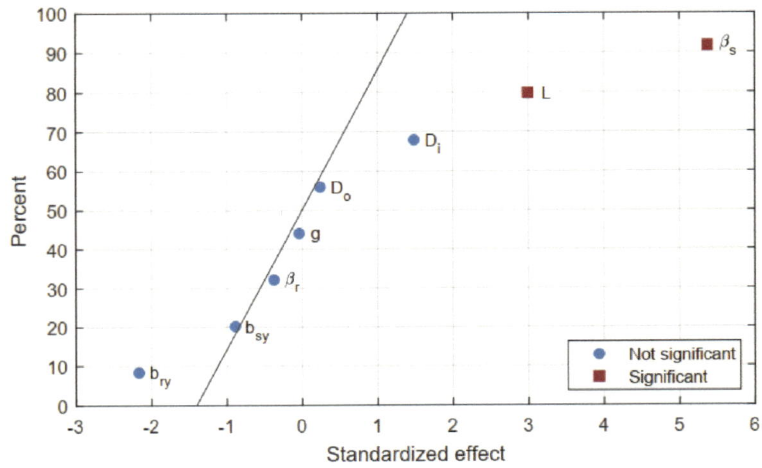

Figure 11.
Normal plot of the standardized effects for total losses (P_{tl}).

5. Conclusions

This chapter focused on detailed explanation of the design procedures of switched reluctance machines, for which the contributions of several authors are summarized and presented to the reader. At the outset, the structure and operation of SRM are presented. Subsequently, the SRM design procedures are presented, and we can observe that the authors sometimes disagree on the values and procedures.

Many dimensions are chosen through intervals, and this choice is often made subjectively and empirically. This fact justified the second part of this chapter, which is an analysis of the effects of dimensions on SRM performance. For this purpose, we used definitive screening design, a class of design of experiments that allows analysis of the effects of each factor with a reduced number of simulations. The impact of eight dimensions on the average torque value and on SRM losses was analyzed, the experiments were conducted through finite element simulations, and the analysis was performed by statistical software.

For the average torque, the most important dimensions are the polar arcs and the core length, since the relationship between the polar arcs determines the torque production interval and, therefore, the average torque value. However, the stator pole arc and core length are the two most significant factors for losses, and this can compromise the machine's efficiency. These results show that changes in dimensions must be made carefully and that effect analysis is important to draw these conclusions.

The results presented by the effect analysis complement the design procedures and become a useful tool for beginner designers. This data provides an insight into how each variable affects machine performance, which is important information on which choices are based at the design stage.

Acknowledgements

The authors thank CAPES (Coordination for the Improvement of Higher Level Personnel Agency) of the Brazilian Ministry of Education for the resources allocated to the development of this research work.

Author details

Ana Camila Ferreira Mamede[1*], José Roberto Camacho[1] and Rui Esteves Araújo[2]

1 Federal University of Uberlandia, Uberlandia, Minas Gerais, Brazil

2 INESC TEC and Faculty of Engineering of the University of Porto, Porto, Portugal

*Address all correspondence to: anacamilamamede@gmail.com

References

[1] Lawrenson P, Stephenson J, Fulton N, Blenkinsop P, Corda J. Variable-speed switched reluctance motors. IEE Proceedings B Electric Power Applications. 1980;**127**(4): 253-265

[2] Pyrhönen J, Jokinen T, Hrabovcová V. Design of Rotating Electrical Machines. 2nd ed. Chichester: John Wiley & Sons; 2014

[3] Somesan L, Padurariu E, Viorel I. Two simple analytical models, direct and inverse, for switched reluctance motors. Progress in Electromagnetics Research. 2013;**29**:279-291

[4] Besmi M. Geometry design of switched reluctance motor to reduce the torque ripple by finite element method and sensitive analysis. Journal of Electric Power and Energy Conversion Systems. 2016;**1**(1):23-31

[5] Teixeira V, Oliveira D, Pontes R, Viana S. Influence of the switched reluctance machines design parameters on its steady-state operation characteristics. In: 2007 International Conference on Electrical Machines and Systems (ICEMS). Seoul; 2007. pp. 1455-1459

[6] Krishnan R. Switched Reluctance Motor Drives: Modeling, Simulation, Analysis, Design, and Applications. Boca Raton, Florida: CRC Press; 2001

[7] Balaji M, Kamaraj V. Evolutionary computation based multi-objective pole shape optimization of switched reluctance machine. International Journal of Electrical Power & Energy Systems. 2012;**43**(1):63-69

[8] Duraković B. Design of experiments application, concepts, examples: State of the art. Periodicals of Engineering and Natural Sciences. 2017;**5**(3):421-439

[9] Duraković B, Bašić H. Textile cutting process optimization model based on six sigma methodology in a medium-sized company. Journal of Trends in the Development of Machinery and Associated Technology. 2012;**16**(1): 107-110

[10] Paulo F, Santos L. Design of experiments for microencapsulation applications: A review. Materials Science and Engineering: C. 2017;**77**: 1327-1340

[11] Yu P, Low M, Zhou W. Design of experiments and regression modelling in food flavour and sensory analysis: A review. Trends in Food Science and Technology. 2018;**71**:202-215

[12] Schlueter A, Geyer P. Linking BIM and design of experiments to balance architectural and technical design factors for energy performance. Automation in Construction. 2018;**86**: 33-43

[13] Hibbert D. Experimental design in chromatography: A tutorial review. Journal of Chromatography B. 2012;**910**: 2-13

[14] Garud S, Karimi I, Kraft M. Design of computer experiments: A review. Computers and Chemical Engineering. 2017;**106**:71-95

[15] Ma C, Qu L. Multiobjective optimization of switched reluctance motors based on design of experiments and particle swarm optimization. IEEE Transactions on Energy Conversion. 2015;**30**(3):1144-1153

[16] Jones B, Nachtsheim C. A class of three-level designs for definitive screening in the presence of second-order effects. Journal of Quality Technology. 2011;**43**(1):1-15

[17] Tang Y. Characterization, numerical analysis, and design of switched reluctance motors. IEEE Transactions on Industry Applications. 1997;**33**(6): 1544-1552

[18] Raminosoa T, Blunier B, Fodorean D, Miraoui A. Design and optimization of a switched reluctance motor driving a compressor for a PEM fuel-cell system for automotive applications. IEEE Transactions on Industrial Electronics. 2010;**57**(9): 2988-2997

[19] Miller T. Switched Reluctance Motors and Their Control. Oxford: Magna Physics; 1993

[20] Shoujun S, Weiguo L, Peitsch D, Schaefer U. Detailed design of a high speed switched reluctance starter/ generator for more/all electric aircraft. Chinese Journal of Aeronautics. 2010; **23**(2):216-226

[21] Elhomdy E, Li G, Liu J, Bukhari S, Cao W. Design and experimental verification of a 72/48 switched reluctance motor for low-speed direct-drive mining applications. Energies. 2018;**11**(1):192

[22] Rafajdus P, Peniak A, Peter D, Makys P, Szabo L. Optimization of switched reluctance motor design procedure for electrical vehicles. In: 2014 International Conference on Optimization of Electrical and Electronic Equipment (OPTIM). Bran, Romania; 2014. pp. 397-404

[23] Wang Y. Switched reluctance motor analysis and design for a lunar roving vehicle [Master thesis]. Montreal, Canada: McGill University; 2013

[24] Vijayraghavan P. Design of switched reluctance motors and development of a universal controller for switched reluctance and permanent magnet brushless DC motor drives [PhD thesis]. Blacksburg, USA: Virginia Polytechnic Institute and State University; 2001

[25] International Electrotechnical Commission. IEC 60072-1:1991. Dimensions and output series for rotating electrical machines - Part 1: Frame numbers 56 to 400 and flange numbers 55 to 1080. 6th ed. 1991

[26] Riba J, Garcia A, Romero I. An educational tool to assist the design process of switched reluctance machines. International Journal of Electrical Engineering Education. 2016; **54**(1):35-56

[27] Naayagi R, Kamaraj V. Optimum pole arcs for switched reluctance machine with reduced ripple. In: 2005 International Conference on Power Electronics and Drives Systems. Kuala Lumpur, Malaysia; 2005. pp. 761-764

[28] Xue X, Cheng K, Ng T, Cheung N. Multi-objective optimization design of in-wheel switched reluctance motors in electric vehicles. IEEE Transactions on Industrial Electronics. 2010;**57**(9): 2980-2987

[29] Sheth N, Rajagopal K. Optimum pole arcs for a switched reluctance motor for higher torque with reduced ripple. In: 2003 IEEE International Magnetics Conference (INTERMAG). Boston, USA. 2003. p. CP-06

[30] Radun A. Design considerations for the switched reluctance motor. IEEE Transactions on Industry Applications. 1995;**31**(5):1079-1087

[31] Radun A. Analytically computing the flux linked by a switched reluctance motor phase when the stator and rotor poles overlap. IEEE Transactions on Magnetics. 2000;**36**(4):1996-2003

[32] Sheth N, Rajagopal K. Estimation of core loss in a switched reluctance motor

based on actual flux variations. In: 2006 International Conference on Power Electronic, Drives and Energy Systems. New Delhi, India: 2006. pp. 1-5

[33] Materu P, Krishnan R. Estimation of switched reluctance motor losses. IEEE Transactions on Industry Applications. 1992;**28**(3):668-679

[34] Hayashi Y, Miller T. A new approach to calculating core losses in the SRM. IEEE Transactions on Industry Applications. 1995;**31**(5):1039-1046

[35] Juran J, Godfrey A. Juran's Quality Handbook. 5th ed. New York, USA: McGraw-Hill Professional Publishing; 1998

Chapter 4

Using Optimization Algorithms in the Design of SRM

Petrushin Alexandr Dmitrievich,
Kashuba Alexandr Viktorovich
and Petrushin Dmitry Alexandrovich

Abstract

Switched reluctance motors (SRM) are increasingly used in various industries and vehicles, in addition, in household appliances and in medical equipment. This chapter presents the results of research on the geometric optimization of the magnetic circuit of the switched reluctance motor (SRM) and optimization of the control algorithm. The developed optimization method of magnetic circuit geometry is based on the existing Monte Carlo method. It also allows finding global maxima and minima of an objective function. The main differences of the developed method are as follows: it includes the normal distribution and searches for the geometric shape of the rotor pole as smooth curves (the air gap is not necessarily constant). The problem of optimal control was solved using the Pontryagin's maximum principle. The initial conditions for the auxiliary functions were determined using the Newton-Raphson method. The recommendations on the practical implementation of the optimal control algorithm are given.

Keywords: switched reluctance motor, optimization, Monte Carlo method, magnetic circuit, random curve generation algorithm, optimal control algorithm, Newton-Raphson iteration method

1. Introduction

The approaches used in the design of electric drives are more and more similar to the laws of nature evolution [1–3]. Thus, during the evolution of living organisms, the problem of reliable life support is solved with minimal expenditure of energy with a constant change in natural environmental conditions. The theory and practice of using modern electric drives increasingly give them the features of an intellectual approach. This makes it possible to maintain a high level of reliability and operate efficiency with changing external influences.

In the process of the electric drive synthesis, there is no ideal technical solution for the same terms of reference and formulated requirements (like the laws of nature evolution, there are many different variations).

When creating an electric drive, it is often necessary to make a compromise between a highly specialized technical solution for one specific device and a universal version for use in various fields.

When creating electric drives, the best technical solutions are obtained using optimization algorithms. It should be noted that a more general formulation of the problem and less restrictions provide better results.

There are significant difficulties for the optimization problem (including optimization of the magnetic circuit geometry and control optimization) for the following reason. The existing methods of designing the magnetic circuit of switched reluctance motor (SRM) do not fully take into account the drive control algorithm, since they are guided either by the nominal mode of operation or by some mode for a particular application. This is caused by uncertainty: if the algorithm is divided into time sections (within which the control actions are constant), many optimal variants of the magnetic circuit are obtained when solving the optimization problem. The number of optimal variants is equal to the number of time sections. That is, when controlling the drive, the extremum of the cost function (CF) will be obtained only if the magnetic circuit of the electric motor (magnetic core and winding) changes its configuration. But this is impossible now and in the near future. We have covered this problem in paper [4]. In the same paper, we proposed an algorithm for determining the only variant of the magnetic circuit of the traction motor when solving the optimization problem with varying control actions.

This chapter (Sections 2–5) presents the results of the optimization of the magnetic circuit of the motor using the Monte Carlo method. The Monte Carlo method was adapted to find the optimal geometric shape of the rotor pole as smooth curves (the air gap is not necessarily constant). To reduce the calculation time and ensure a high probability of obtaining a global extremum when performing an optimization by an improved method, a variable probability density of a random variable is used.

This chapter (Section 6) presents the results of solving a boundary problem of optimal SRM control. The cost function includes the square of the voltage applied to the stator winding.

Recommendations on the practical implementation of the optimal control algorithm are given (using the most common half-bridge power circuit and modern electronic components).

2. Development of an optimization algorithm for the SRM magnetic circuit

There are many different optimization methods: simplex method, Monte Carlo method, ant colony optimization algorithms, gradient descent, evolutionary algorithm, etc. [5, 6]. All of them are adapted for specific aims and make it possible to take into account the specific conditions inherent in a particular task.

As an optimization criterion, there may be different characteristics of the motor: average electromagnetic torque, torque ripple, efficiency, heat loss, and others (depending on the requirements of terms of reference). In multi-objective optimization, the objective function must be set using mathematical equations that take into account the priority of a certain criterion or use multi-objective optimization algorithms.

The developed optimization algorithm is based on the Monte Carlo method. Experience in designing SRM [4] shows that using this method to optimize the geometric dimensions of the magnetic circuit can significantly improve the performance of the motor compared to the results of the design without the use of optimization.

The Monte Carlo method allows finding the global extremum of the several CF variables. The essence of this method is as follows:

1. Setting limits for all optimized parameters.

2. A series of pseudorandom numbers is generated (uniform distribution of a random variable) and is performed within the specified limits for each optimized parameter.

3. CF value calculation.

Repeating steps 2 and 3 provides an extensive basis for establishing various relationships. By analyzing the data obtained, it is possible to know which parameter values provide the extremum of CF.

One of the advantages of the Monte Carlo method is that it guarantees with a fairly high probability that the global (not local) extremum of the cost function is found. But the main drawback is the need to perform a huge amount of CF calculations (hundreds, thousands, tens of thousands) to get an acceptable result. This leads to more time spent on optimization. Hence, SRM design lasts much longer. Even with a few numbers of optimized parameters, this disadvantage is very noticeable.

The algorithm developed on the basis of the Monte Carlo method makes it possible to find the global extremum of the CF with an accuracy sufficient for many practical problems in a relatively short time.

It is possible to reduce the time of optimization by generating random numbers near the coordinates of the global extremum.

The main difference of the developed algorithm is the generation of random numbers performed according to the Gaussian laws. Its probability density function is expressed as

$$f(x) = \frac{e^{-\frac{(x-\mu)^2}{2\sigma^2}}}{\sqrt{2\pi\sigma^2}}, \tag{1}$$

where μ is the expected value, σ is the standard deviation, and σ^2 is the variance.

Before starting the main part of the algorithm, it is necessary to set a certain number of random values of parameters (uniform distribution of a random variable). After that the best point should be selected of all available coordinates (the best point is the coordinates at which the CF has an extreme value). The selected point is the first approach to the extremum. Then the random number generation with Gaussian distribution is used (expected value corresponds to the coordinates of the first approach, and the variance is still large). After each calculation of the CF for the next approach point, it is necessary to take the last point if the value of the CF of this point is the best among all. Thus, self-adaptation occurs by adjusting the expectation and reducing the variance (**Figure 1**). It should be noted that the dependence of the variance on the calculation number of the CF affects the optimization time.

Figure 1 is an example for the case if the limits for the variation by the optimized parameter were from 0 to 1 and the extremum would be located at the point with coordinates 0.55 for this parameter. After all the calculations of the function have expired, the last result (quasi-extremum) is taken as the result of optimization.

In order to make it convenient to depict the graphic dependence of the variance on the CF calculation number, we introduce the parameter T. It is the reciprocal for the variance σ^2

$$T = \frac{1}{\sigma^2}. \tag{2}$$

As stated above, before starting the main part of the algorithm, it is necessary to set a certain number of random values of parameters (uniform distribution of a

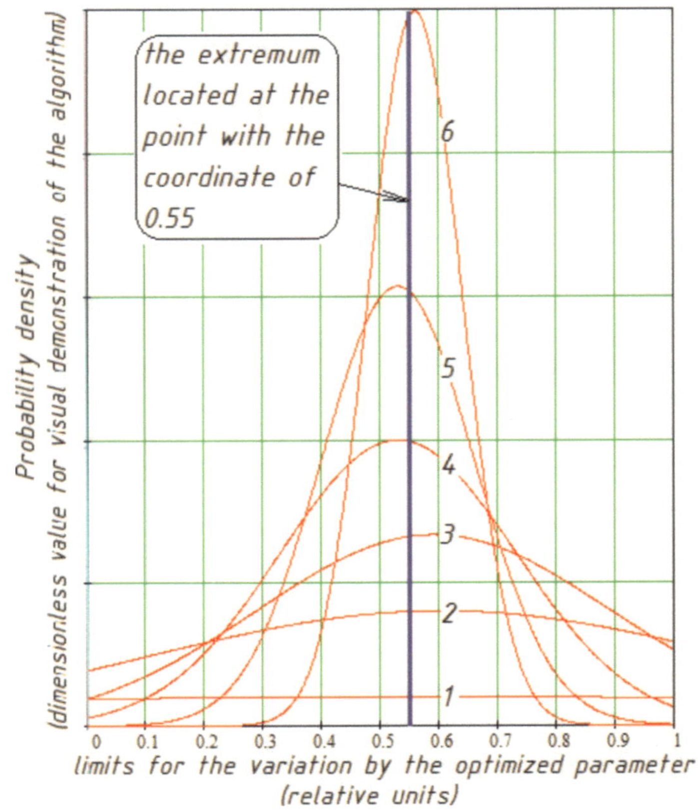

Figure 1.
Examples of probability density graphs (1, start of optimization; 6, completion of optimization).

random variable, variance value is high, T is close to zero). This will suggest that the global extremum will not be missed (familiarization zone, **Figure 2**). Next, self-adaptation should occur, and then a more accurate calculation of the extremum coordinates should follow. The dependence of the T on the CF calculation number is expressed as follows:

$$T(n_0) = k \cdot \frac{arctg(w \cdot n_0 - w \cdot S) + arctg(w \cdot S)}{arctg(w - w \cdot S) + arctg(w \cdot S)}, \tag{3}$$

where n_0 is the indicator of optimization completion, defined by Eq. (4), k is the coefficient that determines the value of the variance at the final stage of optimization (**Figure 3**), w is the parameter determining the duration of self-adaptation (**Figure 4**), and S is the parameter that determines when the self-adaptation will begin (**Figure 5**):

$$n_0 = \frac{n}{n_\Sigma}, \tag{4}$$

where n is the current calculation number of the CF and n_Σ is the total number of calculations of the CF (set before optimization).

Figures 3–5 show the influence of parameters k, S, and w on the general form of the Eq. (3).

The analysis of **Figures 4** and **5** is presented in the form of **Table 1**.

As **Table 1** shows, various combinations of values can have both positive and negative effects. The values of S and w should be selected in such a way that the

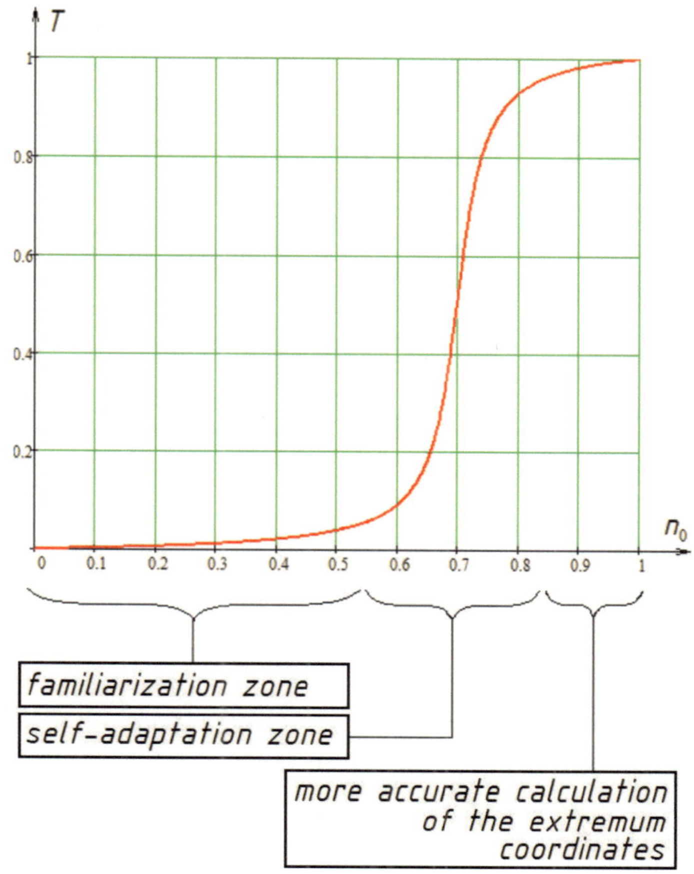

Figure 2.
The dependence of T (reciprocal for the variance σ^2) on the CF calculation number (k = 1; w = 30; S = 0.7).

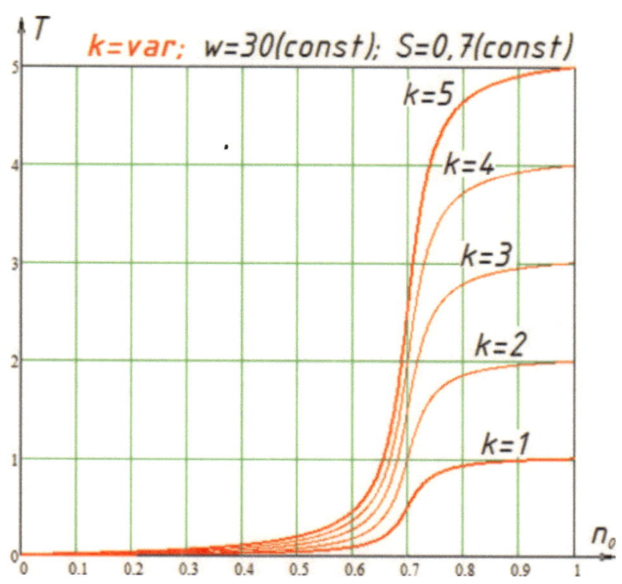

Figure 3.
The effect of parameter k on the general form of the dependence $T(n_o)$.

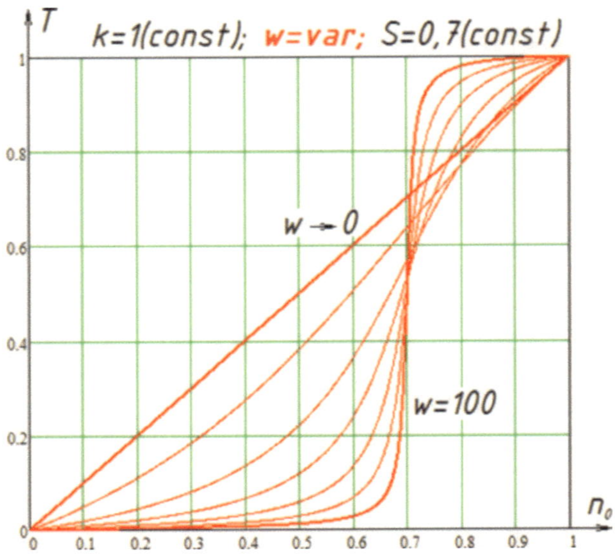

Figure 4.
The effect of parameter w *on the general form of the dependence* $T(n_o)$.

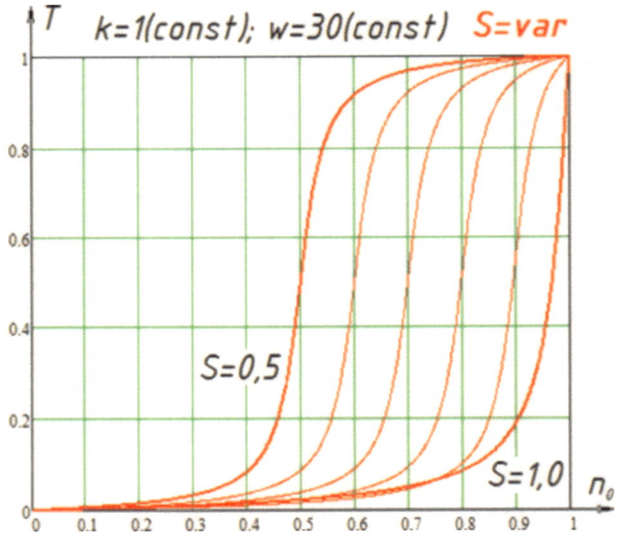

Figure 5.
The effect of parameter S on the general form of the dependence $T(n_o)$.

optimization result gives the greatest probability of finding a global extremum with the necessary accuracy. For different forms of CF, the best result will be given by different combinations of S and w. Therefore, the averaged values of S and w were found: $S = 0.7$ and $w = 30$ (**Figure 2**). These values were chosen empirically by multiple optimizations of test functions. The test functions were specified analytically. The coordinates of their extremum can be calculated without optimizing (to be able to assess the accuracy of finding a global extremum).

The coefficient k in Eq. (3) indicates the variance at the final stage of optimization (the accuracy of the extremum coordinates calculation).

The value of the coefficient k must be determined for each parameter being optimized separately. When optimizing the geometrical forms of the magnetic cores, this coefficient should be determined on the basis of the accuracy of the

	High value	Low value
S	*Positive effect:* • Increases the probability of finding a global extremum, rather than a local extremum	*Positive effect:* • Increases the number of CF calculations for more accurate calculation of the coordinates of the extremum
	Negative effect: • Decreases the number of CF calculations for more accurate calculation of the coordinates of the extremum	*Negative effect:* • Decreases the probability of finding a global extremum, rather than a local extremum
w	*Positive effect:* • Increases the probability of finding a global extremum, rather than a local extremum • Increases the number of CF calculations for more accurate calculation of the coordinates of the extremum	*Positive effect:* • Self-adaptation zone is increasing • Limitations in which the search for extremum which takes place can slowly narrow, so the real extremum will not be outside
	Negative effect: • Self-adaptation zone is reduced • Limitations in which the search for extremum which takes place can quickly narrow, leaving a real extremum outside (find it becomes impossible because the rapidity of variance reduction is large)	*Negative effect:* • Decreases the probability of finding a global extremum, rather than a local extremum • Decreases the number of CF calculations for more accurate calculation of the coordinates of the extremum

Table 1.
Analysis of the effect of parameters S and w on the result of optimization.

manufacturing equipment used in the manufacture of the SRM magnetic core. In addition, when determining the coefficient k, it is necessary to take into account the influence of each parameter value on the CF. Applied to SRM this question deserves special research [7].

When using the coefficient, it should be guided by the "68–95–99.7" rule: a random variable having a normal distribution does not deviate from the expectation in absolute value by more than 3σ with a probability of 99.73%.

The coefficient k should ensure the generation of random numbers at the final stage of optimization within limits equal to the limits of accuracy required (**Figure 6**).

The value of the coefficient should be determined by the following formula

$$k = \frac{1}{\sigma^2} = \frac{1}{\left(\frac{\delta}{3}\right)^2}, \tag{5}$$

where δ is the allowable relative error of extremum coordinates calculation.

Number 3 in the denominator indicates the value of three sigmas.

Eventually, the developed optimization algorithm was based on the distribution function with variable expectation and standard deviation (Eq. (6)). The expectation depends on the intermediate optimization results. The standard deviation and variance depend on the calculation number of the CF:

$$\begin{cases} f(x, \mu, n) = \dfrac{1 + erf\left(\dfrac{x-\mu}{\sigma(n)\sqrt{2}}\right)}{2}; \\ \sigma(n) = \left(\dfrac{2\cdot\delta}{6}\right)\sqrt{\dfrac{arctg(w - w\cdot S) + arctg(w\cdot S)}{arctg\left(w\cdot\dfrac{n}{n_\Sigma} - w\cdot S\right) + arctg(w\cdot S)}}, \end{cases} \tag{6}$$

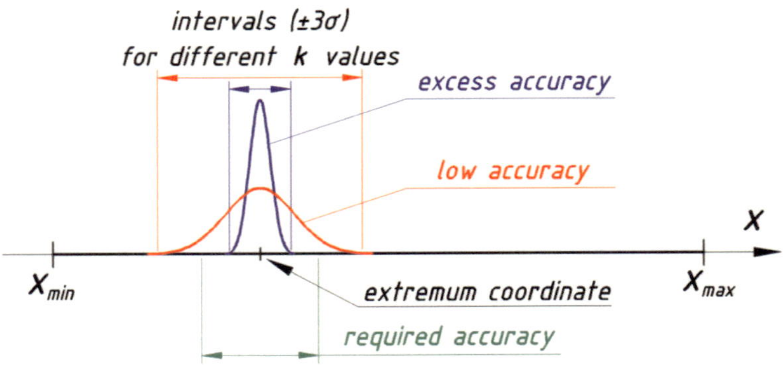

Figure 6.
The influence of the coefficient k on the accuracy of the extremum coordinates calculation.

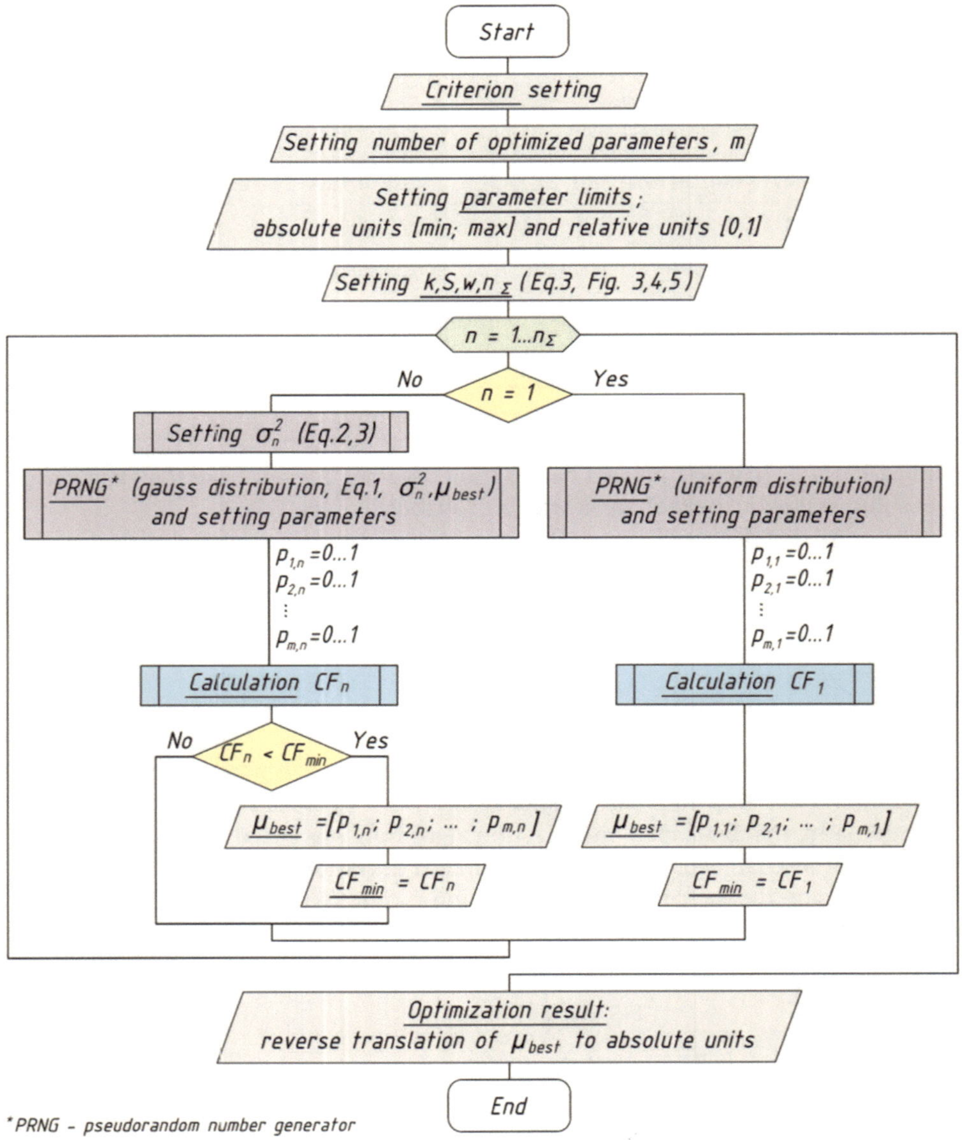

Figure 7.
Flowchart defining the developed optimization algorithm.

where $erf(x)$ is the Gauss error function (Eq. (7)) [8], μ is the expected value, $\sigma(n)$ is the standard deviation, n is the current calculation number of the CF, δ is the allowable relative error of extremum coordinates calculation, w is the parameter determining the duration of self-adaptation (**Figure 4**), S is the parameter that determines when the self-adaptation will begin (**Figure 5**), n_Σ is the total number of calculations of the CF:

$$erf\ (x) = \frac{2}{\sqrt{\pi}} \int_0^x e^{-t^2} dt. \tag{7}$$

Figure 7 summarizes Section 2 and shows the flowchart defining the developed optimization algorithm.

3. Mathematical description of the pole curved shape

When designing SRM, the shape of the poles of the rotor and stator is most often defined by contours similar to an isosceles trapezoid using segments and arcs [2, 9]. This results in a uniform air gap in the motor. **Figure 8** shows an example of the rotor pole contours (uniform air gap).

These restrictions on the shape of the magnetic core do not allow to fully realizing the possibilities for improving the performance of the SRM by changing the geometry of its magnetic circuit.

The scientific novelty of our research is the way of setting restrictions for varying the optimized geometric parameters of the SRM rotor by means of the zone shown in **Figure 9**.

As **Figure 9** shows, the location of the rotor pole contour is limited only by the diameter of the rotor shaft, the maximum approach to the stator, and the sector which depends on the number of poles of the rotor. The left side of the pole is set by mirroring the contours of the right side.

Note that it does not complicate the manufacturing technology, because modern equipment allows to make sheets of magnetic core without restrictions on their shape.

Consider the mathematical description of the rotor pole curved shape. When optimizing the geometry of the curved pole shape, it is necessary to generate a *random curve* at each step (**Figure 9**), in contrast to the optimization of the geometry of the trapezoidal pole shape (**Figure 8**), where it is necessary to generate *random numbers*, to setting specific geometric dimensions (b1r, b2r, etc.).

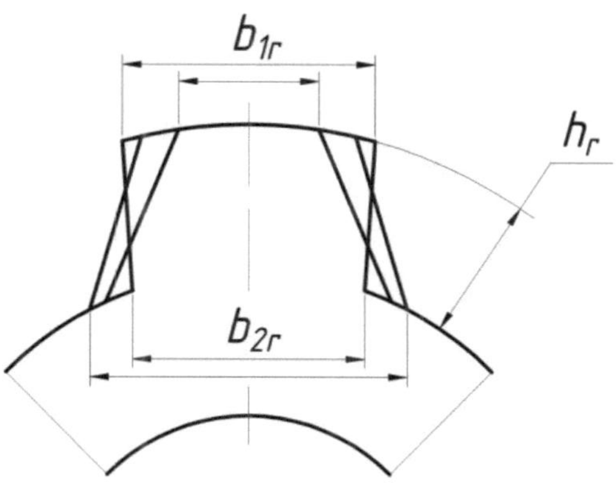

Figure 8.
Example of standard rotor pole defined using segments and arcs.

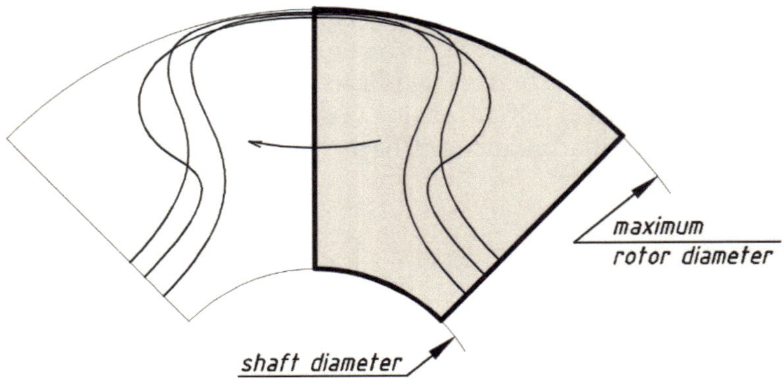

Figure 9.
The area of possible contours of the rotor poles when setting the curved shape of the pole.

To solve this problem, an algorithm for generating a random curve was developed. The design features of the SRM magnetic cores make it necessary to take into account some restrictions on the curve shape that determines the magnetic core at each step of optimization:

1. Point 1 (belongs to the pole contour and is located on the pole axis) must be farther from the rotor axis of rotation than point 2 (belongs to the pole contour and is located on the groove axis). This rule is shown in **Figure 10**.

2. The pole contour should not be in sharp protrusions, peaks, and other irregularities. This may degrade motor performance due to deep local magnetic saturation (**Figure 11**) [10].

3. To fulfill Eq. (2), it is necessary that on the border of the pole zone, the tangents to the magnetic core are perpendicular to the radial direction (**Figure 12**).

It is possible to take into account the above conditions if the random curve is set by means of the Cartesian coordinate system using the analytical formula, then bring the curve to the form required for the SRM pole (**Figure 13**).

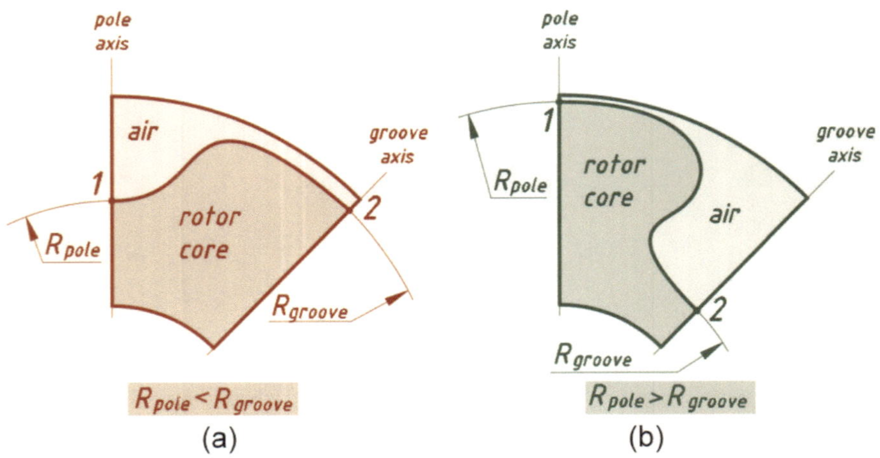

Figure 10.
An example of incorrect (a) and correct (b) setting of the coordinates of the boundary points of the pole contour.

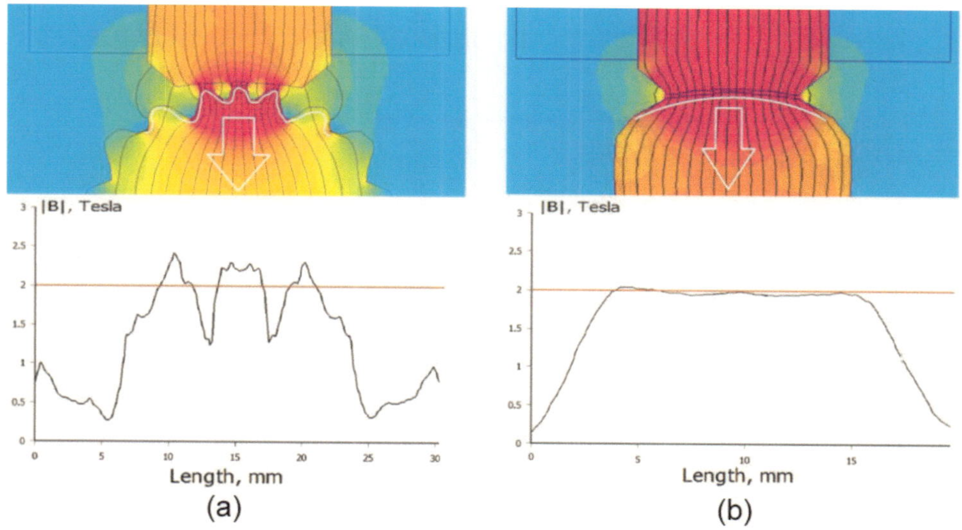

Figure 11.
An example of the influence of irregularities of the pole contour on the magnitude of the flux density in the magnetic core: incorrect (a) and correct (b) setting of the curve describing the pole contour.

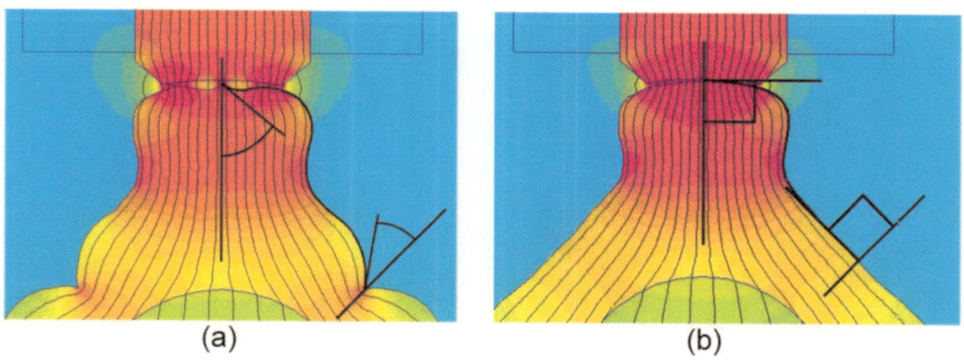

Figure 12.
Examples of incorrect (a) and correct (b) setting of the inclination of the extreme sections of the contour of the rotor pole.

Curve plotting condition: it should be located in the pole sector so that the segments (numbered from 1 to 21, **Figure 13b**) are separated by the curve in the same proportions as the corresponding segments in the Cartesian coordinate system (**Figure 13a**).

The proposed location of the segments (from 1 to 21) in the zone sector pole ensures Eq. (1) ($R_1 > R_2$, **Figure 10**).

To fulfill Eq. (2) (the pole contour should not be in sharp protrusions, **Figure 11**), it is advisable to set a random curve using the sum of sinusoids with different amplitudes, frequencies, and phases:

$$y_1(x) = \sum_{i=1}^{n} A_i \cdot \sin(\omega_i \cdot x + \varphi_i), \tag{8}$$

where n is the number of sinusoids, A is the amplitude, ω is the frequency, and φ is the sine wave phase.

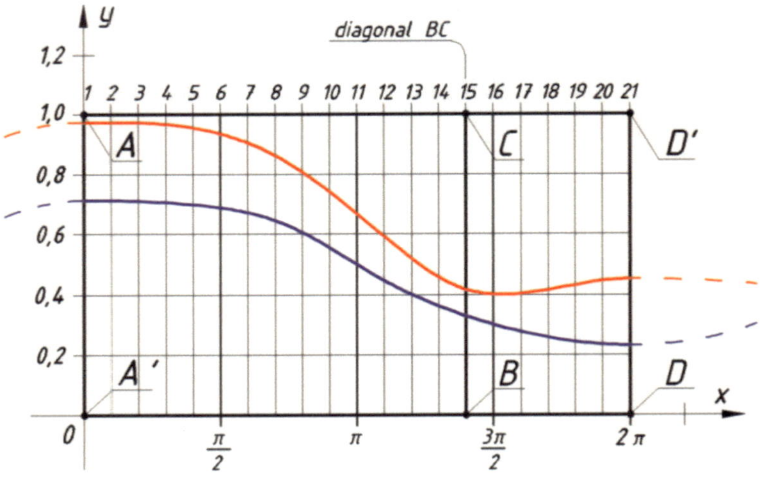

(a)

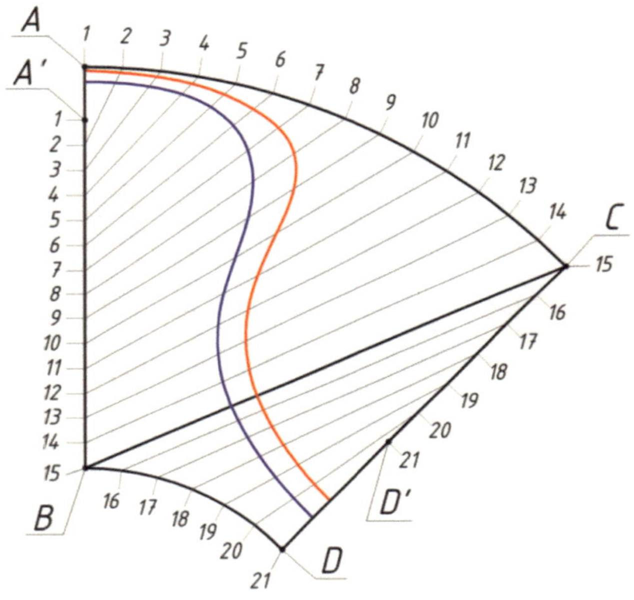

(b)

Figure 13.
The principle plotting of the rotor pole curved shape. The magnetic core curves (b) divide the segments (numbered from 1 to 21) in the same proportions as the corresponding initial curves (a) defined using a mathematical expression and represented in a Cartesian coordinate system.

As shown by the experience of designing SRM [11], in order to avoid protrusions and irregularities, the values n, A, and ω should be set as follows (**Table 2**).

To eliminate the irregularities of the magnetic core at the top of the pole and in the center of the groove (Eq. (3), **Figure 12**), it is necessary to further develop measures that exclude the inclination of the random curve in the abscissas 0 and 2π: the derivative in these coordinates should be zero, and this should not contribute to the appearance of high fluctuations on the curve. We write the equation of the curve as follows:

$$y_2(x) = \begin{cases} y_1(0) \cdot q(x) + y_1(x) \cdot (1 - q(x)), \text{ if } x < \pi; \\ y_1(2\pi) \cdot q(x) + y_1(x) \cdot (1 - q(x)), \text{ if } x \geq \pi, \end{cases} \tag{9}$$

Parameter name	Symbol	limitation		Value
		Min	Max	
Amplitude	A	0	1	0...1
Frequency	ω	0	1	0...1
Sine wave phase	φ	0	2π	0...2π
Number of sinusoids	n	—	—	2

Table 2.
Values of mathematical parameters of a random curve.

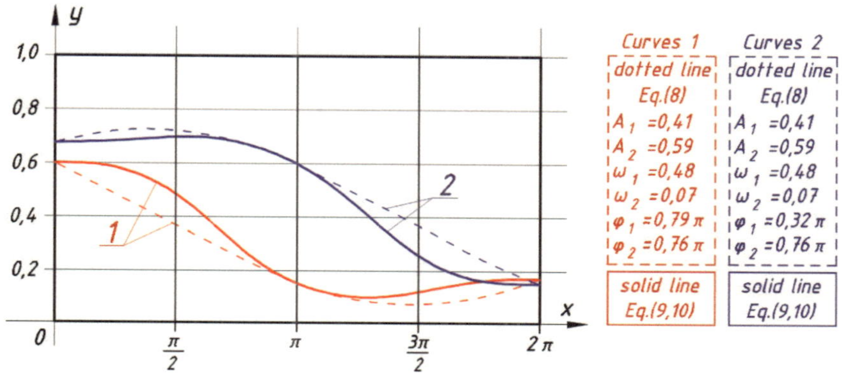

Figure 14.
Examples of changing the shape of curves after conversion (Eqs. (9) and (10)).

where $q(x)$ is the function that has value, $q(0) = 1$, $q(\pi) = 0$, and $q(2\pi) = 1$, and derivatives in these coordinates, $q'(0) = q'(\pi) = q'(2\pi) = 0$.

The most suitable for this is the equation:

$$\begin{cases} q(x) = \dfrac{\cos(x) + 1}{2}; \\ q'(x) = -\dfrac{\sin(x)}{2}. \end{cases} \tag{10}$$

The function values at the three significant coordinates ($x = 0, x = \pi, x = 2\pi$) remain unchanged after the conversion. The derivatives at boundary coordinates ($x = 0$ and $x = 2\pi$) become zero in the middle of this interval (for $x = \pi$); the derivative is equal to the derivative of the original function. This avoids the inclination of the curve in the boundary coordinates with a slight change in its shape compared to the original curve (**Figure 14**).

4. Optimization of the curved shape of the rotor pole

The above optimization algorithm is applicable to optimize the trapezoidal pole shape. To optimize the curved shape, it is necessary to take into account some features of the implementation of this algorithm in computer programming.

When optimizing the trapezoidal pole shape, the optimization parameters are clearly defined: they represent the specific geometric dimensions of the magnetic core. The curved shape of the pole does not have these geometric dimensions. The curved shape of the pole can be defined differently, for example, by the set of

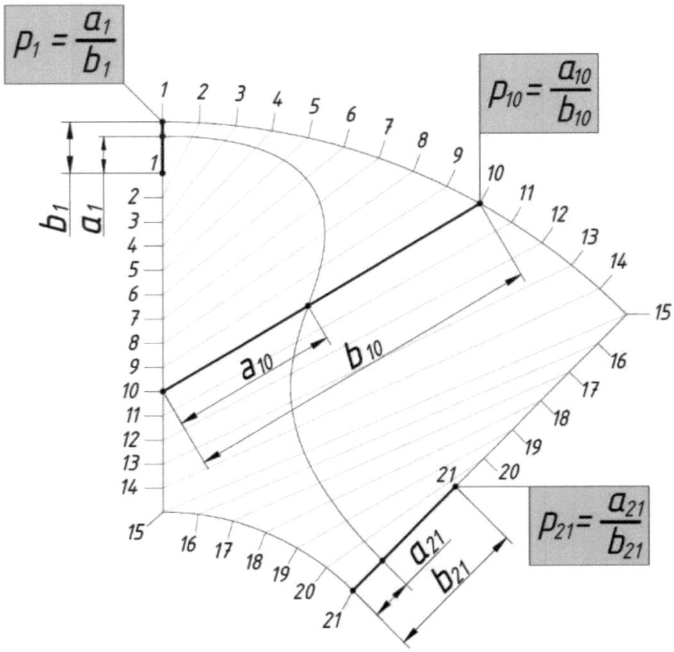

Figure 15.
Formation of optimization parameters and their values.

coordinates of the points belonging to this curve. Therefore, as optimization parameters, it is advisable to set the ratio (Eq. (11)) of the geometric dimensions *a* and *b* for each of the 21 segments (**Figure 15**).

$$p_n = \frac{a_n}{b_n}, \tag{11}$$

where p_n is the optimization parameter and a_n and b_n are the geometrical dimensions that determine the shape of the pole (**Figure 15**).

Thus, the number of optimization parameter is equal to the number of points defining the half-pole contour. It should be noted that an increase in defining points does not entail an increase in optimization time. The curve is set using six random numbers (for two sinusoids, as in our case) that are independent of each other, similarly, optimization by six parameters.

The number of defining points is selected for each design problem singly, depending on the required accuracy and smoothness of the pole contour. The number of defining points is limited only by the features of creating a mesh in a computer program for calculating magnetic fields (e.g., FEMM). The more points that define the pole, the more the number of nodes in the output mesh. This entails a significant increase in the calculation time without increasing their accuracy [12].

5. The example of optimizing the geometric shape of a rotor pole

As an example, consider the optimization of the curved shape of the rotor magnetic core for SRM 6/4; the outer diameter of the stator magnetic core is 131 mm, and the shaft diameter is 30 mm. The number of points defining the half-pole contour is 21.

Optimization criteria: minimum of torque ripple, it is defined as follows [4, 11, 13]:

$$\Delta T_e = \frac{\max(T_e) - \max(T_e)}{T_a} \cdot 100\%, \tag{12}$$

where ΔT_e is the torque ripple value (%), T_e is the instantaneous torque value (N m), and T_a is the average value of electromagnetic torque (N m).

When modeling electromagnetic processes in a SRM, the following assumptions were made:

1. The supply network has infinite power.

2. Losses in power switches and converter conductors are not taken into account.

3. Rotor eccentricity is absent; there is no deflection and shaft skew.

4. Magnetic field was calculated in two-dimensional space, calculated SRM space is stationary, and leakage flux at the end winding is neglected.

5. Electrical steel has isotropic properties for a two-dimensional computational space.

6. The shape of the current was rectangular (ideal case).

The result of the optimization is shown in the **Figure 16**.

In addition to the calculations, the electric motor with two rotors was tested alternately: with a rotor that was designed without the use of optimization methods and with a rotor whose geometric dimensions were obtained by optimization.

The purpose of the experiment was to obtain curves of the dependence of the torque on the rotor position for a motor with different rotor.

The experiment for each electric motor was carried out according to the following algorithm:

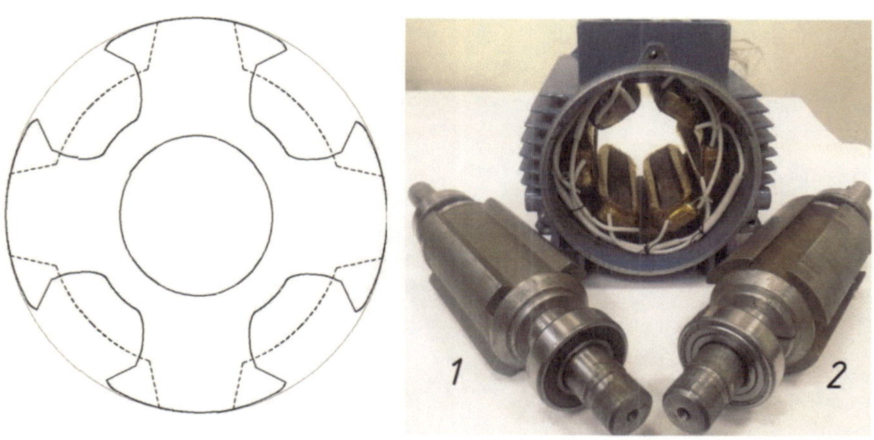

Figure 16.
Experimental SRM sample: 1 (dotted line) is the rotor obtained without optimization algorithm; 2 (continuous line) is the rotor obtained as a result of optimization.

- The values of the torque for the full rotation of the rotor were measured in 0.5° increments with a 0.33 m arm, at the end of which the strain gauge was attached. At the same time, the readings of two measuring instruments were registered: the absolute rotary encoder and the weight terminal (the force occurring in the strain gauge with an accuracy of 0.01 N). The phase windings were supplied with a variable autotransformer (with the help of which a current of 10 A), the diode bridge and a battery of capacitors with a total capacity of 4400 μF.

- According to the data, the curves of the dependence of the torque on the rotor position were obtained.

- The averaging of all sections of the curve, limited by the repeat interval, was performed. For a three-phase SRM, which has a 6/4 configuration, this interval is 30° (120 electric degree). Averaging of 12 sections was performed that correspond to different rotor positions. So, it was possible to reduce the error in measurements caused by the rotor eccentricity.

The results of the experiment are shown in **Figure 17**.

As can be seen from **Figure 17**, the experimental and calculated curves that correspond to one rotor are sufficiently close in form. When comparing the two experimental curves with each other, it follows that the torque ripple with the rotor obtained as a result of optimization is much less than for the motor with the rotor, which was designed without optimization. The torque ripple (Eq. (12)) with the rotor, which was designed without the use of optimization, is 137%. The torque ripple with the rotor, which was designed by means of optimization, is 51%. Thus, it was possible to reduce torque ripple more than 2.5 times.

When designing switched reluctance motors for specific applications, in addition to reducing the pulsations of the electromagnetic torque, other parameters, such as efficiency, thermal conditions, etc., need to be improved. To do this, we need to use a multi-objective optimization, for example, compiled using linear scalarization of cost function: The above optimization method based on Monte Carlo optimization method can also be used in the case of a multi-objective optimization:

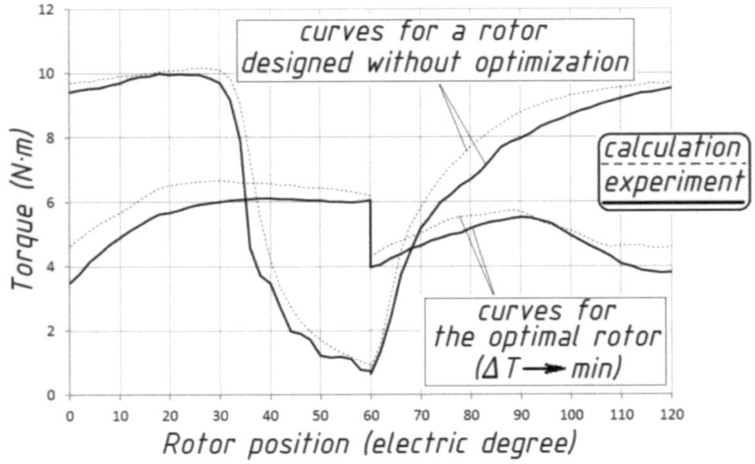

Figure 17.
Comparison of experimental results for two rotor samples.

$$F = \max\left(\sum_{i=1}^{k} w_i \cdot f_i(\overline{x})\right), \tag{13}$$

where F is the cost function of the optimization problem; k—number of optimization criteria; w_i is the weights of the objectives, which indicate the priority of the i-th criterion; and f_i is the function that depends on the values of the optimized parameters and indicates the numerical value of the i-th criterion.

The above optimization method based on Monte Carlo optimization method can also be used in the case of a multi-objective optimization.

Note that the design algorithm developed (allowing to get an uneven air gap and a curved pole shape) is applicable to optimize the stator pole. But for its implementation, it is necessary to take into account the additional restrictions that are created by the stator winding.

6. Development of the SRM optimal control algorithm

The following is one of the variants of the approach to the design of electric motors with regard to the control algorithm using the example of SRM, which are considered promising for applications.

It is known that the wider the range of permissible values of the coordinates and control actions of the control object, the less likely that the optimal dependence will pass along the boundary of the permissible values of coordinates or control actions and the greater the likelihood of obtaining a better final result. We formulate the requirements for the converter, which forms the control algorithm for the SRM on the basis that its functionality should be as wide as possible:

- To power one phase of the SRM, there must be at least one semiconductor key in order to be able to independently switch for this phase. The current must be supplied to the winding during the conduction interval set by the rotor position specified by the control algorithm, in order to ensure the specified operating mode.

- The converter should be able to regulate the phase current in the whole range of frequencies of rotation of the SRM and form a predetermined form of current using pulse-width modulation (PWM) together with frequency-pulse modulation (FPM).

- The converter should be able to simultaneously feed several phases independently of each other, as well as ensure the operation of phases with overlapping.

As a converter for powering SRM, we use the most common half-bridge circuit. Various control laws can be implemented using modulation of the supplying voltage of the SRM and forming a current wave of almost any shape in the stator winding.

Find the power supply parameters of the SRM, which will provide a minimum of cost function, the integrand of which includes the square of the voltage applied to the stator winding:

$$J_{\Phi} = \int_{\theta 1}^{\theta 2} \frac{u^2}{R} d\theta \tag{14}$$

The optimization criterion is selected based on the analysis of variables affecting electrical losses. The integrand (Eq. (14)) determines the electrical losses according to the Ohm's law for linear electrical circuits (containing only active elements). In SRM, the relationship between current and voltage is nonlinear, and the integrand (Eq. (14)) is not identical to electric losses. In spite of this, minimization (Eq. (14)) can significantly reduce the electrical losses in the stator SRM, as shown by calculations.

In solving the problem of optimal control, the maximum principle is used. The equations for the SRM parameters, auxiliary ψ-functions, and the optimal control action u^* are

$$\begin{cases} \dfrac{di_a}{d\theta} = \dfrac{1}{\omega \cdot Z_p \cdot L} \cdot \left(u - i_a R - i_a \cdot \omega \cdot Z_p \cdot \dfrac{dL}{d\theta}\right) \\ \dfrac{d\omega}{d\theta} = \dfrac{1}{J \cdot \omega \cdot Z_p} \cdot \left(\dfrac{Z_p \cdot i_a^2}{2}\dfrac{dL}{d\theta} - T_R\right) \\ \dfrac{d\psi_1}{d\theta} = \psi_1\left(\dfrac{R}{L \cdot \omega \cdot Z_p} + \dfrac{1}{L}\dfrac{dL}{d\theta}\right) - \dfrac{i_a \cdot \psi_2}{J \cdot \omega}\dfrac{dL}{d\theta} \\ \dfrac{d\psi_2}{d\theta} = \dfrac{\psi_1}{L \cdot \omega^2}(u - i_a R) + \dfrac{\psi_2}{J \cdot \omega^2}\left(\dfrac{i_a^2}{2}\dfrac{dL}{d\theta} - T_R\right) \\ u^* = \dfrac{\psi_1 \cdot R}{2 \cdot L \cdot \omega \cdot Z_p} \end{cases} \tag{15}$$

where θ is the rotation angle of the rotor, ω is the angular frequency of shaft rotation, Z_p is the number of salient poles on the rotor, i_a is the stator phase current, R is the resistance of the stator winding, L is the inductance of the stator winding, J is the moment of inertia of the rotor and the mass of the executive mechanism attached to it, and T_R is the torque of resistance on the motor shaft.

The first equation (system of Eq. (15)) is based on the Kirchhoff's voltage law for one phase of the SRM. The second equation is the power balance equation (Cauchy equation): dynamic torque, electromagnetic torque, and torque of resistance. The remaining equations are obtained from the auxiliary function H according to the maximum principle:

$$H = \psi_1\left(\frac{1}{\omega Z_p L}(u - i_a R) - i_a \omega Z_p \cdot \frac{dL}{d\theta}\right) + \psi_2\left(\frac{1}{J\omega Z_p}\left(\frac{Z_p \cdot i_a^2}{2} \cdot \frac{dL}{d\theta} - T_R\right)\right) - \frac{u^2}{R}; \tag{16}$$

$$\frac{d\psi_1}{d\theta} = -\frac{\partial H}{\partial i_a}; \tag{17}$$

$$\frac{d\psi_2}{d\theta} = -\frac{\partial H}{\partial \omega}; \tag{18}$$

$$u^* = \frac{\partial H}{\partial u}. \tag{19}$$

The initial values of the auxiliary functions were found using the Newton-Raphson iteration method based on the following equation:

$$\begin{bmatrix} \psi_1(t_0)^{j+1} \\ \psi_2(t_0)^{j+1} \end{bmatrix} = \begin{bmatrix} \psi_1(t_0)^j \\ \psi_2(t_0)^j \end{bmatrix} - \begin{bmatrix} \dfrac{\partial i_a(t_{end})}{\partial \psi_1(t_0)^j} & \dfrac{\partial i_a(t_{end})}{\partial \psi_2(t_0)^j} \\ \dfrac{\partial \omega(t_{end})}{\partial \psi_1(t_0)^j} & \dfrac{\partial \omega(t_{end})}{\partial \psi_2(t_0)^j} \end{bmatrix}^{-1} \times \begin{bmatrix} i_a(t_{end}) - i_{a/end} \\ \omega(t_{end}) - \omega_{end} \end{bmatrix}, \tag{20}$$

where j = 1, 2... is the iteration number, $\psi_1(t_0)$ and $\psi_2(t_0)$ are the initial values of auxiliary functions, $i_a(t_{end})$ and $\omega(t_{end})$ are the stator winding current and the angular frequency of rotation of the rotor at the j-th iteration at the end of the integration interval at $t = t_{end}$, and $i_{a/end}$ and ω_{end} are the desired values of the stator winding current and the angular frequency of rotation of the rotor at the end of the integration interval (boundary conditions).

More detailed mathematical transformations to obtain the solutions to Eqs. (15) and (20) are given in [3, 14].

The calculations selected SRM with the following initial data: number of salient poles on the stator, 6; number of salient poles on the rotor, 4; phase resistance at 20°C, 0.4 Ω; number of phases, 3; stator package length, mm, 80; outer diameter of the stator, mm, 131; height of salient poles on the stator, mm, 21.2; width on the stator and rotor salient poles on the side of the air gap, mm, 8.8; stator core height, mm, 8.2; height of the rotor core, mm, 4.4; and inner diameter of the rotor core (shaft diameter), mm, 73.70.

In the practical application of SRM, there are always restrictions on the amplitude of the supply voltage. **Table 3** shows the calculation results for the three options. Three different options are selected from the condition of limiting the voltage in the DC link: for the first option, the voltage is not more than 325 V, for the second option, the voltage is not more than 415 V, and for the third option, the voltage is not more than 566 V.

To correctly compare the magnitude of the electrical losses over the switching period, calculations were performed for the same average value of the electromagnetic torque of all three options.

In the calculations, restrictions were imposed on the amplitude of the supply voltage, so that the parameters of the SRM and the supply converter did not exceed the maximum permissible values. In option 1 (**Figure 18**), a voltage pulse of constant amplitude was applied to the SRM stator winding—an option without optimization. Options 2 and 3 are calculated by the optimization criterion (Eq. (14)) with different voltage limits. For the calculation of option 1 and 3, the voltage value is chosen on the basis that the primary source of voltage for the converter is an industrial AC main voltage of 230 and 400 V, respectively. In carrying out the calculations of option 2, an intermediate nonstandard value was taken to illustrate the tendency of the SRM power parameters to change.

As can be seen from the **Table 3**, the electric losses in the stator winding with the optimal form of the supply current are significantly reduced, and the higher the voltage limit, the lower the electrical losses. From **Figure 18** it can be seen that the current I_3, corresponding to the lowest electrical losses, is characterized by a decrease in the amplitude value at the beginning of the switching process and a higher value of the current in the middle of the switching cycle with respect to the rest of the current curves.

Figure 18 shows the power supply interval of the phase winding. The optimal control problem was solved at this interval. The end of this interval corresponds to the moment of power off. The graphical dependencies of the currents I_1 and I_3

No.	U_{max}, V	ΔP (el. loss), %	I_{max}, A	T_a, N·m
1	325	100	31.8	7.7
2	415	84.5	25.9	7.7
3	566	74.1	21.0	7.7

Table 3.
The calculation results for the three options.

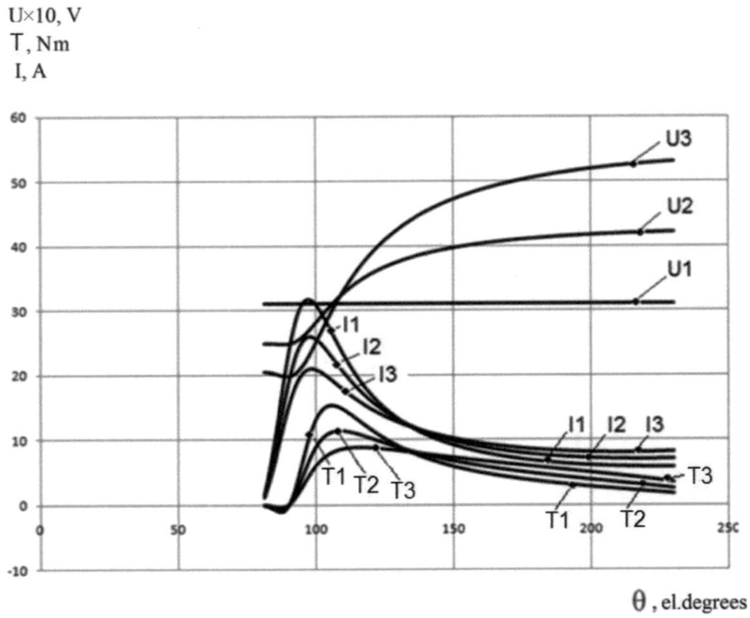

Figure 18.
Curves of changes in the parameter SRM during the controlled switching process, in which the stator winding is under the supply voltage.

versus rotor position (**Figure 19**) show the total current flow interval in the winding, including active interval (voltage applied to the winding) and passive interval (voltage is not applied to the winding, but the energy stored in the winding is returned to the source through the return diodes of the half-bridge power supply circuit). At the optimal variant of the change in current after a power failure, the current decreases faster due to the higher reverse voltage applied to the winding.

Practical implementation of algorithms for optimal control SRM.

Currently, there are a number of options for the formation of the optimal dependence of current on time. It seems rational to use PWM together with FPM, just as for an asynchronous motor, a current form is formed that approximates a sinusoid with PWM. To obtain maximum current repetition of the desired shape of the curve, the number of sampling points of current dependence on time increases, while the maximum pulse duration is reduced, and their frequency is increased with the help of PFM. FPM is used to obtain an optimal SRM current graph, since SRM has a nonlinear current dependence on the rotation angle and rotation frequency, and this in turn imposes limitations on the time of current rise in the SRM phase. From the beginning of each phase SRM to its end, the switching frequency of the half-bridge transistors should constantly change, taking into account the geometric design features of the active part SRM, as well as the specified characteristics. It is also necessary to implement with the help of PWM intermediate current increments that provide a smooth transition from one switching frequency of transistors to another, as well as provide protective and restrictive functions. In practice, all the necessary SRM parameters are read by the control converter as follows:

Phase current is obtained by digitizing an analog signal, a matched level, and certain limits, from a current transformer or a galvanically isolated shunt connected in series with each phase of the SRM.

Voltage is obtained by digitizing the analog signal, the agreed level, and certain limits, from the voltage divider galvanically isolated using linear optocouplers.

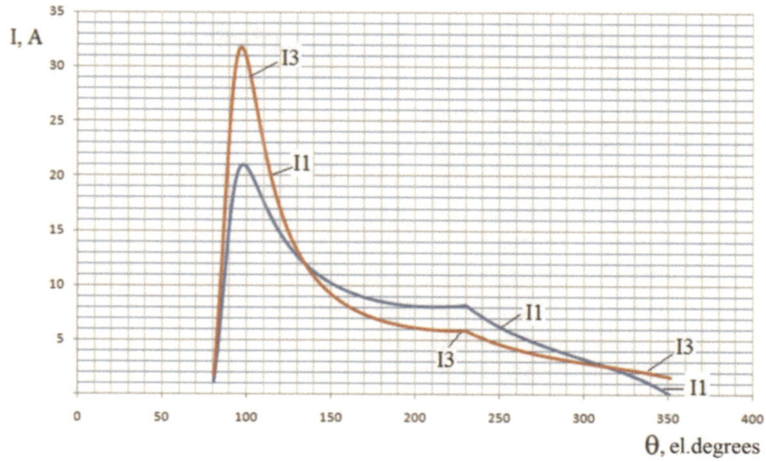

Figure 19.
Chart currents for the full switching cycle of a single-phase SRM.

The rotation angle of the rotor is read from the encoder (both incremental and absolute encoders can be used), and it can also be obtained by processing test pulses sent to the SRM phase (without encoder control system).
The binding of the obtained parameters to the time is carried out by the timers-counters of the microcontroller used to control the SRM.

One of the microcontroller control subroutines solves Eqs. (15) and (20) based on the ADC data obtained using the built-in DMA controller (direct memory access controller). Of course, the microcontroller subprogram solves Eqs. (15) and (20) not in the form presented by the formulas; otherwise it would not have time to process them in time, but in a program-logical form, where the variables of Eqs. (15) and (20) are read and their limits are given. In this case, the entire calculation process takes place in one of the real-time systems such as FreeRTOS or TNKernel. The use of such operating systems helps to properly prioritize and sequence actions when collecting data on the SRM and calculating the sequence of control pulses to obtain the optimal current curve in real time.

Before the microcontroller program issues a sequence of control pulses to the SRM phase control driver, it goes through several stages:

Input of parameters and desired characteristics SRM into the EEPROM of the microcontroller from the interface panel of the configuration
Obtaining parameters SRM using ADC and encoder
Binding of the obtained data to time using timers-counters in the real-time operating system
Setting event flags and reading automatically changing event registers
Simultaneous reading of the time intervals of the components of the process of measuring all parameters and calculating the characteristics SRM over a period of time due to the preset sampling of the angle of rotation and the current rotor speed calculated by the difference in time between the encoder pulses relative to the set reference flags or registers
After receiving and calculating all unknowns in Eqs. (15) and (20), the microcontroller program selects the desired sequence and duration of control pulses to realize the desired current shape depending on the preset limits of parameters and desired characteristics of the SRM.

Even with the achievement of a correctly written program for the microcontroller and the optimization calculation of the SRM geometry, there will always be restrictions for their implementation related to the component base of the converter control. First of all, this refers to the semiconductor elements IGBT and MOSFET. Until recently, IGBT for SRM phase switching was the most popular, because they have small gate capacitances and a short recovery time of the free-wheel diode when switching high currents and voltages, as compared with MOSFET of the same price category. Their only drawback is that the switching frequency is up to 100 kHz, which imposes a restriction on the discretization of pulses forming a SRM current curve. At the moment, the markets appeared are MOSFET, Wolfspeed, and CREE [15]. MOSFET data have zero tail current. Along with this, Wolfspeed produces Schottky SiC diodes with zero reverse recovery current. By using a Z-FET™ and MOSFET™ bundle with Z-Rec™ Schottky diodes, you can get "all-SiC" high-power switching circuits. The switching frequency of such a MOSFET bundle is limited by the maximum operating frequency of the driver and, that, in turn, is limited by the response time of the galvanic isolation circuit. Although these limitations exist, using drivers with a large inverse gate-source current and implementing a driver power supply circuit with a bipolar power source, as well as minimizing parasitic capacitances and inductances in the control circuit of the power transistors during PCB layout, they can be removed completely. The microcontroller operates at large switching frequencies and is not taken into account as a limiting factor for the implementation of the program for the formation of optimal SRM current curves.

The use of optimal SRM control by optimization criterion (Eq. (14)) allowed to significantly reduce electrical losses in the winding.

In practice, the optimal control of the SRM can be realized with the help of shared pulse-width and pulse-frequency modulation, supplemented by a high-speed data acquisition system for the microcontroller.

The derivation of control laws and formulas for calculating the driving pulses that form the optimal SRM current graph for each of its possible operation modes and performance ranges is achieved in practice by a microcontroller control program generating these pulses according to conditions that are initially determined based on the optimal control problem, by means of collecting and calculating all unknown equations of the mathematical model of SRM in real-time systems.

7. Conclusions

Optimization algorithms used at different stages of the SRM life cycle (design, manufacturing, exploitation, recycling) can significantly improve all technical and economic performance of the SRM including environmental issues, loss reduction, and reduction of energy costs in the performance of certain work.

Further development of this line of research will consist in the fact that the optimization problem will be solved in a more general formulation and eventually will cover the entire life cycle of SRM, as a converter of one type of energy into another.

Author details

Petrushin Alexandr Dmitrievich*, Kashuba Alexandr Viktorovich
and Petrushin Dmitry Alexandrovich
Rostov State Transport University, Rostov-on-Don, Russia

*Address all correspondence to: alex331685@yandex.ru

References

[1] Krishnan R. Switched Reluctance Motor Drives. Modeling, Simulation, Analysis, Design and Applications. Boca Raton: CRC Press; 2001. DOI: 10.1201/9781420041644

[2] Miller TJE. Switched Reluctance Motors and their Control. Oxford: Magna Physics Publishing and Clarendon Press; 1993. 205p

[3] Petrushin AD. Energy Saving Switched Reluctance Drives and Asynchronous Drives for an Electric Rolling Stock. Rostov-on-Don: Publishing House of the North Caucasus Scientific Center of Higher Education; 1999. 72p

[4] Petrushin AD, Shcherbakov VG, Kashuba AV. Magnetic system optimization of switched reluctance motor. Russian Electromechanics. 2017; **1**:20-27. DOI: 10.17213/0136-3360-2017-1-20-27

[5] Rastrygin LA. Adaptation of Complex Systems. Riga: Zinatne; 1981. 375p

[6] Himmelblau DM. Applied Nonlinear Programming. New York: McGraw-Hill; 1972. 498p

[7] Shevkunova AV. Study of influence of fragments of magnetic system of a switched-reluctance motor on the average value of electromagnetic torque. Vestnik RGUPS. 2016;**3**:116-123

[8] Bronshtein IN, Semendyayev KA. Math Handbook. Moscow: Nauka; 1986. 544p

[9] Radimov IN, Rymsha VV, Procina ZP. Pole overlaying coefficients of a three-phase switched-reluctance motor. Electrical and Electrical Engineering. 2009;**73**:63-67

[10] Kopylov IP. Electric Motor Design. 4th ed. Moscow: Urait; 2012. 767p

[11] Petrushin AD, Kashuba AV. Improvement of switched reluctance motor performance using optimization algorithms. In: Proceedings of 10th International Conference on Electrical Power Drive Systems ICEPDS 2018; October 3–6, 2018; Novocherkassk: IEEE; 2018. pp. 3-6

[12] Meeker D. Finite Elements Method Magnetics. Version 4.2. User's Manual (FEMM4.2). California, USA; 2013. 161p

[13] Petrushin AD, Kashuba AV. Optimization of the magnetic system of switched reluctance motor. Vestnik RGUPS. 2016;**1**:61-65

[14] Petrushin AD. Energy-saving drives of electric rolling stock based on switched reluctance and induction electric motor [thesis]. Rostov State Transport University, Rostov-on-Don, Russia

[15] SiC and GaN Power and RF Solutions. Wolfspeed [Internet]. 2019. Available from: https://www.wolfspeed.com/ [Accessed: 26 May 2019]

Chapter 5

Scaling Laws in Low-Speed Switched Reluctance Machines

Pedro Lobato, Joaquim A. Dente and Armando J. Pires

Abstract

In low-speed applications, switched reluctance machine (SRM) design methodologies are generally driven by criteria related to the number of pole pair combinations in the stator and rotor of regular machines. In this chapter, a set of scaling laws are developed in order to compare magnetic topologies of regular and non-regular machines. Through the introduction of constraints, the proposed methodology integrates thermal and magnetic saturation phenomena. Within a selected set of short flux-path topologies and using finite element analysis, it is possible to verify which topology is the most appropriate. As an illustrative application, the chosen topology—a modular one—is compared with a regular prototype of a switched reluctance generator designed for a wind energy turbine by using a linear model in conjunction with the scaling laws. The evaluation of the two topologies shows the significant increase of power per unit of mass and lower losses of the modular topology than the regular one. The application of scale model methodology is extensive to the design of higher-speed machines. Valuing the dimensional and similarity arguments, certain assumptions of the machine design gain a special emphasis in this work, bringing the discussion of these machines to another paradigm.

Keywords: machine design, scale model methodology, similarity laws, low-speed energy converters, switched reluctance machine

1. Introduction

Over the last decades, various technical aspects such as design, modelling, analysis, and control of switched reluctance machines (SRM), turned research attention to further improvements in the machine operation. Topics of investigation, including control techniques, loss optimization, torque ripple, and acoustic noise, illustrate the progress in knowledge of motoring mode operation.

The SRM behaves, intrinsically, as a variable speed drive that can be controlled, according to the load, through the energizing and de-energizing moments of the stator phases [1]. Robustness and simple construction (stator-concentrated windings only), control flexibility, and high tolerance over a high-speed range [2] are attributes that qualify it even for the harshest environments [3]. Short flux-path topologies make use of design techniques that facilitate maintenance and improve the machine performance.

The motor operation of this machine has often been the target of research, in detriment of the generator operation. The switched reluctance generator (SRG)

research has been primarily focused on high-speed application design [4]. In these cases, the magnetic topologies are generally of the regular type [5], with two poles diametrically opposed in each stator winding phase and a minimum number of rotor poles. The applications of the SRM as a SRG are in the fields of aerospace, automotive, and wind energy [6].

Research work has been published, for more than a decade, on the application of SRG to wind power systems. Despite the efforts and different possibilities of designing the magnetic circuit of the SRG, regular topologies are usually presented for this type of applications, with an appropriate combination of stator and rotor poles [7]. The laws of scale, also called laws of similarity [8], formulated here, are a useful tool to compare magnetic topologies. This proposed scale model methodology allows to evaluate the effect of thermal phenomena and magnetic saturation of the materials that constitute the electrical and magnetic circuits of the machine. The scale relationships that are established between these circuits differ from topology to topology, by the specific characteristics of these circuits and the relative position they occupy in the structure of the machine.

Thus, in this work, dimensional and symmetry arguments are invoked in the choice of magnetic topologies, leading the discussion about the SRG project to another paradigm.

In the field of renewable energy conversion, high-torque and low-speed machines have gained market space, namely, in wind power applications. The trend towards offshore resource exploitation and consequently the increasing capacity of wind turbines impose robust and reliable solutions, which are critical attributes for wind generators, for an economically and environmentally sustainable operation. In this sense, gearboxes are a component of the system to be avoided in system design, because of the high weight, cost, and unpredictable lifetime.

These are the main reasons that manufacturers in conjunction with research teams have been pursuing direct drive wind energy systems, where the permanent magnet synchronous generator (PMSG) stands out. However, the fluctuating cost of the neodymium (NdFeB), constituting the magnets, has raised some reservations on the consensus opinion about these generators [9]. In addition, harsh on-site installation and operation conditions require special attention to thermal and mechanical stresses; otherwise they will degauss and reduce overall system performance [10]. At this delicate issue, the SRM presents itself as a simple, robust, and cost-effective structure as it does not require permanent magnets or rotor windings. Combined with control flexibility, high reliability, and fault tolerance, it makes this machine attractive for direct drive energy converters (**Figure 1**).

The remainder of this chapter is organized as follows: Section 2 examines some design issues and proposes similarity laws for a SRG design methodology. Section 3

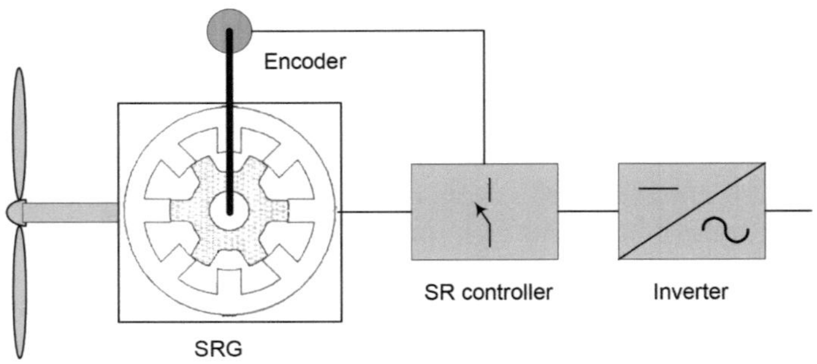

Figure 1.
Schematic representation of a SRG applied to a direct drive wind energy converter (the stator structure is shown without the coils, for an easier understanding).

discusses the features of short-flux path magnetic topologies, using finite element analysis, and a linear field-based model is proposed. Section 4 presents results from the comparison of a modular short-flux path topology, with a 20 kW regular SRG prototype operating at a rated speed in the region of 100 rpm, for use in a direct drive wind turbine [11]. Finally, Section 5 offers some conclusions and points out some impacts of the scale model methodology in the design procedure.

The work presented in this chapter introduces a simple and already known dimensional similarity-based methodology, built however on another paradigm, ready to be used as a design tool for estimation of SRG characteristics and comparison of SRM topologies.

2. Formulation of similarity laws

The SRM design flexibility is consistent with a large number of poles, and there are various magnetic topologies that can support low-speed applications. In spite of that diversity of feasible topologies, there are some intrinsic constraints affecting their dimensions and performance. The characterization and differentiation of constraints are considered priorities in order to adopt a suitable methodology to help in the selection of the magnetic topology and geometry for each application. A constraint can be distinguished by its nature, physical and material. Since the operation of the SRM is sustained by the magnetic field, a first type of constraints can be a result of the physical phenomena involved in the electromagnetic conversion. A second type of constraints emerges from the domain of the material properties, depending on the limits of insulators and magnetic materials used. As examples, the practical constraint of fixing the flux density results from the saturation limit inherent to the magnetic material, or fixing the temperature limits prevents premature deterioration of insulators and the shortening of the equipment lifetime. Additionally, variables such as power, losses, and weight should be taken into consideration as they also play an important role in the machine design and construction. The estimation of certain characteristics of the machine can be facilitated by the relationship between these variables. Models based on similarity laws were chosen, bearing in mind a machine design oriented towards a change of scale.

2.1 Rated power and losses

In the machines based on magnetic field properties, the displacement current term in Maxwell's equations can be neglected, taking into account the frequencies and sizes involved. By neglecting this term, Maxwell's equations will be in the magneto-quasi-static form. There is a physical constraint due to the electromagnetic linkage of electrical and magnetic circuits in both cases, the motoring and the generating mode operations. This linkage is represented in Eq. (1), where an implicit relationship of characteristic dimensions for each type of circuit is also shown:

$$rot\ H = J \tag{1}$$

Considering a change of scale in the construction-oriented machine design, Eq. (1) can be written in the form of (2), where $\boldsymbol{l}$ represents a characteristic linear dimension:

$$B \propto Jl \tag{2}$$

This proportional relationship highlights that magnetic circuits, excited by a current, impose a scale factor in their linear dimensions, which is echoed in the relationship $B/\mathbf{J}$. Therefore, by modifying that characteristic linear dimension, important changes of performance and machine features can be achieved.

Other variables, like power, losses, and material weight, are also critical in the machine design. So, it is important to find the scale relationships involving some of the mentioned variables and the machine dimensions, with the purpose of comparing and predicting certain SRG dimensional characteristics and parameters, based on scale models.

For that aim, the understanding of the energy conversion principles of the SRM is a first step and can be easily described using the generating mode operation, represented in **Figure 2**.

Neglecting saturation phenomena at this point, **Figure 2** shows the profiles of idealized inductance $\boldsymbol{L}$, phase flux-linkage $\boldsymbol{\psi}$, phase current $\boldsymbol{i}$, and voltage $\boldsymbol{u}$ for a SRG for single-pulse operation, where V_S is the dc-link voltage and L_a and L_{na} are, respectively, the aligned and unaligned inductances. In a regular SRM, with $\boldsymbol{N_R}$ rotor poles, those waveforms are periodic in $\boldsymbol{\tau_R}$, which is the rotor pole pitch $(2\pi/N_R)$. During the excitation period, from the rotor position θ_{on} toθ_{off}, energy is stored in the machine magnetic field, with the current supplied by the electrical source. After commutation, at θ_{off}, the excitation energy is returned to the source, and simultaneously, the energy provided by the prime mover is converted to electrical energy.

At θ_e, all the flux from excitation period is extinguish, and no more electrical energy is returned to the source. Neglecting the losses, the output energy over each stroke exceeds the excitation energy by the mechanical energy supplied.

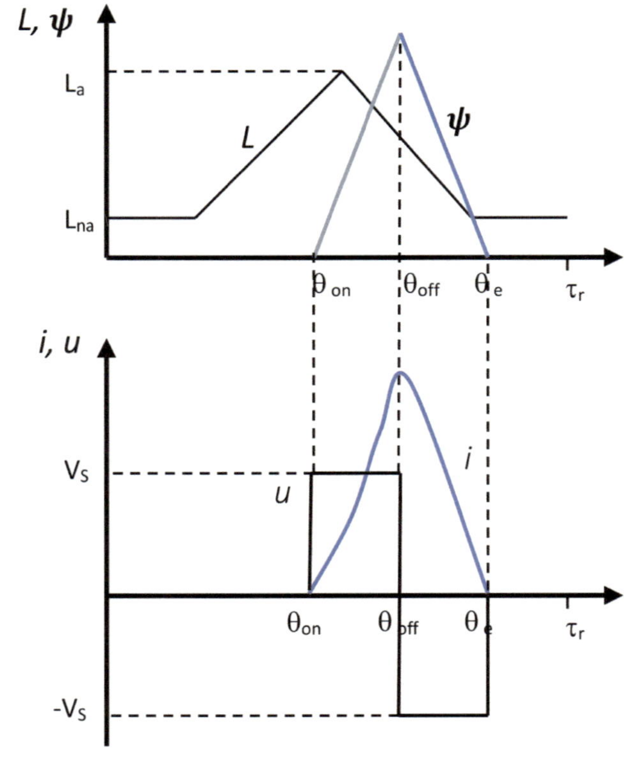

Figure 2.
Idealized inductance profile and waveforms of a single-pulse controlled SRG.

The SRM can operate continuously as a generator, if the electrical output energy exceeds the excitation energy, i.e., maintaining the excitation period so that the bulk of the winding conduction period comes after the aligned position $(dL/d\theta < 0)$. It is considered that the SRG can operate in single-pulse mode, at all speeds, which results in low switching losses.

To better clarify the issues related to the excitation of the SRG and to extract energy from the phase winding, it is useful to look at the electrical equations of an SRG phase as follows:

$$u = \frac{d\psi}{dt} + Ri \tag{3}$$

Multiplying both members of Eq. (3) with the phase current, neglecting the ohmic losses and assuming a non-saturated machine ($\boldsymbol{\psi} = \boldsymbol{L}(\boldsymbol{\theta})\boldsymbol{i}$), the instantaneous power is given as

$$ui \cong i\frac{d\psi}{dt} \cong iL\frac{di}{dt} + i^2\frac{dL}{d\theta}\omega \tag{4}$$

where $\boldsymbol{\omega}$ is the angular speed of the rotor ($\boldsymbol{\omega} = \boldsymbol{d\theta}/\boldsymbol{dt}$).

The SRM energy converter can be represented as a lossless magnetic energy storage system, with electrical and mechanical terminals, as shown in **Figure 3**. In this type of system, the magnetic field serves as the coupling channel between the electric and mechanical terminals. This kind of representation is valid in situations where the loss aspects can be separated from the storage aspects and is also useful in generating mode operation, where the average torque is negative.

For the lossless magnetic energy storage system, the instantaneous power flow in the system can be expressed by

$$ui = \frac{dW_m}{dt} + T_e\omega \tag{5}$$

where $\boldsymbol{W_m}$ is the stored energy in the magnetic field expressed by (6) and $\boldsymbol{T_e}$ is the electromagnetic torque. For a linear system, the magnetic energy $\boldsymbol{W_m}$ and coenergy $\boldsymbol{W_c}$ are numerically equal. Thus, the time rate of the magnetic energy change is given by (7), and $\boldsymbol{T_e}$ can be obtained from the coenergy as indicated in (8):

$$W_m = \frac{1}{2}Li^2 \tag{6}$$

$$\frac{dW_m}{dt} = iL\frac{di}{dt} + \frac{1}{2}i^2\frac{dL}{d\theta}\omega \tag{7}$$

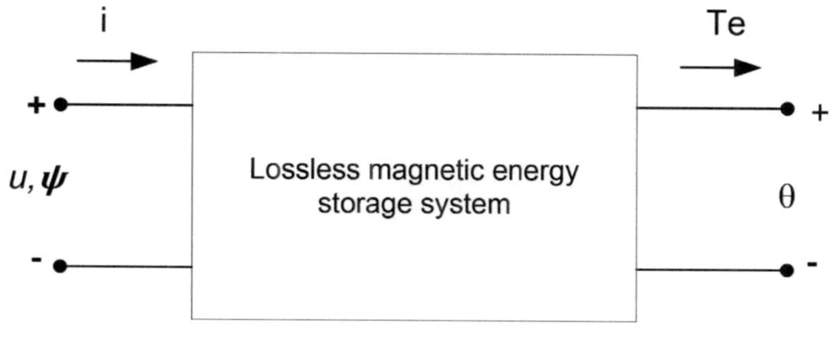

Figure 3.
Schematic magnetic field-based SRM energy converter.

$$T_e = \frac{\partial W_c}{\partial \theta} = \frac{1}{2} i^2 \frac{dL}{d\theta} \tag{8}$$

After replacing $\boldsymbol{T_e}$ in Eq. (5), comparing it with Eq. (4) allows to conclude that less than half of the electrical power involved in the system is converted from the mechanical power of the prime mover. In fact, part of it is being stored in the magnetic field. It should be noted that this conclusion has been established in terms of instantaneous power and for a non-saturated machine. However, for scaling design, based on similarity laws, the average power and the torque are considered fundamental. The average power $\langle \boldsymbol{p} \rangle$ is given by (9), where the rotor angular speed is considered constant. Both current and phase inductance can be written as functions of the rotor position and the maximum values, $\boldsymbol{I}$ and $\boldsymbol{L_a}$, respectively, as represented in Eqs. (10) and (11). Multiplying these functions, the phase flux-linkage is represented by (12). Thus the average power per phase, expressed in (9), is transformed in by (13), in such way that the integral $\int_0^{\tau_r} \boldsymbol{\xi}_{(\theta)}^2 \frac{d\chi_{(\theta)}}{d\theta} \boldsymbol{d\theta}$ should result in a constant value.

$$\langle p \rangle = \langle ui \rangle = \langle T_e \rangle \omega \tag{9}$$

$$i = I \cdot \xi_{(\theta)} \tag{10}$$

$$L = L_a \cdot \chi_{(\theta)} \tag{11}$$

$$\psi = Li = L_a I \cdot \xi_{(\theta)} \chi_{(\theta)} \tag{12}$$

$$\langle p \rangle = \frac{\omega}{2\tau_r} \int_0^{\tau_r} i^2 \frac{dL}{d\theta} d\theta = \frac{\omega}{2\tau_R} I^2 L_a \int_0^{\tau_r} \xi_{(\theta)}^2 \frac{d\chi_{(\theta)}}{d\theta} d\theta \tag{13}$$

Considering the dimensional analysis, the similarity laws for $\boldsymbol{\psi}$, $\boldsymbol{I}$, and $\boldsymbol{L_a}$ can be expressed by (14), (15), and (16), where the scale factor naturally takes place. Having in mind Eq. (13), a proportional relationship for $\langle \boldsymbol{p} \rangle$ can be written as (17):

$$\psi \propto B l^2 \tag{14}$$

$$I \propto J l^2 \tag{15}$$

$$L_a \propto \frac{\psi}{I} \propto \frac{B}{J} \tag{16}$$

$$\langle p \rangle \propto N_R \omega \left(J l^2 \right)^2 \frac{B}{J} \propto N_R \omega B J l^4 \tag{17}$$

Considering an **m**-phase SRM, and in terms of similarity laws, the rated power can be expressed by (18)

$$P \propto m N_R \omega J B l^4 \tag{18}$$

Identical similarity laws can be written for motor mode since the physical fundamental concepts are similar. It should also be highlighted that the criteria of selecting the number of variables, greater or smaller, to be explicit in the scale laws, depend only on the parameters and characteristics considered relevant for the comparison of topologies that is proposed. For instance, the number of rotor poles is a significant parameter in evaluating low-speed SRG designs, since J should be kept under certain limits due to copper losses and the preservation of the insulating material and B is limited by magnetic saturation of the iron. The influence of the commutation of phases on the torque, for this dimensional analysis, was neglected.

In fact, assuming similar levels of saturation (of B), it is assured that the abovementioned comparison of topologies, in relative terms, is valid. Therefore, a compromise involving the number of poles, the dimensions related with both the magnetic and electrical circuits, and the phase current is required. For that purpose, particular design details of SRM topologies related with the magnetic circuit drawing and the windings location will be investigated, in order to increase the available inner space of the machine, taking advantage of an improved air circulation. These aspects will be addressed in Section 3.

2.2 Multi-machine topology

Regarding to SRM design, there is a high number of feasible topologies, which are differentiated by the properties of the electrical and magnetic circuits, and their relative location. A possible SRG topology for an energy converter can include a series of n-SRM assembled on a common axis. The mono-axial multi-machine topology presents a higher fault tolerance and a simpler maintenance, compared with a monolithic SRM.

The previous relationships, applied in SRG design, are suitable to observe certain limitations and some effects of the scale laws. The methodology will be applied considering a SRG that is characterized by the rated power, the copper and iron losses, the maximum value of flux and the current densities, the temperature, and the efficiency. These values are assumed as reference values, allowing the comparison with similar machines, but built in a different scale.

In the methodology of scale models, applied to an electromagnetic device, there is one freedom degree to select the variables B and J as both are interconnected by (2). The saturation levels should be kept constant in all regions of the magnetic circuit, although the flux density is significantly different depending on the regions and instants. Therefore, the practical limitation imposed by the saturation of the magnetic materials can be expressed, in terms of scale laws, by (19):

$$J \propto \frac{B}{l} \tag{19}$$

According to scale laws, it is possible to compare the relative losses of a hypothetical SRG topology, composed of a set of n machines assembled in the same shaft, with a similar monolithic SRG, even though much larger but with the same rated power of the previous one. Assuming a constant rotor speed and equal number of phases and rotor poles of both topologies, the scale law for rated power P_N can be rewritten simply by

$$P_N \propto B^2 l^3 \tag{20}$$

For one SRM module of the multi-machine topology, the rated power can be expressed by

$$P_m \propto B_m^2 l_m^3 \propto B^2 l_m^3 \tag{21}$$

where B and B_m are the flux densities, considered equal once both topologies use the same core material, and l_m is a linear dimension of the multi-modular topology. However, this multi-modular topology is characterized by the total rated power shared by n SRM modules as follows:

$$P_m = \frac{P_N}{n} \tag{22}$$

Moreover, for comparison purposes of topologies, one searches a relation that involves linear dimensions of both topologies and number of modules. Using (20), (21), and (22), a scale law can be expressed by

$$\frac{l_m}{l} \propto \left(\frac{1}{n}\right)^{1/3} \tag{23}$$

This result states, as expected, that as the number of modules is higher, smaller will be the relative linear dimension. Using the previous proportionality law, a relationship that allows the comparison of the relative copper losses in both topologies can be established:

$$\frac{P_{JmT}}{P_J} \propto \frac{nmJ_m{}^2 l_m{}^3}{mJ^2 l^3} \propto n\left(\frac{l_m}{l}\right) \propto n^{2/3} \tag{24}$$

Using (23) and (24) the scale law concerning the relative copper losses can be expressed by (25) and represented in **Figure 4**, with the condition of having a constant flux density:

$$p_{Jr} \propto \frac{P_{JmT}}{P_N} \propto n^{2/3} P_N^{-2/3} \tag{25}$$

So, regarding copper losses, it is possible to conclude about a scale gain for larger monolithic SRM, rather than a modular system composed by SRM units. However, the multi-machine topology could be elected for certain applications where other aspects are a priority like fault tolerance, low cost, and easy maintenance.

2.2.1 Constraint imposed by temperature

In the machine design, at some stage, the heating phenomena have to be considered. In effect, the compromise settled the temperature increase, and the proper temperature levels of insulating materials are critical in the machine lifetime. In order to obtain a scale law, under the thermal stability conditions, it is assumed that the temperature variation $\boldsymbol{\Delta\vartheta}$ can be expressed by (26). It is introduced $\boldsymbol{h_e}$ as the

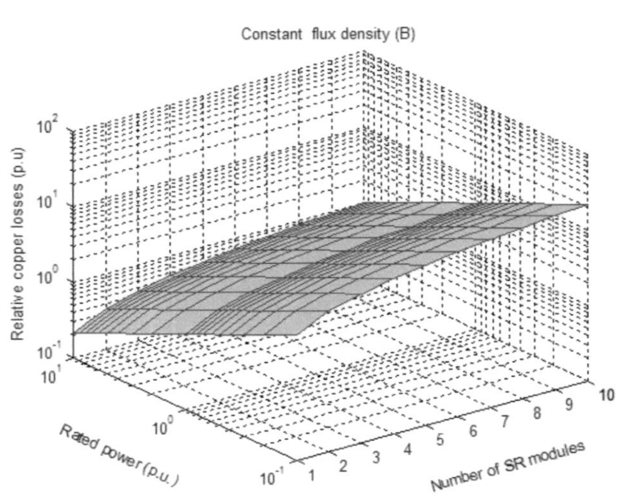

Figure 4.
Relative copper losses for scale models of SRM considering magnetic saturation constraint.

equivalent heat transfer coefficient for all heat exchanges between a fluid and a solid, where $\boldsymbol{A_S}$ is the area of the boundary surface of the material:

$$\Delta\vartheta \propto \frac{P_J}{h_e A_S} \propto \frac{mJ^2 l^3}{l^2} \propto J^2 l \tag{26}$$

Once the constraint of temperature as an influential criterion for comparison purposes is introduced, the current density is expressed by (27), which points out the lower current density in larger machines:

$$J \propto \frac{1}{\sqrt{l}} \tag{27}$$

Therefore, the former scale relationship indicated for rated power (28), and the relationship between linear dimensions (23), is replaced respectively by (28) and (29):

$$P_N \propto mN_R \omega l^4 \tag{28}$$

$$\frac{l_m}{l} \propto \left(\frac{1}{n}\right)^{1/4} \tag{29}$$

Taking into account these two relationships and following the steps presented in Subsection 2.2, the relative copper losses in such temperature conditions are ruled by the scale laws (30) and (31). **Figure 5A** shows the behavior of those losses considering the rated power of the reference machine and the number of modules:

$$\frac{P_{JmT}}{P_J} \propto n^{1/2} \tag{30}$$

$$P_{Jr} = \frac{P_{JmT}}{P_N} \propto n^{1/2} P_N{}^{-1/2} \tag{31}$$

It is observed that the scale gain effect persists for larger monolithic SRM systems, rather than multi-machine systems, despite the temperature constraint and due to lower relative losses in the previous case. Indeed, the relative copper losses are smaller for larger machines and higher for smaller machines due to the current density, which assume, in relative terms, smaller values in large SRM, as shown in **Figure 5B**. This dimensional analysis makes possible to forecast that the efficiency tends to increase in large SRM systems and decrease in small ones. To counteract that trend in smaller machines, it is considered a good practice to increase, in relative terms, the linear dimension of the copper and decrease the dimension of the magnetic circuit.

From **Figure 5B** it can be inferred that for larger machines it is easy to drive into saturation once the flux density can achieve higher levels. Unlike the smaller machines design, in larger machines emphasis should be placed on the circuit magnetic dimension combined with the decrease of the relative dimension of the conductive material. These results suggest a split of scales, thereby benefiting from the behaviors of B and J with regard to the rated power. That split of scales reflects in practice by making use of two distinct scales, one for copper and another for iron. Next steps of the scale analysis will follow in that direction.

2.2.2 Differentiated scales for copper and iron

Until now, the design study has been based on a geometrical scale factor with additional constraints to fix B and J densities. It becomes clear, from the last results,

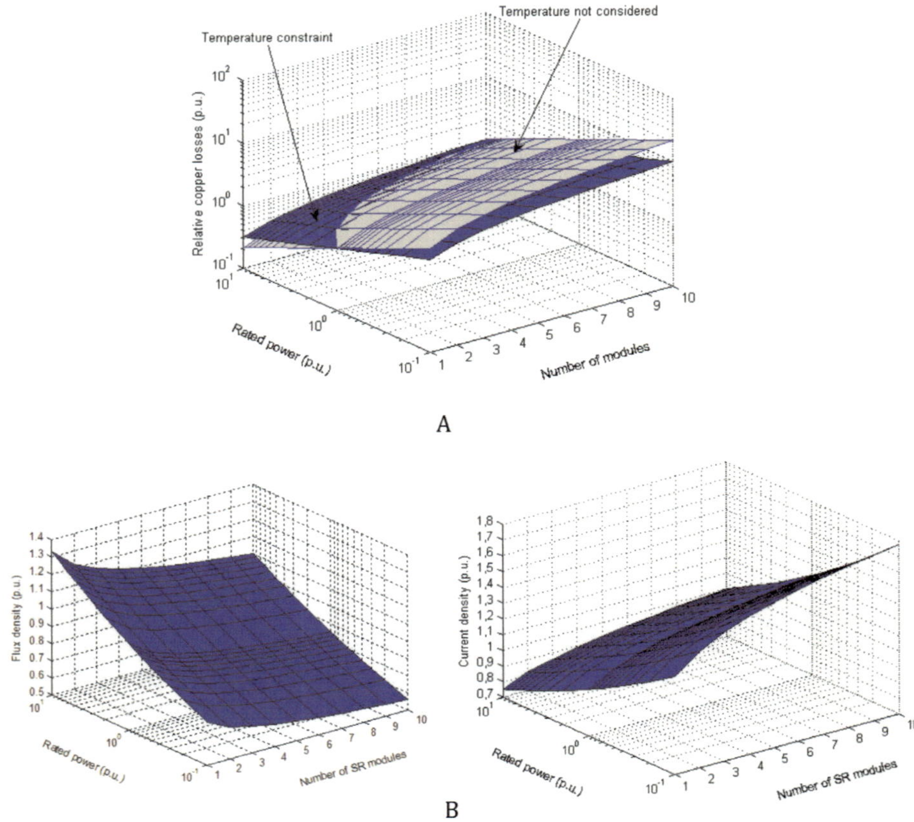

Figure 5.
(A) Relative copper losses for scale models of SRM considering temperature constraint. (B) Normalized current density and flux density for scale models of SRM considering temperature constraint.

that a degradation of the overall system efficiency, when the SRG is composed of several SRM units, is expected. In fact, the adoption of the scale criteria has shown an increase of modules current density, which in turn causes higher copper losses. This increase of copper losses is not balanced by the flux density decrease.

Trying to improve the use of the conductive material and the core iron, it is of interest to introduce a modification into the linear scales. This option is accomplished through a structural change, in which two specific dimensions of each material will be used, l_{Cu} and l_F. Therefore, the relationship (18) assumes the form of (32), and the relationship (2) is rewritten as (33). These two degrees of freedom correspond to an equal number of constraints in order to fix the flux and current densities:

$$P_N \propto m N_R \omega \left(B l_F^{\,2}\right)\left(J l_{Cu}^{\,2}\right) \tag{32}$$

$$B l_F \propto J l_{Cu}^{\,2} \tag{33}$$

Taking as reference the regular monolithic machine, each SRM unit provides only a part of the rated power and enables to infer a new relationship between the relative dimensions of the SRM and the number of modules.

The two available degrees of freedom are used to keep the density flux, as well as the temperature variation as a constant. In such case, the copper characteristic dimension plays a key role in limiting the temperature increase. The relationship (27) will be replaced by (34):

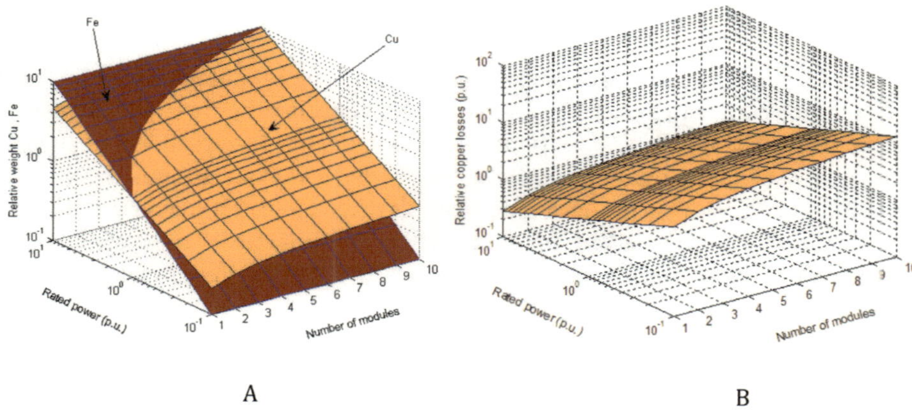

Figure 6.
(A) Relative weights of iron and copper, keeping flux density and temperature increase both constant. (B) Relative copper losses with temperature increase and density flux both constant.

$$J \propto \frac{1}{\sqrt{l_{Cu}}} \tag{34}$$

The materials weight is proportional to the cube of the respective specific dimensions. So, the relationships (35) and (36) are considered representative of the copper and iron weights of the modules set, $Wt_{F_{m_T}}$ and $Wt_{Cu_{m_T}}$.

Observing **Figure 6A** it can be highlighted that the copper weight increases with the number of modules and decreases with the rated power, in relative terms. The iron weight is proportional directly to the rated power and is independent from the number of SRM modules:

$$Wt_{F_{m_T}} \propto P_N \tag{35}$$

$$Wt_{Cu_{m_T}} \propto n^{1/3} {P_N}^{2/3} \tag{36}$$

The variation of temperature, within a restricted range, and the constant flux density lies on preventing the limits of temperature and the materials magnetic saturation to be exceeded. Under these constraints, the scale relationships reporting Joule losses at the modular SRM system are formulated and thereby indicated by (37) and (38). Observing **Figure 6B** it can be concluded that the relative losses increase, as the number of modules increase. That increase is steeper for lower rated powers:

$$P_{Jm_T} \propto n^{5/9} {P_N}^{4/9} \tag{37}$$

$$P_{Jr} \propto n^{5/9} {P_N}^{-5/9} \tag{38}$$

3. Machine design

According to the scale criteria, the previous evaluation of the copper weight and copper losses clearly support the preference for a monolithic topology, instead of a multi-machine topology. The selection of a magnetic structure that considers a shorter flux-path may reduce the magnetomotive force (MMF) absorbed in the iron core.

3.1 Short flux-path topology

Examples of a long flux-path of the regular SRG is presented in **Figure 7A**, and of a short flux-path is presented in **Figure 7B**, highlighting the flux-path closing between two adjacent rotor poles. In order to compare the flux-path length $\boldsymbol{l_{Fsf}}$ of a short flux-path (SFP) topology with the length of the regular machine $\boldsymbol{l_F}$, an external radius $\boldsymbol{R_2}$, similar for both topologies, is considered. The circular arc length of the shaft contour is neglected, for simplicity of analysis. Using the scale laws, it is possible to evaluate the effect on copper losses and weight, in relative terms. The relationship of the flux-paths is given by (39):

$$\frac{l_{Fsf}}{l_F} \cong \frac{\frac{\pi}{m}(R_1 + R_2) + 2(R_2 - R_1)}{\pi R_2 + 2R_2} \tag{39}$$

As inferred from the geometry of both topologies, in the short flux-path case, it is considered a relationship between the exterior and interior radius $\boldsymbol{R_1}$ given by $R_2 = 3R_1$. Both machines have four phases (m' = m = 4), but the SFP topology has a higher number of rotor poles than the reference machine, $N'_R = 14$. With the above assumptions, the ratio of flux-path lengths $\boldsymbol{l_{Fsf}}/\boldsymbol{l_F}$ is approximately 1/2.

Choosing a modular SFP topology, as a magnetic structure, with eight stator modules separated by nonmagnetic spacers, part of the coil wound on the base of each module is exposed, making the copper cooling process and the heat removal more effective. Other favorable arguments to choose this SFP topology are the high fault tolerance, the easy maintenance, and the simplicity of the manufacturing [12].

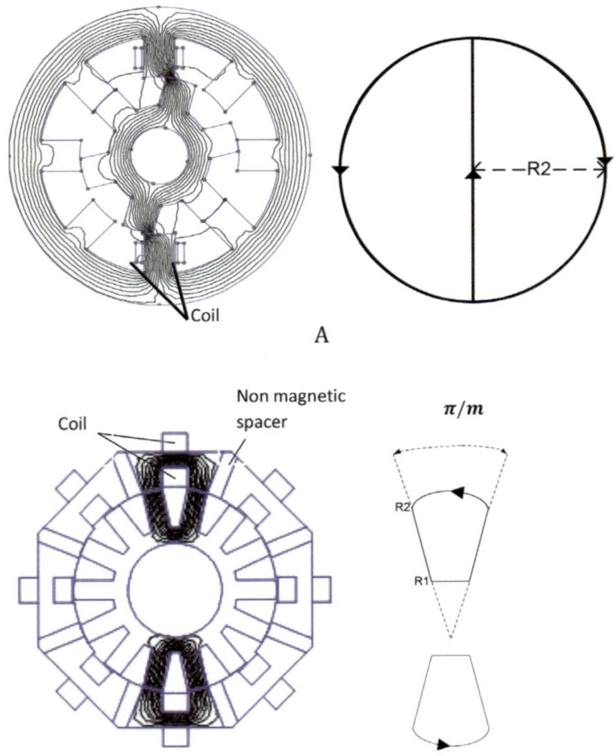

A

B

Figure 7.
(A) Long flux-path for a regular topology (m = 4; N_R=6). (B) Short flux-path for a SFP topology (m = 4; N'_R=14) with stator modules separated by nonmagnetic spacers [1].

The laws related to differentiated scales for copper and iron allow to track the course of the copper losses of the SFP topology, with respect to the regular 8/6 topology. With a limited increase of temperature and a constant flux density, the relationship between the iron and copper characteristic dimensions is expressed by (40) and the copper losses by (41).

$$Bl_F \propto {l_{Cu}}^{3/2} \tag{40}$$

$$P_J \propto mJ^2{l_{Cu}}^3 \propto m{l_{Cu}}^2 \propto m(Bl_F)^{4/3} \tag{41}$$

The copper losses depend on the length of the flux path, becoming lower in the SFP topology. For identical rated power, the copper losses decrease, considering a wide power range suggests a rescaling operation, i.e., a reduction of the dimensions of the SFP machine compared to the dimensions of the regular machine. For that purpose, together with scale laws, a lumped-parameters model based on field theory will be developed. Previously, in order to build the lumped-parameters model as simple as possible, a finite element analysis is used to evaluate some starting hypothesis. In **Figure 8B** it can be seen that the mutual flux-linkage is very small compared with the flux-linkage of each phase. Therefore, in a similar way to regular topologies, the mutual flux-linkage between two phase windings of the SFP magnetic topology is negligible. The magnetic saturation has a significant effect in this modular SFP topology, as can be seen in **Figure 8A**. However, a steep saturation can also appear in regular machines as it is presented in [13].

3.2 Field-based model for dimensional analysis

Under the scope of scale comparisons and regarding electromagnetic rotating systems, it is suitable to work with simple models representing the distribution of the flux density and the magnetic energy. The torque and power relationships can be estimated with those field-based models, which are supported in real system dimensions. An evaluation of relevant topology characteristics and parameters is expected to be achieved making use of scale laws. Thus, a model of a basic reluctance rotating system will be developed. The core is built with identical and isotopic material, assuming very high magnetic permeability and magnetic linear behavior.

The idea of constructing field-based models, assuming linearity of the magnetic circuit, in such SRG topologies, which are characterized by operating into the

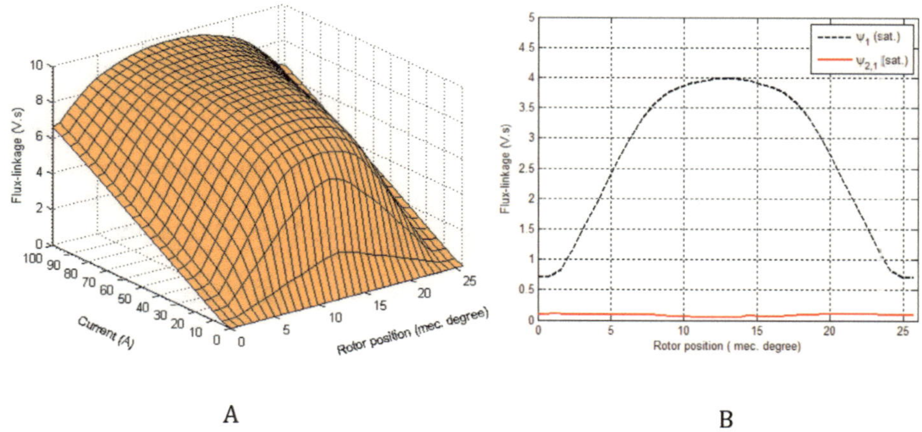

Figure 8.
(A) Flux-linkage characteristic for the SFP topology. (B) Mutual flux-linkage effect when SFP topology was driven hard into saturation.

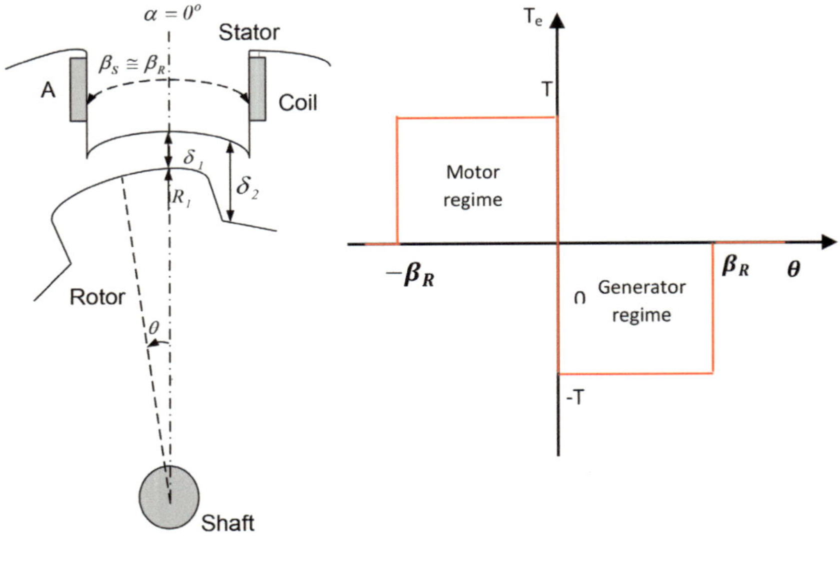

Figure 9.
(A) Schematic view of part (two poles) of a basic reluctance rotating system [1]. (B) Torque-position profile and operation regimes of the switched reluctance system [1].

saturation region, may seem contradictory. However, the saturation effect being extended to both topologies, it is possible to compare the characteristics (e.g., having different numbers of poles), preserving certain dimensions of the magnetic circuit where the flux paths lie on. Therefore, the stator external diameter; the air gap length, δ_1; the radius of the air gap, R_g; as well as the core length, L, will be fixed and kept constant.

A part of a basic rotating reluctance system is shown in **Figure 9A**. It is composed of two poles of equal dimensions, one (rotor pole) having the capability of movement with respect to the other (stator pole) which is in a fixed position. One of these poles is confined to an area $\boldsymbol{A}$ and magnetized by a coil with current density J. Thus, a torque will be produced in order to reduce the reluctance of the system magnetic circuit, i.e., by varying the relative position of the poles.

Two angular coordinates are sufficient to determine the position of the rotor pole and the quantities involved in the system. One is the absolute coordinate associated to the inertial fixed referential, $\boldsymbol{\alpha}$. The other is the coordinate that indicates the relative position of the rotor pole regarding the stator pole, $\boldsymbol{\theta}$. The pole arcs of the rotor and the stator, $\boldsymbol{\beta_R}$ and $\boldsymbol{\beta_S}$, respectively, are approximately equal. It is also assumed that the stator winding comprises another coil, wound on a pole diametrically opposed to the first one, through which the flux-path closes by itself. A last assumption to mention is that the fringing and leakage fields in the air gap will be neglected.

Applying Eq. (1) to this reluctance system yields Eq. (42). Thus, the flux density, B, at the air gap and the magnetic energy stored in the system, W_δ, are expressed by (43) and (44).

$$H_{(\alpha,\theta)}\delta_{(\alpha,\theta)} = JA \tag{42}$$

$$B_{(\alpha,\theta)} = \frac{\mu_o JA}{\delta_{(\alpha,\theta)}} \tag{43}$$

$$W_\delta = 2\int_{V_\delta}\left(\int_0^B H\cdot dB'\right)dV = \int_{V_\delta}\frac{B^2}{\mu_0}dV \tag{44}$$

It should be noted that (44) includes the presence of two volumes of air gap V_δ in the magnetic circuit. Regarding the air gap lengths, it is assumed that $\delta_1 \ll \delta_2$. In these terms, the electromagnetic torque T_e is given by the derivative of the magnetic coenergy W_C with respect to the rotor position (45), and the maximum torque is given by (46). The electromagnetic power, in the generator regime, as well as in the motor regime, shown in **Figure 9B**, is calculated using the average torque $\langle T_e\rangle$ as presented in (47).

$$T_e = \frac{\partial W_C(J,\theta)}{\partial\theta} = \frac{\partial W_\delta(J,\theta)}{\partial\theta} \tag{45}$$

$$T = \mu_o(JA)^2LR_g\left(\frac{1}{\delta_1}-\frac{1}{\delta_2}\right) \cong \frac{\mu_o(JA)^2LR_g}{\delta_1} \tag{46}$$

$$P = \langle T_e\rangle\omega = m\frac{\beta_R}{\tau_R}\left(\frac{\mu_o A^2 LR_g}{\delta_1}\right)\omega J_{rms}^2 \tag{47}$$

4. Design study results and discussion

The scale law methodology and the field models are used now to compare the SFP topology, represented in **Figure 10A**, with a 12/16 laboratory prototype (three-phase with regular structure), built by M. A. Mueller for a direct drive wind turbine (**Figure 10B**). This topology, with 12 stator and 16 rotor poles, was elected based on torque density criteria [11]. However, the only relevance of that work for the present study lies on the similar parameters of the prototype that will support the scale comparison of topologies. Both magnetic circuits have four stator poles involved in the flux-path when one phase is excited. Identical air gap dimensions and core length $(\boldsymbol{\delta}, \boldsymbol{R_g}, \boldsymbol{L})$ are assumed, and an equal MMF per stator pole is imposed.

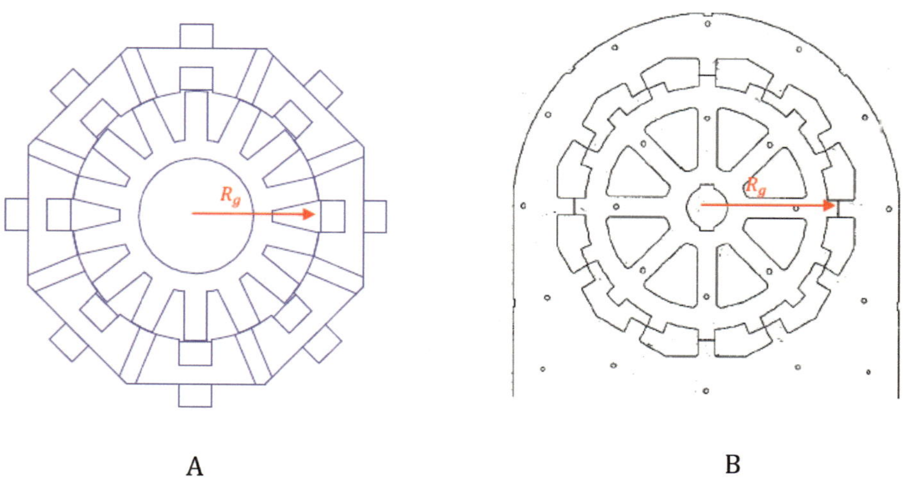

Figure 10.
(A) Modular SFP topology with identical airgap radius, $\mathbf{R_g}$*. (B) SRG prototype (three phases, 12/16).*

Thus, Eq. (48) can be used. Concerning the obtained results, the modular SFP topology presents a torque 56% higher than the regular machine. This is an added value that lies on a greater number of phases, $\boldsymbol{m}$ (one more than the prototype) and a larger section of rotor poles, $\boldsymbol{\beta_R}$ (pole arc of the rotor), as observed in **Table 1**, where the rotor pole pitch is given by $\boldsymbol{\tau_R} = \frac{2\pi}{N_R}$.

The results, in terms of power, allow enough flexibility to perform a rescaling operation of the modular magnetic structure.

Adopting differentiated scales and keeping constant the flux density and the temperature variation, the rated power is expressed by the relationship (48). As illustrated in the diagram of **Figure 11**, and keeping in mind a modular machine and a standard one with equal power $(\boldsymbol{P'_N} = \boldsymbol{P_N})$ to the standard machine $(\boldsymbol{P_N})$, the proportion of power values presented in **Table 1** enables to infer the following rescaling relationships, (49) and (50):

$$P_N \propto m N_R \omega B^2 l_F^{\,3} \tag{48}$$

$$\frac{P'_N}{P_N} \propto \frac{m' N_R'}{m N_R}\left(\frac{l_F'}{l_F}\right)^3 \tag{49}$$

$$l_F' \propto \left(\frac{1}{1,56}\frac{m N_R}{m' N_R'}\right)^{1/3} l_F \propto 0,82\, l_F \tag{50}$$

The rescaling operation shown in **Figure 11** consists on the reduction for identical rated power, in proportional terms, of the modular machine (M_{MOD})

	m	N_R	τ_R [rad]	β_R [rad]	*P (p.u)*
Regular 12/16 prototype	3	16	$\pi/8$	$\pi/24$	1
SFP modular topology	4	14	$\pi/7$	$\pi/18$	1,56

Table 1.
Parameters of the compared topologies (the base power is peak power), m (number of phases), N_R (number of rotor poles), τ_R (rotor pole-pitch), β_R (rotor pole-arc), P (rated power).

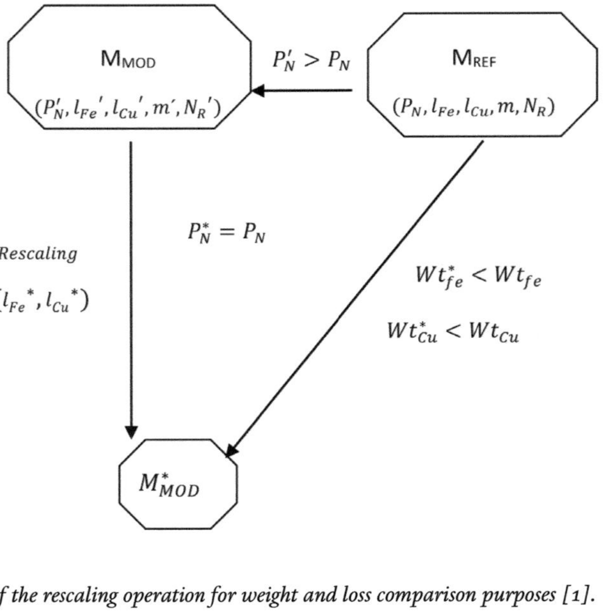

Figure 11.
Schematic diagram of the rescaling operation for weight and loss comparison purposes [1].

characteristic dimensions compared to the standard machine (M_{REF}). After rescaling, it is well-timed to establish relationships for both the weight of iron Wt_F^* and the weight of copper Wt_{Cu}^* as well as to evaluate the copper losses P_{Jr}^*, in relative terms. The iron weight and the copper weight of the modular topology are compared with the SRG prototype, as indicated by (51) and (52):

$$Wt_F^* \propto \frac{(l_F{}^*)^3}{(l_F)^3} t \propto 0,55\ Wt_F \tag{51}$$

$$\mathbf{Wt}_{\mathbf{Cu}}^* \propto \frac{(\mathbf{l_{Cu}}{}^*)^3}{(\mathbf{l_{Cu}})^3}\mathbf{Wt_{Cu}} \propto \frac{(\mathbf{l_F}{}^*)^2}{(\mathbf{l_F})^2}\mathbf{Wt_{Cu}} \propto \mathbf{0,67\ Wt_{Cu}} \tag{52}$$

Regarding the modular topology, the iron weight is approximately 55% when compared to the standard topology, reducing the volume taken by the iron by 45%. Concerning the relative weight of the copper, it allows a reduction of 33% in respect to the regular 12/16 SRG.

In terms of specific power, expressed in W/Kg, it is predicted that there will be an increase of power of 80% at the modular SRG per unit of iron mass, and an increase of close to 50% per unit of copper mass, when compared to the standard SRG. According to the relationships (53) and (54), the relative losses of both machines show the same proportionality, even after the rescaling operation and the resultant reduction of the volume of the modular machine. The rescaling operation is performed on an identical rated power basis:

$$\frac{P_{Jr}^*}{P_{Jr}} = \frac{\frac{P_J^*}{P_N^*}}{\frac{P_J}{P_N}} \propto \frac{m^*(l_{Cu}{}^*)^2}{m l_{Cu}{}^2} \propto 1,\ P_N^* = P_N \tag{53}$$

$$P_{Jr}^* \propto P_{Jr} \tag{54}$$

The weight of the materials is an important factor in choosing the equipment to be used in a generator, due to the general high-altitude location of the wind turbines, whether onshore or offshore. Also, the cost of material and maintenance comes lower, together with lesser weight. Furthermore, when the modular topology is selected, higher fault tolerance is expected. The heat removal and the temperature distributed within the machine take benefit from the modular configuration and the winding location on the stator modules. In fact, the heat transfer is at least so relevant as the electromagnetic design and therefore will be treated in a future work.

Finally, the good performance is preserved, and it is not expected that the inclusion of iron losses in the previous calculations will degrade the results in a significant way.

5. Conclusions

A comparative analysis based on scale models has been presented for low-speed SRM. The comparison and evaluation of magnetic structures play an important role in the SRM design. General design methodologies are usually oriented towards the evaluation of stator/rotor poles combinations for regular SRM. Besides covering that feature, the proposed formulation of scale laws is also suitable to compare other SRM topologies, distinguished by different characteristics of electric and magnetic circuits and their own relative location. As shown by the authors, this methodology can be extended to other physical phenomena, like thermal changes and magnetic saturation, by introducing some constraints. The study performed in this paper compares a modular short flux-path topology versus a low-speed 20kW prototype SRG, at a rated speed in the region of 100 rpm, designed for a direct drive wind turbine. The modular topology can optimize the efficiency and weight, taking benefits from the significant gain of power per unit of mass and lower losses.

The comparison results achieved in this dimensional analysis indicate that additional work should be developed concerning a detailed design of the modular SFP topology, and so, the work should be seen as a guideline, and not as an end. Considering the real dimensions and the material characteristics, a full-scale machine design will be able to compare initial costs of the modular SFP SRG topology with classical generators, using gearbox.

This work should not be seen as attempting to address a detailed design of a novel SRG, nor an original methodology, but rather highlight the usefulness and effectiveness of the similarity law formulation, as an assistant tool for the machine design. Furthermore, its application to regular and non-regular SRM topologies clearly emphasizes some design details of magnetic structures in machine.

Acknowledgements

The authors thank the Polytechnic Institute of Setúbal and FCT/Nova University of Lisbon for providing facilities as part of an existing cooperation protocol. This work was supported by national funds through FCT under contract UID/EEA/00066/2019.

Author details

Pedro Lobato[1]*, Joaquim A. Dente[2] and Armando J. Pires[1]

1 ESTSetúbal, Polytechnic Institute of Setúbal, Setúbal, Portugal

2 Instituto Superior Técnico, University of Lisbon, Lisbon, Portugal

*Address all correspondence to: pedro.lobato@estsetubal.ips.pt

References

[1] Lobato P, Martins J, Dente JA, Pires AJ. Scale models formulation of switched reluctance generators for low speed energy converters. IET Electric Power Applications. 2015;**9**(9):652-659

[2] Kioskeridis I, Mademlis C. Optimal efficiency control of switched reluctance generators. IEEE Transactions on Power Electronics. 2006;**21**(4):1062-1071

[3] Bilgin B, Emadi A, Krishnamurthy M. Design considerations for switched reluctance machines with a higher number of rotor poles. IEEE Transactions on Industrial Electronics. 2012;**59**(10):3745-3756

[4] Chancharoensook P, Rahman MF. Control of a four-phase switched reluctance generator: experimental investigations. Proceedings of the IEEE International Electric Machines and Drives Conference. 2003;**2**:842-848

[5] Miller TJE. Switched Reluctance Motors and Their Control. Oxford, UK: Magna Physics Publishing and Clarendon Press; 1993

[6] Bao YJ, Cheng KWE, Cheung NC, Ho SL. Experimental examination on a new switched reluctance wind power generator system for electric vehicles. IET Power Electronics. 2012;**5**(8): 1262-1269

[7] Mueller MA. Design and performance of a 20 kW, 100 rpm, switched reluctance generator for a direct drive wind energy converter. In: Proceedings of the IEEE International Conference on Electric Machines and Drives. San Antonio, Texas, U.S.A.; 2005. pp. 56-63

[8] Jufer M. Electric Drive: Design Methodology. UK: ISTE Ltd and John Wiley & Sons Inc; 2010

[9] Boldea I, Tutelea L, Blaabjerg F. High power wind generator designs with less or no PMs: An overview. In: Proceedings of 17th International Conference on Electrical Machines and Systems (ICEMS). Hangzhou, China; 2014

[10] Liu X, Park K, Chen Z. A novel excitation assistance switched reluctance wind power generator. IEEE Transactions on Magnetics. 2014; **50**(11):1-4

[11] Mueller MA. Design and performance of a 20 kW, 100 rpm, switched reluctance generator for a direct drive wind energy converter. In: Proceedings of the IEEE International Conference on Electric Machines and Drives. San Antonio, Texas, U.S.A.; 2005. pp. 56-63

[12] Ruba M, Viorel I, Szabó L. Modular stator switched reluctance motor for fault tolerant drive systems. IET Electric Power Applications. 2013;**7**(3):159-169

[13] Parreira B, Rafael S, Pires AJ, Costa Branco PJ. Obtaining the magnetic characteristics of an 8/6 switched reluctance machine: FEM analysis and experimental tests. IEEE Transactions on Industrial Electronics. 2005;**52**(6): 1635-1643

Chapter 6

Linear Switched Reluctance Motors

Jordi Garcia-Amoros, Pere Andrada and Baldui Blanque

Abstract

This chapter deals with linear switched reluctance machines (LSRMs). Linear switched reluctance machines are the counterpart of the rotary switched reluctance machine (SRM), and now they have aroused great interest in the field of electrical machines and drives. In this chapter, first, a mathematical model is presented, and then a procedure for the design of this kind of machines is proposed. Next, a linear switched reluctance force actuator, based on the before designed procedure, is simulated. In addition, experimental proofs of the goodness of the design process and of the accuracy of the simulation of the linear switched reluctance force actuator are given.

Keywords: linear switched reluctance machines (LSRMs), mathematical model, finite element analysis (FEA), design procedure, simulation

1. Introduction

Nowadays, there is a great interest in linear electric machines and especially in linear switched reluctance machines (LSRMs). LSRMs are an attractive alternative to permanent magnet linear motors (PMLM), despite the fact that the force/volume ratio is about 60% lower for LSRMs [1]. On the other hand, the absence of permanent magnet makes them less expensive and easy to assemble and provides a greater robustness and a good fault tolerance capability. LSRMs have been proposed for a wide range of applications such as precise motion control [2], propulsion railway transportation systems [3], vertical translation [4], active vehicle's suspension system [5], life-support applications [6], and in direct-drive wave energy conversion [7].

The LSRMs consist of two parts: the active part or primary part and the passive or secondary. The active part contains the windings and defines two main types of LSRMs: transverse and longitudinal. It is longitudinal when the plane that contains the flux lines is parallel to the line of movement and transverse when it is perpendicular. Other classifications are considering the windings totally concentrated in one coil per phase [2] or partially concentrated in two poles per phase (i.e., single-sided) or four poles per phase (double-sided) [3, 4]. **Figure 1** shows all the possible configurations belonging to this classification. The simplest structure is the single-sided flat LSRM shown in **Figure 1a**, in which the number of stator active poles is $2 \cdot m$, and the number of poles per phase (N_{pp}) is 2. A conventional double-sided flat LSRM (see **Figure 1b**) is created by joining two single-sided structures; in this case $N_{pp} = 4$ and the number of stator active poles is $N_p = N_{pp} \cdot m$, where m is the number of phases. The double-sided structure (**Figure 1b** and **c**) balances the normal force over the mover, and therefore, the linear bearing does not have to support it. This

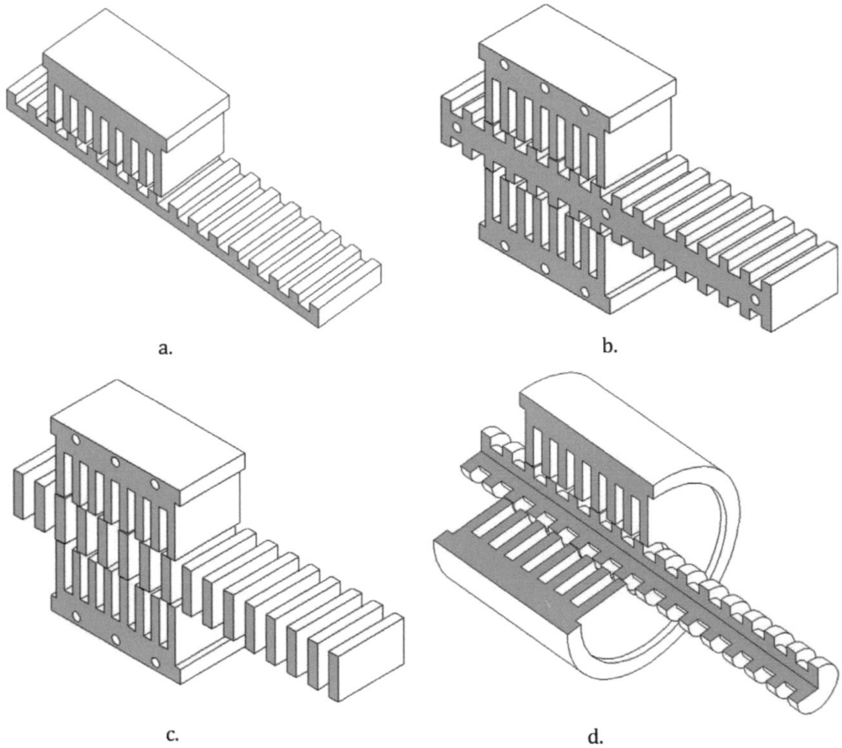

Figure 1.
Longitudinal flux LSRM topologies. (a) Single-sided; (b) conventional double-sided; (c) modified double-sided; and (d) tubular.

configuration has twice air gaps and coils than single-sided, which means a double translation force. Conventional double-sided (**Figure 1b**) can operate with one flux loop or two flux loops due to the magnetic connection between secondary poles. In the modified double-sided LSRM (**Figure 1c**), the secondary, the mover, is comprised of rectangular poles without connecting iron yokes between them but are mechanically joined by nonmagnetic mounting parts [8]. This arrangement reduces the mass of the mover, giving a higher translation force/mass ratio than conventional double-sided flat LSRM, which reduce the mover weight and its inertia although only allows operating with one flux loop. The tubular structure is shown in **Figure 1d**.

It is important to note that in an LSRM, thrust or translation force is produced by the tendency of its secondary or mover to translate to a position where the inductance of the excited phase is maximized, i.e., to reach the alignment of primary and secondary poles. Therefore, as in its rotary counterpart (SRM), a power converter with solid-state switches, usually an asymmetric bridge (with two switches and two diodes per phase), is needed to generate the right sequence of phase commutation. Thus, it is necessary to know, in every instant, the position of the secondary part or mover, for which a linear encoder is generally used.

2. Mathematical model of LSRM

The mathematical model of the LSRM consists on the voltage phase equation, the internal electromechanical force, and the mechanical equation, balance between internal electromagnetic force and load, friction, and dynamic forces.

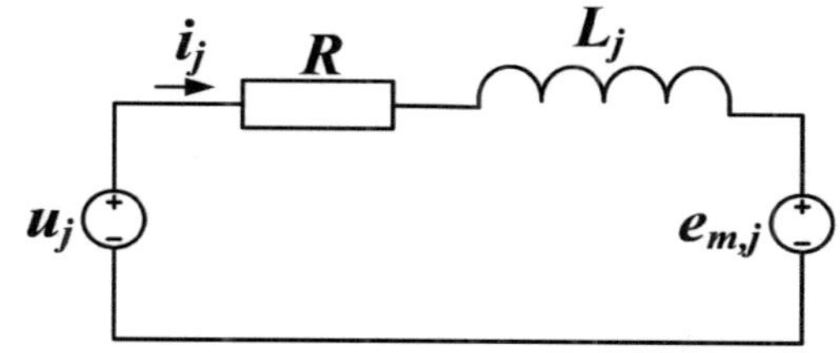

Figure 2.
Single-phase equivalent electric circuit of an LSRM.

The voltage equation of j-phase is equal to the resistive voltage drop plus the partial derivative of the j-phase flux-linkage respect time. This equation that can be written as (1) where in its second member the first term is the resistive voltage drop, the second term involves the voltage induced by the current variation, and the third is the induced voltage due to the relative movement of the primary and secondary parts at the speed u_b:

$$u_j = R \cdot i_j + \frac{\partial \psi_j(x,i)}{\partial i} \cdot \frac{di}{dt} + u_b \cdot \frac{\partial \psi_j(x,i)}{\partial x} \tag{1}$$

The respective derivatives in x (position) and i (current) of the *phase*-flux linkage $\left(\psi_j\right)$ give the j-phase incremental inductance, L_j (2), and the j-phase back electromotive force, $e_{m,j}$ (3).

$$L_j = \frac{\partial \psi_j(x,i)}{\partial i} \tag{2}$$

$$e_{m,j} = u_b \cdot \frac{\partial \psi_j(x,i)}{\partial x} \tag{3}$$

Rewriting (1)

$$u_j = R \cdot i_j + L_j \cdot \frac{di_j}{dx} + e_{m,j} \tag{4}$$

Then, the electrical equivalent circuit per phase of the LSRM is shown in **Figure 2**.

The total internal electromagnetic force (F_X) summing the force contribution of each phase is given by

$$F_X = \sum_{j=1}^{m} \left(\frac{\partial}{\partial x} \left(\int_0^I \psi_j(x,i) \cdot di \right) \bigg|_{I=ctn} \right) \tag{5}$$

The total internal electromagnetic force (5) is balanced by the dynamic force, product of mass by the acceleration, the friction force, and the applied mechanical load:

$$F_X = M \cdot \frac{du_b}{dt} + F_r + F_L \tag{6}$$

Rearranging Eqs. (1) and (6), we obtain the state-space equations (7), which define the dynamical model of an LSRM per phase:

$$\left.\begin{array}{r} I_j = \int \dfrac{1}{\dfrac{\partial \psi_j(x, i_j)}{\partial i_j}} \cdot \left(u_j - R.i_j - \dfrac{\partial \psi_j(x, i_j)}{\partial x} \cdot u_b \right) \cdot dt \\ u_b = \int \dfrac{1}{M} \cdot (F_X - F_L - F_r) \cdot dt \end{array}\right\} \quad (7)$$

The simulation of the dynamic mathematical model (7) requires the flux-linkage characteristics $\psi_j(x, i)$ (see **Figure 3a**) and its partial derivatives $\frac{\partial \psi_j}{\partial i}$ (see **Figure 3c**) and $\frac{\partial \psi_j}{\partial x}$ (see **Figure 3d**), as well as the internal electromagnetic force F_X (see **Figure 3b**), whose values are obtained from FEM analysis.

3. LSRM design procedure

The design of electric rotating motors usually starts with the output equation. This equation relates the main dimensions (bore diameter and length), magnetic loading, and electric loading to the torque output. In this case, it introduced a similar development for the output equation of LSRM, in which the average translation force (output equation) depends on geometric parameters, magnetic loading, and current density [9].

Although the present study is focused in the longitudinal double-sided LSRM and the longitudinal modified double-sided LSRM (**Figure 1b** and **c**), the main dimensions describing its geometry are the same to that single-sided LSRM shown in **Figure 4**.

3.1 Design specification

The first step is to define the design specifications. These specifications affect not only the electromagnetic structure but also the power converter and the control

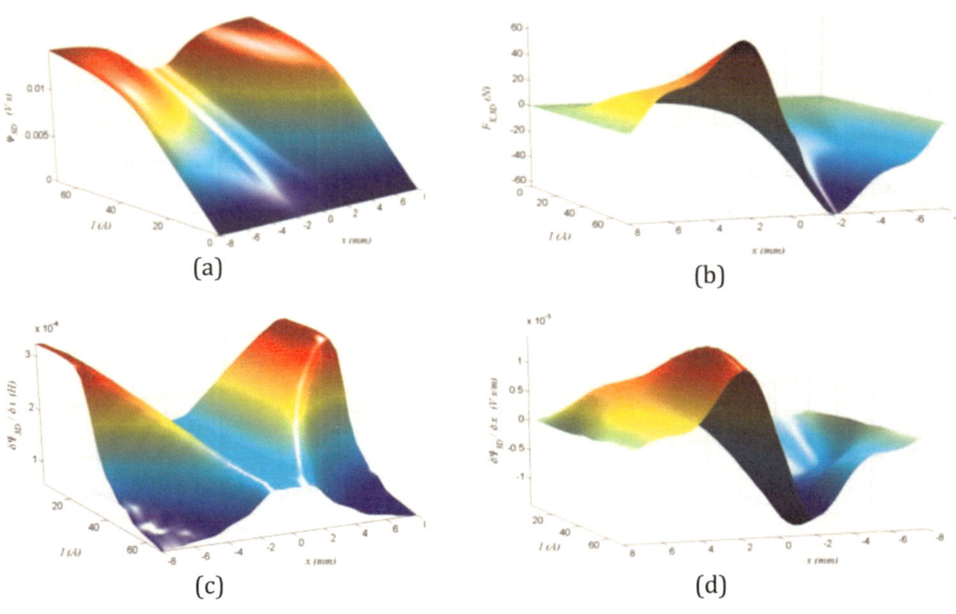

Figure 3.
FEM results plots for position x∈[−8,8] mm and current I∈[0,69] A. (a) Flux-linkage. (b) Internal electromagnetic force. (c) Partial derivative of flux-linkage with respect to current. (d) Partial derivative of flux-linkage with respect to position.

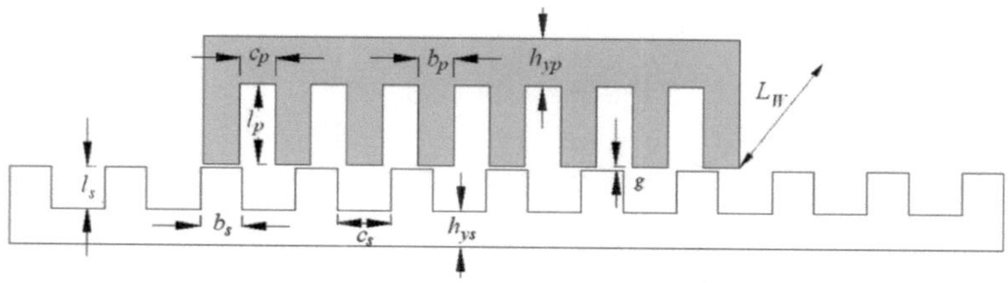

Figure 4.
Single-sided LSRM main dimensions.

Requirements	Constraints
• LSRM type • Power converter topology • Control strategy • Translation force (F_x) • Velocity (u_b) • Acceleration/deceleration (a) • Thermal duty cycle • Number of phases (m) • Pole Stroke (PS) • Mover stroke (TS)	• DC bus voltage (V_b) • Magnetic material • Temperature rise • Some critical dimensions (i.e., air-gap length)

Table 1.
Requirements and constrains.

strategy. They are also different in nature (mechanical, electrical, thermal) and can be classified into two general areas, requirements and constraints; the most usual are listed in **Table 1**.

Figure 5 shows a flowchart with the different steps of the design process [9]. These steps begin with the definition of the specifications. Then, the main dimensions are obtained using the output equation. In the next step, the number of turns and the wire gauge are determined following an internal iterative process. Then a first performance FEM-computation is performed. Finite element analysis and thermal analysis are used in order to check whether the motor parameters meet the expected specifications. The design steps are repeated in an iterative process until the design specifications are obtained.

3.2 Output equation

The LSRM design is addressed using two different approaches. In the first approach, it is performed in the rotary domain which is then transformed back into the linear domain [10, 11]. In the second approach, the LSRM design is carried out by using an analytical formulation of the average translation force determined by means of an idealized energy conversion loop [12, 13]. A design procedure for longitudinal flux flat LSRMs, based on this second approach, is proposed according the flowchart of shown in **Figure 5**, in which the average translation force, is determined in terms of magnetic loading, current density and geometrical relationships derived from a sensitivity analysis reported in [14]. Once the main dimensions are obtained, the number of turns per phase is determined by means of an iterative process.

The number of phases (m) and the pole stroke (PS) can be used to determine T_p, T_s, N_p, and N_s, by means of the following equations:

$$\left.\begin{aligned} N_P &= 2\cdot m \\ N_S &= 2\cdot(m\pm 1) \end{aligned}\right\} \tag{8}$$

$$\left.\begin{aligned} T_P &= \frac{1}{2}\cdot N_S\cdot PS = b_p + c_p \\ T_S &= \frac{1}{2}\cdot N_P\cdot PS = b_s + c_s \end{aligned}\right\} \tag{9}$$

The average internal electromagnetic force or translation force ($F_{x,avg}$) is calculated using an idealized nonlinear energy conversion loop in which the unaligned magnetization curve is assumed to be a straight line and the aligned magnetization curve is represented by two straight lines [13, 14]. This simplified model accounts for the saturation effect and is described in **Figure 6**. Assuming a flat-topped current waveform (hysteresis control), the area OACFO is the energy conversion area (W). Excluding iron and friction losses, the average translation force per phase ($F_{X,avg}$) is then obtained by

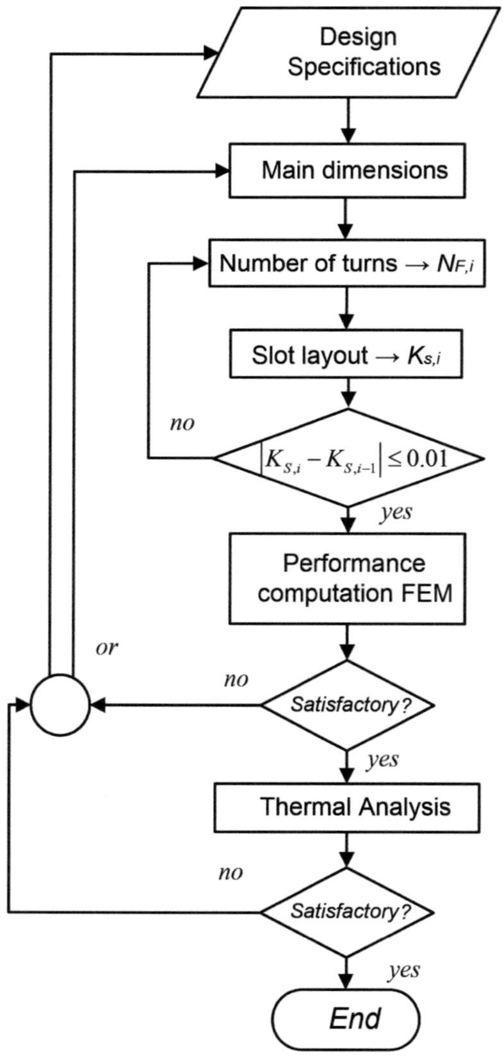

Figure 5.
Flowchart of the overall design procedure.

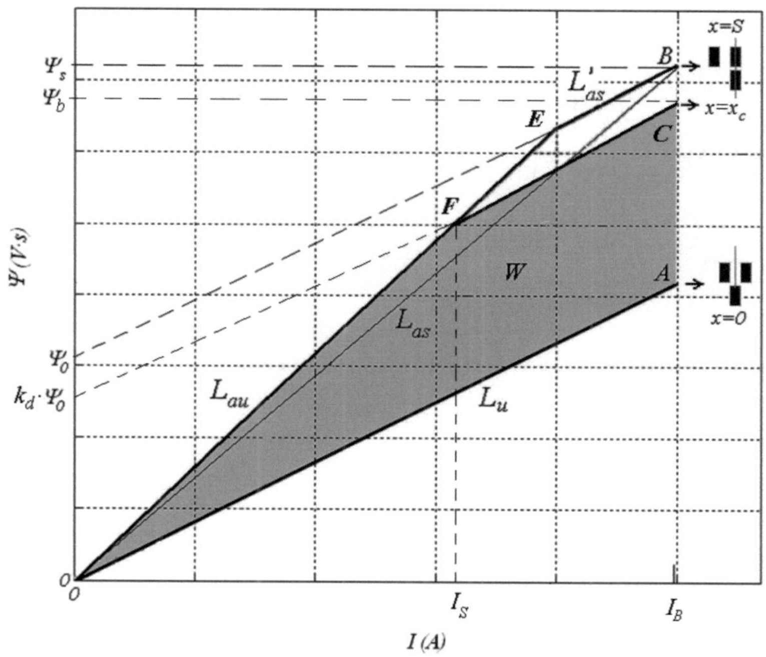

Figure 6.
Idealized nonlinear energy conversion loop.

$$F_{X,\,avg} = \frac{W}{S \cdot k_d} \tag{10}$$

where k_d is the magnetic duty cycle factor defined as $k_d = x_c/S$ (see **Figure 6**) and S is the distance between aligned and unaligned positions given by

$$S = \frac{T_s}{2} = \frac{N_p}{N_s} \cdot \frac{T_p}{2} \tag{11}$$

From **Figure 6** the following expressions can be derived:

$$W = I_B^2 \cdot L_{as} \cdot K_L \cdot k_d \tag{12}$$

where K_L is a dimensionless coefficient defined from the inductances depicted in **Figure 6**, by

$$K_L = \left(1 - \frac{L_u}{L_{as}}\right) \cdot \left(1 - \frac{1}{2} \cdot \frac{L_{as} - L_u}{L_{au} - L_u} \cdot k_d\right) \tag{13}$$

At point B (see **Figure 6**), the poles are fully aligned and therefore

$$\psi_s = L_{as} \cdot I_B = B_p \cdot N_1 \cdot N_{pp} \cdot b_p \cdot L_W \tag{14}$$

The total ampere-turns per slot ($N_1 \cdot I_B$) can be expressed, considering the slot fill factor (K_s) by means of the current density peak (J_B) by

$$N_1 \cdot I_B = \frac{1}{2} \cdot c_p \cdot l_p \cdot K_s \cdot J_B \tag{15}$$

Combining (15) and (16) into (13)

$$W = \frac{1}{2} \cdot (K_L \cdot K_s \cdot k_d) \cdot (c_p \cdot b_p \cdot l_p \cdot L_W \cdot N_{pp}) \cdot (B_p \cdot J_B) \tag{16}$$

Therefore, the average translation force per phase is

$$F_{X,\,avg} = N_{pp} \cdot \left(\frac{N_s}{N_p}\right) \cdot (K_L \cdot K_s) \cdot \left(\frac{c_p \cdot b_p \cdot l_p \cdot L_W}{T_P}\right) \cdot (B_p \cdot J_B) \tag{17}$$

In order to obtain dimensionless variables, the stator pole pitch (T_P) normalizes the geometric variables depicted in **Figure 4**, obtaining

$$\alpha_p = b_p / T_P \tag{18}$$

$$\alpha_s = b_s / T_P \tag{19}$$

$$\beta_p = l_p / T_P \tag{20}$$

$$\beta_s = l_s / T_P \tag{21}$$

$$\gamma_W = L_W / T_P \tag{22}$$

$$\delta_y = h_y / T_P \tag{23}$$

Rewriting (18) by considering (19)–(24)

$$F_{X,\,avg} = N_{pp} \cdot \left(\frac{N_s}{N_p}\right) \cdot (K_L \cdot K_s) \cdot \left(\left(\alpha_p - \alpha_p^2\right) \cdot \beta_p \cdot \gamma_W\right) \cdot T_P^3 \cdot (B_p \cdot J_B) \tag{24}$$

The output Eq. (25) is applicable to all the types of LSRMs considered in **Figure 1**, just considering N_{pp} = 2 for single-sided flat and tubular LSRMs and N_{pp} = 4 for conventional double-sided LSRMs and for modified double-sided LSRMs.

3.3 Selection of magnetic loading, current density and normalized geometric variables

The magnetic flux density in the stator pole (B_p) depends on the chosen magnetic lamination material; a good choice is to take a value slightly lower than the value at which laminations reach magnetic saturation. The current density, J_B, strongly depends on operation conditions and cooling facilities. The current density should be kept within reasonable margins if the temperature rise should not exceed a specified value. For high force LSRMs with air natural/forced convection and continuous duty cycle, J_B = 5 A/mm^2 is a good value while for the same conditions but following a short time intermittent duty cycle, J_B = 15 A/mm^2 could be more advisable.

The K_L coefficient depends on the geometrical parameters (see **Table 2**) and the current density (J_B). For values of current density between 5 and 20 A/mm^2, a good initial choice is K_L = 0.3.

The influence of the normalized geometric variables involved in the output equation as well as the current density has been investigated in [14]. **Table 2** shows the set of values of α_p and β_p, for different values of current density, recommended to obtain high values of average force [9].

The average force is proportional to L_W. However, an excessive stack length increases mass and iron losses. The air-gap length (g) should be as small as possible

J_B (A/mm²)	α_p	β_p
5	[0.333, 0.417]	≤3.5
10	[0.375, 0.5]	≤3
15	[0.417, 0.542]	≤2.5
20	[0.458, 0.542]	≤2

Table 2.
Recommended values of α_p and β_p to obtain high average force.

to maximize the average force compatible with tolerances and manufacturing facilities; it is advisable to avoid air-gap lengths under 0.3 mm. In the case of double-sided, LSRM is very important in the assembly process ensure that the upper and the lower air gaps have the same length.

3.4 Number of turns and wire gauge

The maximum flux linkage at point B (see **Figure 6**), at a constant velocity, u_b, with a flat-topped current waveform and disregarding resistance, is related to the DC voltage V_b by means of

$$\psi_0 = \frac{V_b}{u_b} \cdot S \tag{25}$$

Thus

$$\psi_0 = \psi_s \cdot (1 - L_u/L_{as}) \tag{26}$$

Combining (26) and (27) into (15), the number of turns per pole is given by

$$N_1 = \frac{V_b \cdot S}{N_{pp} \cdot b_p \cdot L_W \cdot u_b \cdot B_p \cdot (1 - L_u/L_{as})} \tag{27}$$

The number of turns per phase is

$$N_F = N_{pp} \cdot N_1 \tag{28}$$

The way to obtain the number of turns is by means of an iterative process. This iterative process is shown in **Figure 7**, in which K_L = 0.3 and $N_1 = V_b \cdot S/(N_{pp} \cdot b_p \cdot L_W \cdot u_b \cdot B_p)$ are taken as initial conditions. Aligned (L_{au}) and unaligned (L_u) inductances can be computed by 2D FEM or by using classical magnetic circuit analysis based on lumped parameters, in both cases considering leakage pole flux and end-effects [15].

Initially, the slot fill factor (K_s) is unknown, so $K_{s,0}$ = 0.4 is a good starting point. Once the number of turns per pole and the wire gauge have been obtained, K_s should be recalculated again by means of

$$K_s = \frac{2 \cdot S_c \cdot N_1}{c_p \cdot l_p} \tag{29}$$

If the slot fill factor (K_S) obtained from (30) differs from its initial value plus an accepted error (see **Figure 5**), then the number of turns should be recalculated again taking as initial slot fill factor the last value obtained.

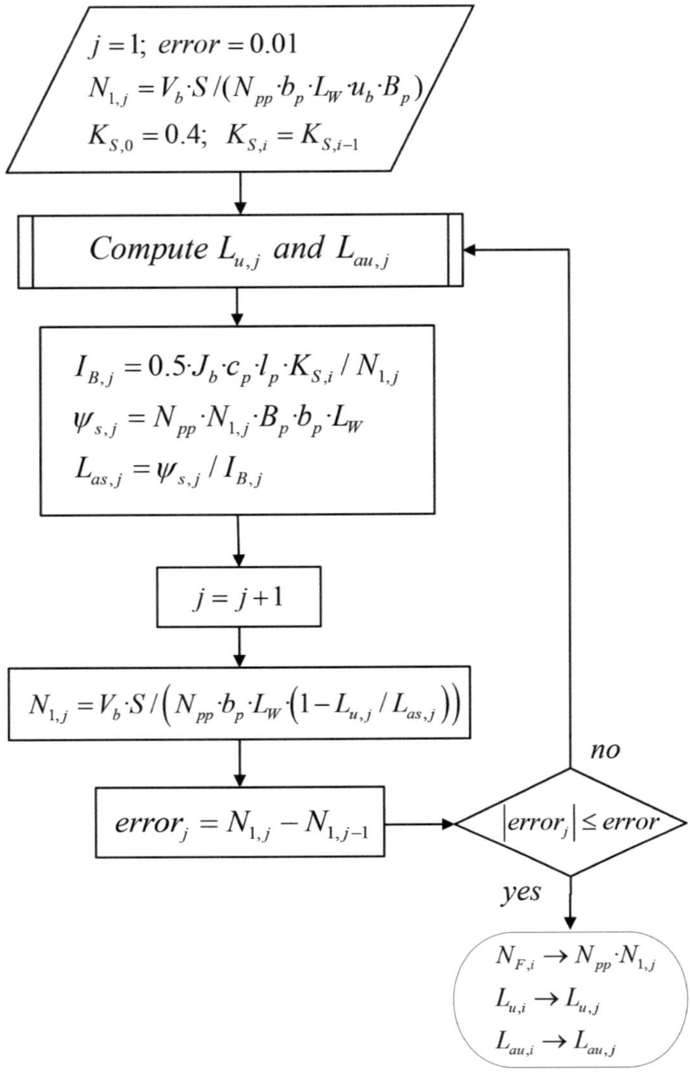

Figure 7.
Iterative process to obtain the number of turns per coil, N_1.

3.5 Finite element analysis

The best option to address the finite element analysis (FEA) process is to use three-dimensional 3D-FEA, but its use is discouraged because of the large computing time it could take. In order to overcome that handicap, it used a 2D-FEA adjusted in accordance with the end-effects. The end-effects in 2D FEA are considered by means of the end-effects coefficient, K_{ee} [16], given by:

$$\psi_{3D} = K_{ee} \cdot \psi_{2D} \tag{30}$$

$$L_{3D} = K_{ee} \cdot L_{2D} \tag{31}$$

where Ψ_{2D} and L_{2D} are the flux linkage and the inductance obtained by 2D-FEA; and Ψ_{3D} and L_{3D} are the 3D flux linkage and the inductance approach that account for the end-effects and are closer to the measured values. The correction factor K_{ee} is defined as [16, 17]

$$K_{ee} = \left(1 + \frac{L_{end} \cdot K_{si}}{L_{2D}}\right) \cdot K_f \tag{32}$$

where L_{end} is the end-winding inductance, K_{si} is a factor that affects L_{end} due to the steel imaging effect [17], and K_f is the axial fringing factor. K_{si} can usually be omitted (K_{si} = 1) since its effect on L_{end} is generally less than 2%. End-winding inductance, L_{end}, can be analytically deduced from end-winding geometry or can be computed by means of an axis-symmetrical 2D finite element model.

The co-energy (W'_{3D}), knowing (Ψ_{3D}), is calculated using the well-known expression:

$$W'_{3D}(x_i, I) = \int_o^I \psi_{3D}(x, i) \cdot di \Big|_{x_i = Ctn} \tag{33}$$

Then, the translation force, including end-effects, is obtained by

$$F_{x,\,3D}(x, I) = \frac{\partial W'_{3D}(x, I)}{\partial x}\Big|_{I = Ctn} \tag{34}$$

In order to offer a practical formulation of (34), it can be rewritten in (35)

$$F_{x,\,3D}(x, I_B) \approx \frac{\Delta W'_{3D}}{\Delta x} = \frac{\Delta I}{\Delta x} \cdot \left[\sum_0^{I_B} \psi_{3D}(x + \Delta x, I) - \sum_0^{I_B} \psi_{3D}(x, I)\right] \tag{35}$$

3.6 Thermal analysis

The objective of this analysis is to check that, within the specified conditions of operation, the temperature rise in the different parts of the LSRM does not surpass the limit value of the chosen insulation class. Thermal analyses of electric rotating machines have been extensively described in the literature [18–26], but up to now little attention has been paid to the thermal analysis of LSRMs [22]. Thermal analyses can be conducted by means of analytical or numerical methods. The analytical method based on lumped parameters is faster, but its accuracy depends on the level of refinement of the thermal network and on the knowledge of the heat transfer coefficients. In this paper a lumped parameter thermal model adapted to the LSRM is used in which the heat transfer coefficients are estimated taking into account previous studies in rotating machines [23–25].

3.7 Design verification

In order to verify the described design procedure, a four-phase double-sided LSRM prototype has been designed, built, and tested. Its main design specifications and its main dimensions, obtained following the proposed design procedure, are shown in the Appendix (**Table 4**).

3.7.1 Finite element verification

The finite element analysis is carried out by means of a 2D-FEM solver. The magnitudes computed are the 2D linked flux $\psi_{2D}(x, i)$, for a set of evenly distributed current ($0 \div I_B$), and positions between alignment and nonalignment, in **Figure 8** the flux density plots for the aligned (see **Figure 8a**) and nonaligned positions (see **Figure 8b**) for the LSRM prototype are shown.

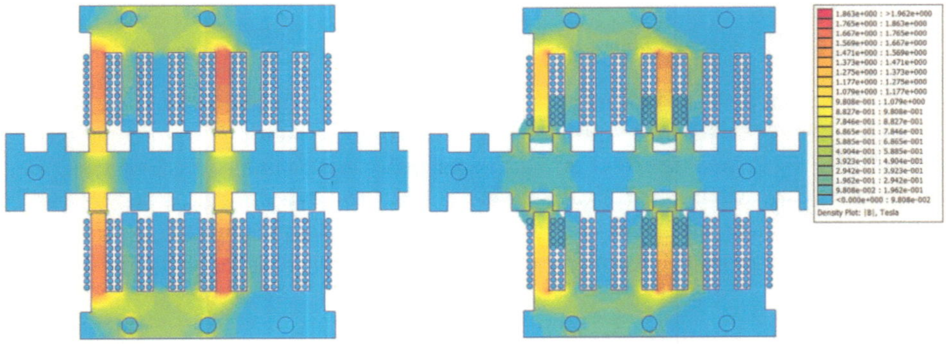

Figure 8.
Flux density plots from 2D FEA of the four-phase LSRM (a) aligned x = 0. (b) Unaligned x = S.

In order to verify and compare the results, the prototype was analyzed by means of the finite element method (FEM) described in Section 3.5 adapted to account for end-effects. The values of static force were also obtained experimentally using a load cell UTICELL 240. The measured and FEM computed force results are shown in **Figure 9**. Finally, the results for the average static force obtained by experimental means and by FEM are compiled in **Table 3**.

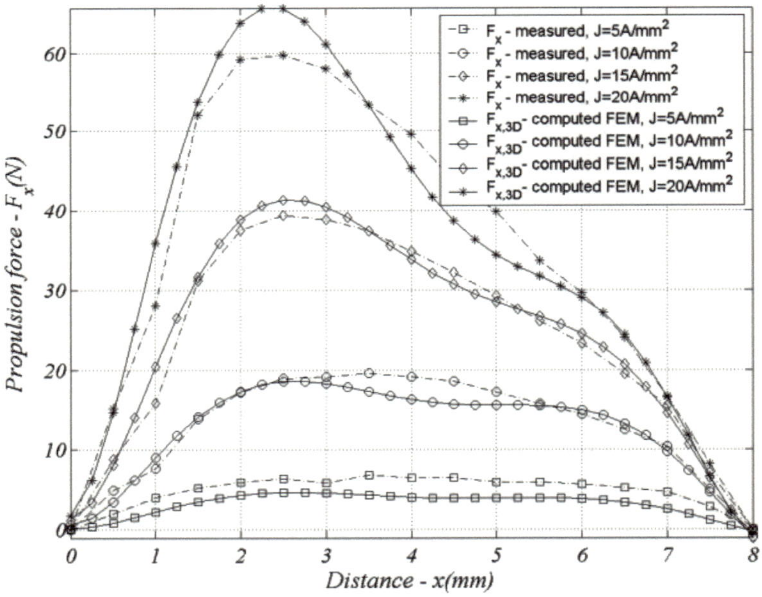

Figure 9.
Static force $F_x(J,x)$. Comparison of results.

J_B = 15 A/mm^2	$F_{x,avg}$ (N)
Measured	23.3
FEM	24.5

Table 3.
Average static force comparison of results.

3.7.2 Thermal verification

The lumped parameter thermal model mentioned in Section 3.6 and explained in depth in [26] was applied to our case study. The location of the nodes in the cross section of the double-sided flat LSRM prototype is shown in **Figure 10**, and the completed lumped thermal model is depicted in the circuit of **Figure 11**. The temperature rise over ambient temperature in each node was obtained solving the thermal network with *MATLAB-Simulink*. **Figure 12** shows the simulated and experimental results of a heating test consisting on feeding a phase with DC current at 15 A/mm^2 for a period of 1800 s and after that a cooling period of 1800 s by natural convention. The time evolution of temperature in node 4 (critical node) is compared with a platinum resistance thermometer sensor (PT100) placed in the same point.

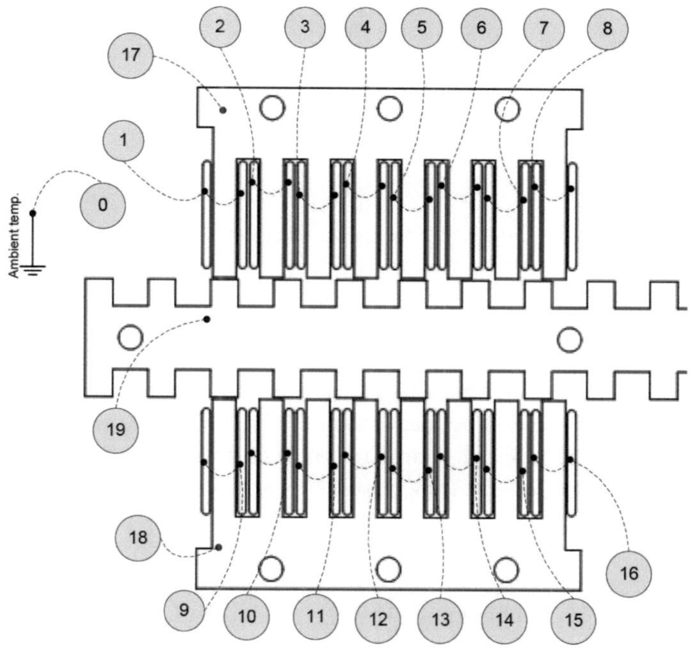

Figure 10.
Cross section of double-sided LSRM prototype showing node location.

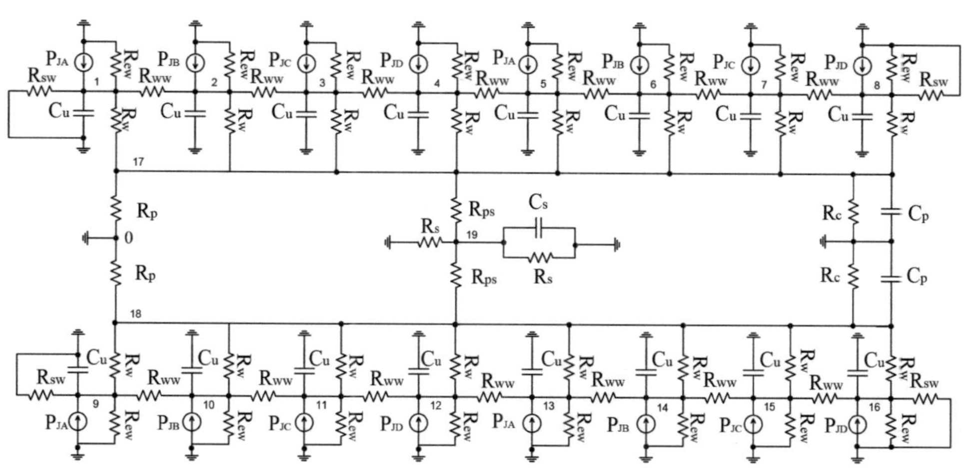

Figure 11.
Lumped-parameter thermal model for the double-sided LSRM prototype.

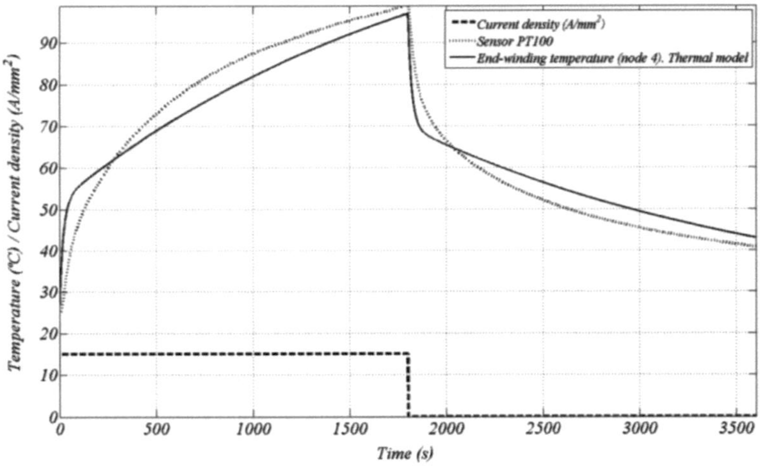

Figure 12.
Comparison of temperature rise results in node 4, for the LSRM prototype.

3.8 Discussion of results

Once built and tested, the prototype of double-sided LSRM is appropriate to proceed with a discussion of the results. It can be observed (**Figure 9**) that the static force results obtained by 2D-FEA, adjusted to take into account end-effects, are in good agreement with those measured experimentally except for those corresponding to high values of current density (20 A/mm^2), values that are outside the scope of application of the designed LSRM. The average static force values, for a current density of 15 A/mm^2, obtained by measurements are very close with those results of simulation by FEA (**Table 3**). The comparison of temperature raises results for node 4, in which a sensor of temperature was placed, between the values obtained using the proposed thermal model, and the experimental values measured by means of a sensor PT100 are quite good, but they also show that it would be advisable to improve the model, increasing the level of refinement of the thermal network. Anyway, the comparison between computed and experimental results is enough and good to validate the proposed design procedure.

4. Simulation model and experimental results of an LSRM actuator

The simulation of an LSRM force actuator is presented [27]. This linear actuator is formed by a longitudinal flux double-sided LSRM of four phases that has been designed following the design procedure before being described, and of which the main characteristics are given in the Appendix (**Table 4**). It is fed by an electronic power converter, an asymmetric bridge with two power MOSFETs switches and two diodes per phase, which incorporate drivers, snubbers, and current transducer for each phase. An optical linear encoder designed for this purpose, composed by four optical switches (S1, S2, S3, S4) is used in order to know the position at any time. The actuator is controlled by a digital force controller. The LSRM force actuator simulation model has been implemented in *MATLAB-Simulink*.

The simulation block diagram is shown in **Figure 13**, and it consists of three blocs: the power converter block, the LSRM motor block, and the digital control block.

The electronic power converter is implemented in *MATLAB-Simulink* by means of the *SimPowerSystems* toolbox. This block needs the previous knowledge of the gate signals which are generated by the switching signals module of the digital control block.

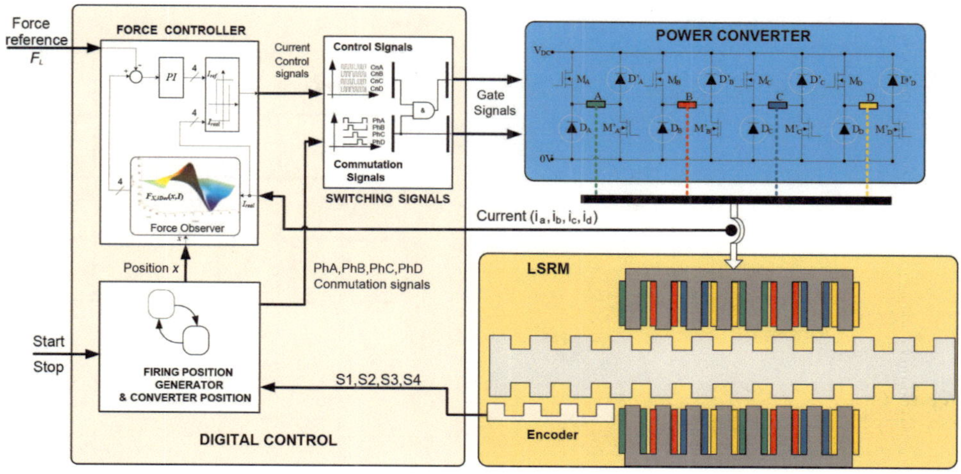

Figure 13.
LSRM force actuator simulation block diagram.

The LSRM block has to solve the mathematical model of the SRM, i.e., the space-state equations (7). To solve the instantaneous phase current (8), it is needed to know the phase voltages, the partial derivatives of the flux (lookup tables), and the phase resistance (**Figure 14**). The optical switch signals are obtained from integrating the speed of the mechanical equation (8), generating a Boolean set of digital signals in order to produce the phase activation sequence shown in **Figure 15**.

The force control block implements a PI controller and a hysteresis loop for generating the current control signals. The force is estimated using a force-observer which consists in a lookup table (static force curves of LSRM), previously computed using the 2D FE procedure described in Section 3.5, and therefore, the knowledge of phase currents (i_a, i_b, i_c, i_d) and of the position (x) is required. The phase currents are directly obtained from the electronic power converter output, and the position is from the firing position generator module of the digital control block. A hysteresis

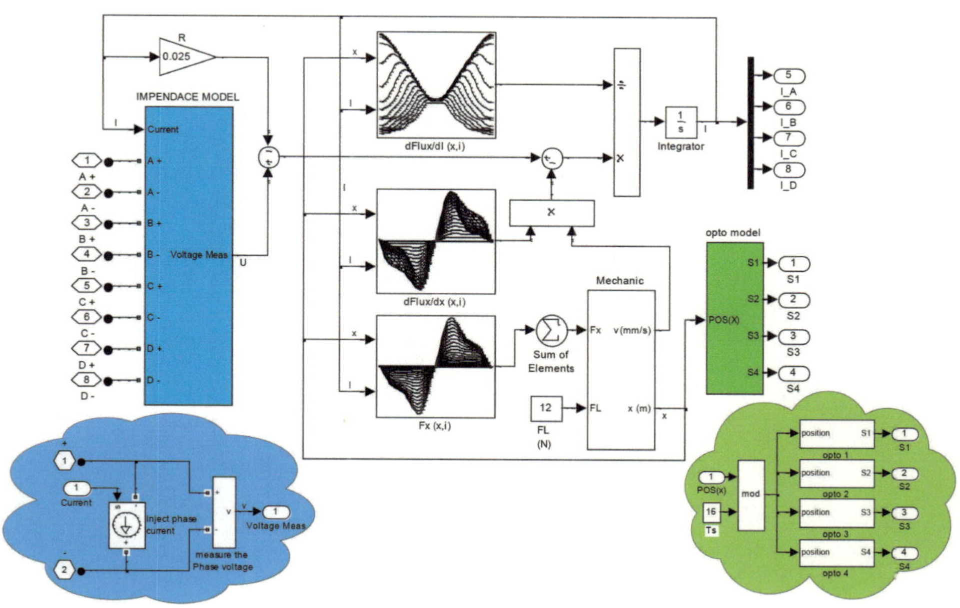

Figure 14.
LSRM, load and opto-switches Simulink model.

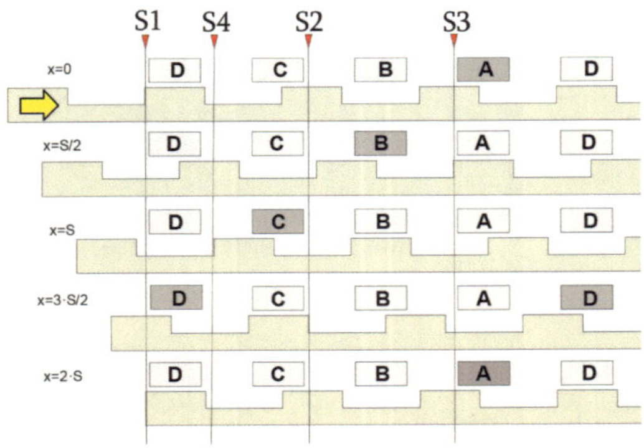

	S1	S4	S2	S3
x=0	1↓0	0	1	1
x=S/2	0	1	0	1↓0
x=S	0	1↓0	0	0
x=3·S/2	1	0	0↑1	0

Figure 15.
Encoder: phase activation sequence.

control adjusts the translation force to a given reference force (F_L). The program adjusts the frequency of the gate control signals of the MOSFET in order to match to the required force. The force control is implemented by means of a DSPACE ACE kit 1006 (**Figure 16**).

The simulation results are presented in **Figure 17**. The conducting interval is equal to the pole stroke, which is 4 mm in all the cases. In **Figure 17**, it can also be shown the influence of the firing position (x_1) over the current waveform. When firing at x_1 = 0 mm, the electromagnetic force $F_{x,\,3D}(x, i)$ is zero at the beginning of the conduction interval (see **Figure 17a**), and a current peak appears near this position. Firing at x_1 = 1 mm and x_1 = 2 mm, the resulting conduction intervals are from 1 to 5 mm and from 2 to 6 mm, respectively. In these intervals is where the force reaches the maximum values, which produces a current waveform almost flat. When the firing position is at x_1 = 3 mm, the conduction interval is from 3 to 7 mm, and a current peak appears at the end of the period because the electromagnetic force at x = 7 mm is quite low (is 0 at x = 8 mm), and therefore, a high increasing in current is required to maintain the force constant. In conclusion, firing near aligned (x = 0 mm) and unaligned (x = 8 mm) positions at low speed produces high current peaks, which is not advisable.

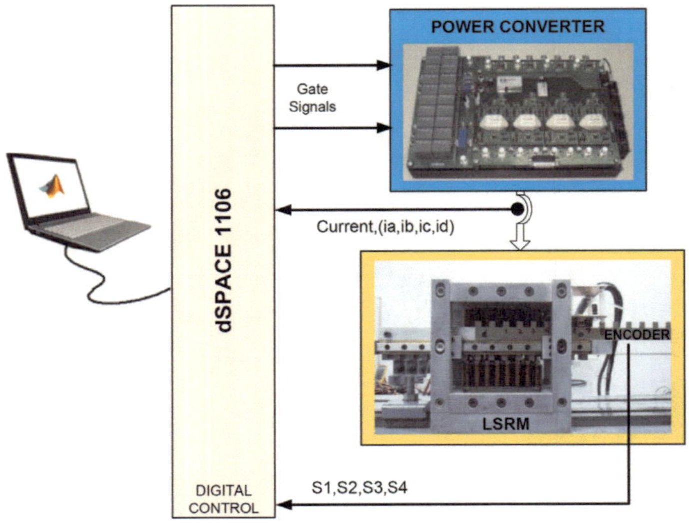

Figure 16.
LSRM force actuator hardware implementation.

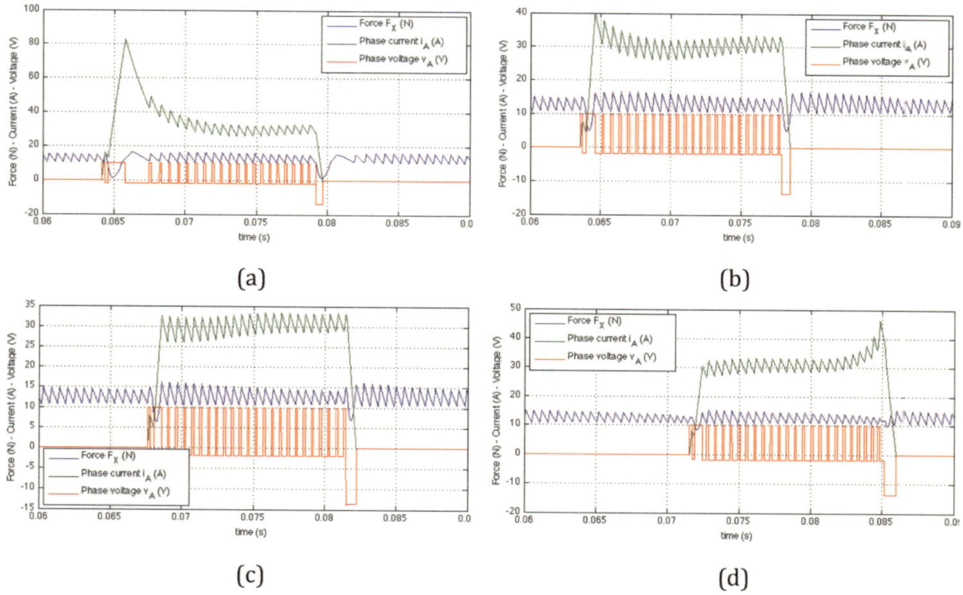

Figure 17.
Simulation results for different turn on (x_1) positions. (a) x_1 = 0 mm; (b) x_1 = 1 mm; (c) x_1 = 2 mm; and (d) x_1 = 3 mm.

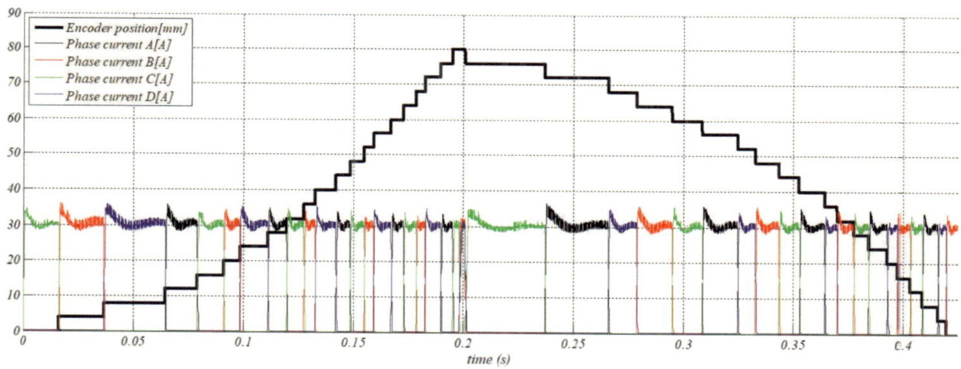

Figure 18.
Measured position and measured phase currents at turn on x_1 = 1 mm.

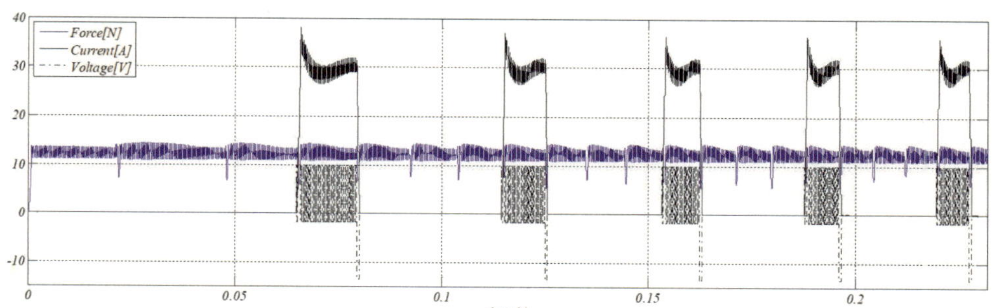

Figure 19.
Measured phase-a results during acceleration at turn on x_1 = 1 mm.

Figures 18 and **19** show the experimental results obtained from the four-phase LSRM which are shown in **Figure 16**. The experimental results display a good agreement with simulation results. The mover stroke is 80 mm, and this distance is covered in 0.2 s, which gives an average speed of 0.4 m/s.

5. Conclusion

In this chapter, after the presentation of a mathematical model of the LSRM, a design methodology for LSRM is proposed. This methodology is based on an analytical formulation of the average translation force determined using a nonlinear energy conversion loop. The main dimensions of the LSRM were determined from machine specifications, the aforementioned average translation force formula, and geometric relationships. 2D finite element analysis, corrected to take into account end-effects, and lumped parameter thermal analysis were used to refine and/or to validate the proposed design. An LSRM prototype was built following the described design approach that was validated by experimental results. Then, modeling and simulation of an LSRM force actuator are presented. This linear actuator is formed by the LSRM prototype, by an electronic power converter with two power MOSFET switches and two diodes per phase, incorporating drivers, snubbers, and current transducers for each phase and by an original, simple, and low-cost optical linear encoder designed for this purpose, composed of four optical switches. The actuator is controlled by a digital force controller that is implemented by means of a PI controller and a hysteresis loop for generating the current control signals. The force is estimated using a force-observer which consists in a lookup table previously computed using the 2D finite element analysis. The LSRM force actuator simulation model was implemented in MATLAB-Simulink. Experimental results were in good agreement with simulations and confirmed that the proposed LSRM actuator as an alternative to pneumatic actuators or of the assembly of AC servomotors coupled to a timing belt or a ball screw for injection molding machines.

Conflict of interest

The authors declare no conflict of interest.

Nomenclature

a	acceleration (m/s^2)
b_p	primary pole width (m)
B_p	magnetic flux density in the active pole (T)
b_s	secondary pole width (m)
c_p	primary slot width (m)
c_s	secondary slot width (m)
$e_{m,j}$	back electromotive force (V)
F_x	translation force, internal electromagnetic force (N)
F_r	friction force (N)
F_L	load force (N)
g	air-gap length (m)
h_{yp}	primary yoke height (m)
h_{ys}	secondary yoke height (m)
I_B	flat-topped current peak (A)

i_j	current phase j (A)
J_B	current density peak (A/m^2)
k_d	magnetic duty cycle factor
K_s	slot fill factor
L_{au}	unsaturated aligned inductance (H)
L_{as}	saturated aligned inductance (H)
L'_{as}	saturated aligned incremental inductance (H)
L_{end}	end-winding inductance (H)
l_p	primary pole length (m)
l_s	secondary pole length (m)
L_j	incremental inductance (H)
L_u	unaligned inductance (H)
L_W	stack length (m)
m	number of phases
M	mass of the mover (m_t) plus the payload (m_l)
n	number of switching devices per phase
N_1	number of coils per pole
N_p	number of active poles per side (primary)
N_{pp}	number of active poles per phase
N_s	number of passive poles per side (secondary)
PS	stroke (m)
R	phase resistance (Ω)
S	distance between aligned and unaligned positions (m)
S_c	cross section of the wire (m^2)
T_p	primary pole pitch (m)
T_s	secondary pole pitch (m)
TS	mover stroke (m)
u_b	velocity (m/s)
u_j	voltage phase j (V)
V_b	DC bus voltage (V)
W	energy conversion loop (J)
x	mover position (m)
x_c	turn-off current position (m)
ψ	flux linkage (Wb)

Appendix

Specifications		
Rated force	F_X	25 N
Number of phases	m	4
Lamination steel	M-19 (B_{sat} = 1.8 T)	
DC Bus voltage	V_b	12 V
Temperature rise (class F)	ΔT	100°C
Dimensions		
Pole stroke	PS	4 mm
Primary pole width	b_p	6 mm
Primary slot width	c_p	6 mm
Primary pole pitch	T_p	12 mm

Number of active poles per side	N_p	8
Primary pole length	l_p	30 mm
Secondary pole width	b_s	7 mm
Secondary slot width	c_s	9 mm
Secondary pole pitch	T_s	16 mm
Number of passive poles per side	N_s	6
Secondary pole length	l_s	7 mm
Yoke length	h_y	8 mm
Stack length	L_W	30 mm
Number of turns per pole	N_1	11
Wire diameter	d_c	2.1 mm
Air-gap length	g	0.5 mm

Table 4.
LSRM prototype main dimensions.

Author details

Jordi Garcia-Amoros[1*], Pere Andrada[2] and Baldui Blanque[2]

1 Electrical Engineering Department (DEEEA), Universitat Rovira i Virgili, Tarragona, Spain

2 GAECE, Electric Engineering Department, Universitat Politècnica de Catalunya, BARCELONATECH, EPSEVG, Vilanova i la Geltrú, Spain

*Address all correspondence to: jordi.garcia-amoros@urv.cat

References

[1] Bianchi N, Bolognani S, Corda J. Tubular linear motors: A comparison of brushless PM and SR motors. In: International Conference on Power Electronics, Machines and Drives (Conf. Publ. No. 487); June 4–7, 2002; pp. 626-631. DOI: 10.1049/cp:20020189

[2] Zhao SW, Cheung NC, Gan W-C, Yang JM, Pan JF. A self-tuning regulator for the high-precision position control of a linear switched reluctance motor. IEEE Transactions on Industrial Electronics. 2007;**54**(5):2425-2434. DOI: 10.1109/TIE.2007.900348

[3] Kolomeitsev L, Kraynov D, Pakhomin F, Kallenbach E, Kireev V, Schneider T, et al. Linear switched reluctance motor as high efficiency propulsion system for railway vehicles. In: SPEEDAM 2008; 2008. pp 155-160. DOI: 10.1109/EPEPEMC.2008.4635495

[4] Lobo NS, Lim HS, Krishnan R. Comparison of linear switched reluctance machines for vertical propulsion application: Analysis, design, and experimental correlation. IEEE Transactions on Industry Applications. 2008;**44**(4):1134-1142. DOI: 10.1109/TIA.2008.926294

[5] Zhu Z, Cheung NC, Cheng KWE, Xue X, Lin J. Direct instantaneous force control with improved efficiency for four-quadrant operation of linear switched reluctance actuator in active suspension system. IEEE Transactions on Vehicular Technology. 2012;**61**(4): 1567-1576. DOI: 10.1109/IWECMS.2011.5952367

[6] Llibre J-F, Martinez N, Nogarede B, Leprince P. Linear tubular switched reluctance motor for heart assistance circulatory: Analytical and finite element modeling. In: 10th International Workshop on Electronics, Control, Measurement and Signals (ECMS); June 1–3, 2011. pp. 1-6

[7] Du J, Liang D, Xu L, et al. Modeling of a linear switched reluctance machine and drive for wave energy conversion using matrix and tensor approach. IEEE Transactions on Magnetics. 2010;**46**(6): 1334-1337. DOI: 10.1109/TMAG.2010.2041041

[8] Garcia Amorós J. Aportaciones al diseño y caracterización del motor lineal de reluctancia autoconmutada [PhD thesis]. Vilanova i la Geltrú (Barcelona): UPC BarcelonaTech; 2010. Available from: http://hdl.handle.net/2117/94579

[9] Garcia-Amoros J, Andrada P, Blanqué B. Design procedure for a longitudinal flux flat linear switched reluctance motor. Electric Power Components & Systems. 2010;**40**(2): 161-178. DOI: 10.1080/15325008.2011.629333.

[10] Byeong-Seok L, Han-Kyung B, Praveen V, Krishnan R. Design of a linear switched reluctance machine. IEEE Transactions on Industry Applications. 2000;**36**(6):1571-1580. DOI: 10.1109/28.887208

[11] Krishnan R. Switched Reluctance Motor Drives. Modeling, Simulation, Analysis, Design and Applications. Boca Raton: CRC Press; 2001. DOI: 10.1201/9781420041644

[12] Boldea I, Nasar SA. Linear Electric Actuators and Generators. Cambridge: Cambridge University Press; 1997. DOI: 10.1017/CBO9780511529641

[13] Miller TJE. Converter volt-ampere requirements of the switched reluctance motor drive. IEEE Transactions on Industry Applications. 1985;**21**(5): 1136-1144. DOI: 10.1109/TIA.1985.349516

[14] Amorós JG, Andrada P. Sensitivity analysis of geometrical parameters on a double-sided linear switched reluctance

motor. IEEE Transactions on Industrial Electronics. 2010;**57**(1):311-319. DOI: 10.1109/TIE.2009.2032208

[15] Amorós JG, Andrada P. Magnetic circuit analysis of a linear switched reluctance motor. In: Proceedings of the 13th European Conference on Power Electronics and Applications (EPE 2009); Barcelona, Spain; 2009

[16] D'hulster F, Stockman K, Belmans RJM. Modelling of switched reluctance machines: State of the art. International Journal of Modelling and Simulation. 2004;**24**(4):216-223. DOI: 10.1080/02286203.2004.11442306

[17] Matveev A, Kuzmichev V, Lomonova E. A new comprehensive approach to estimation of end-effects in switched reluctance motors. In: Proceedings of XV International Conference on Electrical Machines (ICEM 2002); Bruges, Belgium; 2002

[18] Srinivas KN, Arumugam R. Thermal characterization through finite element analysis of the switched reluctance motor. In: Proceedings of IEEE Region 10 International Conference on Electrical and Electronic Technology (TENCON 2001). Vol. 2; Singapore; 2001. pp. 819-823. DOI: 10.1109/TENCON.2001.949707

[19] Mellor PH, Roberts D, Turner DR. Lumped parameter thermal model for electrical machines of TEFC design. Proceedings of the Institution of Electrical Engineers - Part B: Electric Power Applications. 1991;**138**(5): 205-218. DOI: 10.1049/ip-b.1991.0025

[20] Nerg M, Rilla M, Pyrhönen J. Thermal analysis of a radial-flux electrical machines with a high power density. IEEE Transactions on Industrial Electronics. 2008;**55**(10):3543-3554. DOI: 10.1109/TIE.2008.927403

[21] Staton D, Boglietti A, Cavagnino A. Solving the more difficult aspects of electric motor thermal analysis in small and medium size industrial induction motors. IEEE Transactions on Energy Conversion. 2005;**20**(3):620-628. DOI: 10.1109/IEMDC.2003.1210320

[22] Chayopitak N, Taylor DG. Thermal analysis of linear variable reluctance motor for manufacturing automation applications. In: Proceedings of IEEE International Conference on Electric Machines and Drives (IEMDC 2005); San Antonio, USA; 2005. pp. 866-873. DOI: 10.1109/IEMDC.2005.195824

[23] Boglietti A, Cavagnino A, Staton D. Determination of critical parameters in electrical machine thermal models. In: Conference Record of the Proceedings of 42nd IAS Annual Meeting; New Orleans, USA; 2007. pp. 73-80. DOI 10.1109/07IAS.2007.60

[24] Boglietti A, Cavagnino A, Staton D. Determination of critical parameters in electrical machine thermal models. IEEE Transactions on Industry Applications. 2008;**44**(4):1150-1159. DOI: 10.1109/TIA.2008.926233

[25] Staton D, Cavagnino A. Convection heat transfer and flow calculations suitable for electric machines thermal models. IEEE Transactions on Industrial Electronics. 2008;**55**(10):3509-3516. DOI: 10.1109/TIE.2008.922604

[26] Amorós JG, Andrada P, Blanqué B. An analytical approach to the thermal design of a double-sided linear switched reluctance motor. In: Proceedings of XIX International Conference on Electrical Machines (ICEM 2010); Rome, Italy; 2010. DOI: 10.1109/ICELMACH.2010.5608298

[27] Amorós JG, Blanque B, Andrada P. Modelling and simulation of a linear switched reluctance force actuator. IET Electric Power Applications. 2013;**7**(5): 350-359. DOI: 10.1049/iet-epa.2012.0391

Chapter 7

Design of Ultrahigh-Speed Switched Reluctance Machines

Cheng Gong and Thomas Habetler

Abstract

High-speed electric machines have been a timely topic in recent years. Switched reluctance machines (SRMs) are very competitive candidates for high-speed applications in high-temperature environments, due to the single material rotor without permanent magnets (PM). In this chapter, recent research on ultrahigh-speed SRMs (UHSSRMs) for applications from 100,000 rpm to 1,000,000 rpm is reviewed regarding the design and control. First, the proposed control methods in the literature for ultrahigh speeds over 100,000 rpms are introduced. A direct position control using low-cost nonintrusive reflective sensors is described in detail. Experiments are conducted to validate the method on a 4/2 SRM at 100,000 rpm. Next, for even higher speeds up to 1,000,000 rpm, a high-strength, high-torque-density, and high-efficiency rotor structure is introduced. Finally, the complete design of a 1,000,000 rpm SRM is proposed and prototyped using aerostatic air bearings with the help of electromagnetic finite element analysis (FEA) tools.

Keywords: finite element analysis, optical sensors, optimization, switched reluctance machine, ultrahigh speed

1. Introduction

Electric machines have been widely used in various applications such as metro systems [1, 2], electric power generation [3], etc. The size of an electric machine depends on its power level, which can vary from several 100 megawatts [4, 5] to several watts, according to the following equation [6]:

$$S = 11K_{w1} \cdot \overline{B} \cdot ac \cdot D^2 \cdot L \cdot n \tag{1}$$

where S is the motor rating in watts, B is the magnetic loading in Teslas, ac is the electric loading in A/m, D is the stator bore diameter in meters, L is the machine active length in meters, K_{w1} is the winding factor of fundamental frequency, and n is the rated speed in rps.

From Eq. (1), it can be seen that for a certain power rating, since the magnetic and electric loading do not vary too much for certain materials and cooling, the machine size (D^2L) scales down with the increased rotational speed [7], which saves the materials and costs, and would lead to a more compact system.

High-speed drives have become more and more interesting to researchers in both academia and industry in recent years. With the development of modern power electronics, the top speed of electric motors has increased significantly. For ultrahigh-speed drives above 100,000 rpm, there are several research conducted

IntechOpen

for permanent magnet (PM) machines from 200,000 rpm [8, 9] up to 1,000,000 rpm [10]. Besides PM machines, switched reluctance machines (SRMs) [11] are competitive candidates due to their intrinsically simple and robust rotor geometry. They are suitable for ultrahigh-speed applications under high ambient temperatures such as turbochargers. In this chapter, the state-of-the-art ultrahigh-speed switched reluctance drives are presented in detail regarding their control and design.

2. Control of ultrahigh-speed SRMs (UHSSRMs) over 100,000 rpm

Extensive research has been conducted in the literature for high-speed SRMs up to 60,000 rpm. One of the many reasons that prevent an SRM from reaching even higher speeds is the speed sensing. One can find various rotary encoders with a maximum speed less than 60,000 rpm. However, the internal mechanical stability issues of the code disc put a limitation on the maximum rotational speed a rotary encoder can achieve. Other kinds of shaft-connected encoders, for example, resolvers, suffer from the same issue, while a custom-designed encoder is far too expensive for a practical application [12]. To break this mechanical speed limit, either sensor-less control or noncontact sensing control must be applied.

Table 1 concludes different sensing methods for high-speed SRMs over 60 krpm reported in the literature, among which most of them are below the maximum speed limit of commercially available rotary encoders on the market. Only five SRMs are reported to achieve a max speed beyond 60 krpm. Bateman uses the current gradient sensor-less (CGS) method to control a 4/2 high-speed SRM up to 80 krpm, although the target was 100 krpm [24]. Calverley uses a Hall sensor to detect the rotor pole saliency in order to control the motor. But the Hall sensor itself must be located very close (1.5 mm) to the rotor laminations. This implies that this method is not suitable for off-the-shelf machines. Morel uses another sensor-less method and achieved 110 krpm, although the target speed was 200 krpm. The method applies the resonant characteristic of the RLC circuit to detect the rotor position, and it requires a sophisticated design of a complex external circuitry and a carefully selected resonant frequency. Kozuka reports a 6/4 high-speed SRM at 150 krpm with test results. It uses an unknown type of special sensor that generates a series of square waveform with four times the frequency and a duty cycle of 50% of the shaft mechanical speed. However, in order to implement the method, a complex

Max speed (krpm)	Sensor type	Pole no.	Refs.	Max speed (krpm)	Sensor type	Pole no.	Refs.
20	Sensor-less	6/4	[13]	50	Resolver	6/4	[14]
20	Optical	8/6	[15]	50	Optical	6/4	[16]
25	Resolver	6/4	[17]	50	Optical	4/2	[18]
30	Sensor-less	8/6	[19]	52	Resolver	6/4	[20]
30	Hall Sensor	4/2	[21]	60	Unknown	4/2	[22]
40	Unknown	6/6	[23]	80	Sensor-less	4/2	[24]
48	Unknown	4/2	[25]	100	Hall sensor	6/4	[26]
50	Resolver	6/4	[27]	110	Sensor-less	6/2	[28]
50	Hall sensor	6/4	[29]	150	Unknown	6/4	[30]

Table 1.
Different sensor types for high-speed SRMs.

drive with FPGA must be used to provide the exact rotor position, which is also far too complicated and computationally intensive for practical use.

A simple and fast control system using low-cost noncontact optical sensors for ultrahigh-speed SRMs is proposed in [31, 32] and is followed by a multi-physics acoustic analysis [33]. **Figure 1(i)** shows the settings of the method. The photons emitted from the LED are reflected by the surface of the shaft and are then received by the photodiode when they are detecting a white mark. A current signal is thus generated, and a trigger signal is output after an amplifier, a Schmitt trigger, and an output transistor. On the other hand, when the photons are emitted to a black mark, all of them will be absorbed and no signal will be generated. Thus, as the shaft rotates, an output square wave signal is generated to control the motor.

Figure 1(iii) describes a simple and fast direct position control for ultrahigh-speed SRMs. A SRM with four stator slots and two rotor poles is used here for illustration purpose, and other types are similar. The goal is to reduce the cycle per resolution (CPR) to the minimum value, which is equal to the number of rotor poles, and then apply single pulse control for each phase [34]. The black and white marks on the shaft determine the switching-on and switching-off angles (dencted as θ_1 and θ_2, respectively). Note that from this point the photodiodes are assumed to be light on mode, which means that the trigger signal is HIGH when a dark surface is detected and is LOW when a white surface is detected. As shown in **Figure 1(iii)** (a), when the rotor is θ_1 degrees from the unaligned position ($\theta = 0°$, which is also the aligned position for the other phase), the optical sensor detects the black mark and gives a rising edge of the output signal. As the rotor keeps rotating clockwisely, the sensor keeps detecting the black mark and outputs a high voltage. When the rotor is θ_2 degrees from the unaligned position, the black mark ends and makes the trigger signal low voltage again. The switching-on and switching-off angles can be easily modified by changing the position and span of the black mark. A logic AND is

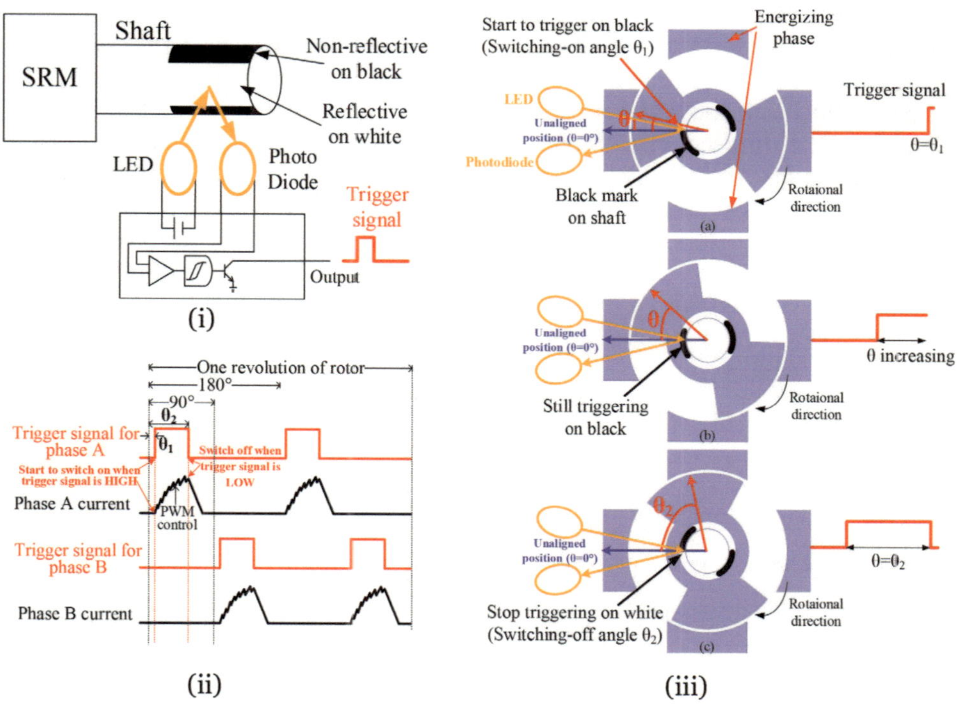

Figure 1.
(i) Setup of the noncontact optical sensor (up left). (ii) PWM current control with the optical sensors (down left). (iii) direct position control (right).

performed between the PWM signal and the trigger signal to control the gate signals of the MOSFETs of the corresponding phase. **Figure 1(ii)** shows a typical waveform of the phase currents and trigger signals during one revolution.

The advantages of the proposed sensing method are as follows:

- **Versatility.** Because it uses noncontact sensors, it can be fitted to any kind of off-the-shelf SRMs without changing the shaft geometry or opening the motor. A simple painting of the shaft is enough.

- **No virtual speed limit.** Thanks to the noncontact setup, there is no mechanical speed limit to the motor. The actual speed limit is determined only by the time delay of the signal processing circuit, which can be designed to be a very small number.

- **Simplicity.** The proposed method does not require speed/position transformation or an extra sampling frequency of DSP. This is because that the trigger signal is an analog type and is used to control the phases directly without and digital/analog speed/position manipulations.

- **Low cost.** Conventional rotary encoders are very expensive. A typical high-speed rotary encoder (i.e., US digital E5) costs several 100 dollars. But the proposed method only costs several dollars for very cheap components such as LEDs, photodiodes, and operational amplifiers.

Table 2 Gives a comparison of the proposed method and different sensing methods proposed in the literature.

The method was validated on a high-speed 4/2 SRM at 100,000 rpm. **Figure 2** shows the experimental setup. Two optical sensors were built for sensing the rotor position for each phase. The optical sensors are the light on mode, which means that the output voltage is 5 V for black and 0 V for white. The mark patterns are drawn in such a way that the two optical sensors detect the two different phases, as is shown by the black and white marks in **Figure 2**. The switching-on and switching-off angles are optimized to be 0° and 70° using a fast equivalent model based on finite element analysis (FEA) [32]. **Figure 3** shows the block diagram of the closed loop control. A video reference can be found in [35]. **Figure 4** shows a screenshot of the oscilloscope at 100,000 rpm. Channel 1 (yellow) is the voltage waveform of phase A with a DC link voltage of 140 V. Chanel 2 (blue) is the optical trigger signal of phase A. Channels 3 and 4 (magenta and green) are the current profiles of phase

	Hall sensor	Optical encoder	Resolver	Sensor-less	Proposed
Speed limit	Low	Low	Low	Medium	High
Nonintrusive?	No	No	No	Yes	Yes
Implementation simplicity	Medium	Medium	Medium	Low	High
Direct control without signal processing?	No	No	No	No	Yes
Accuracy	High	High	High	Medium	Low
Typical cost	\$300	\$200	\$300	\$0	\$5

Table 2.
Comparison of different sensing methods.

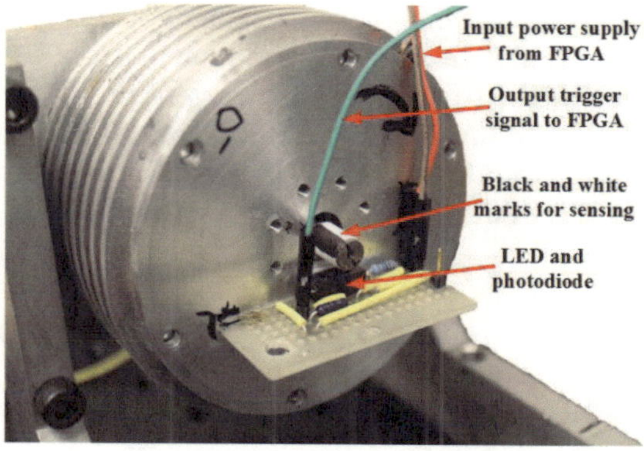

Figure 2.
Picture of the 4/2 SRM and the optical sensor setup.

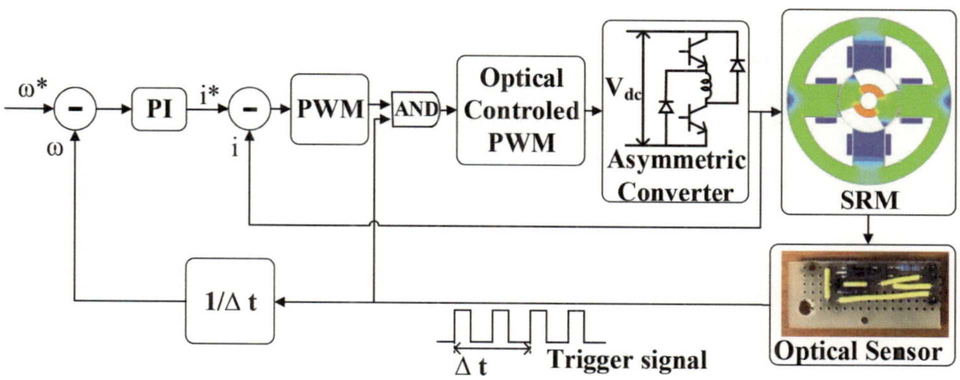

Figure 3.
Block diagram of the closed loop control.

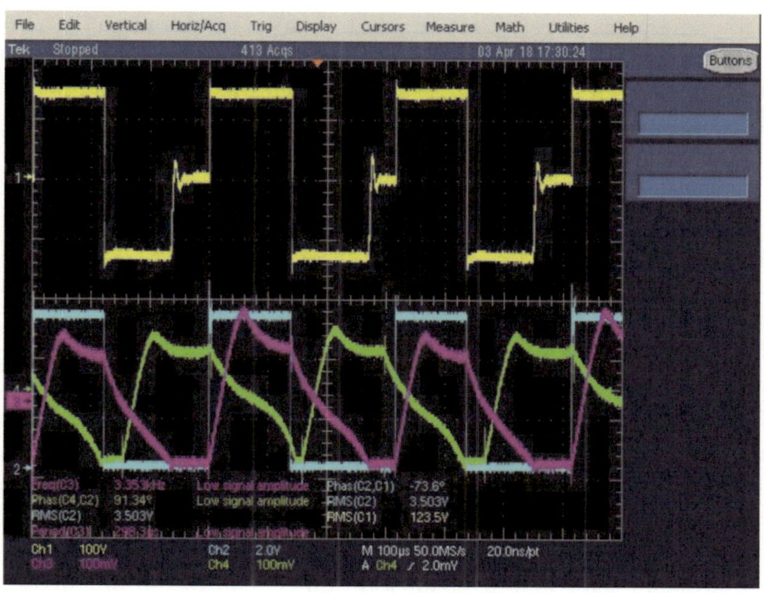

Figure 4.
Experimental result at 100,000 rpm.

A and phase B, respectively. It can be seen that the optical sensor controls the current very well. The current sensor conversion rate is 1A/100 mV. It can be read from the oscilloscope that the frequency of the current profile of phase A (magenta curve wave) is 3.353 kHz. So the speed is 3353/2•60 ≈ 100,000 rpm. Higher speeds could have been achieved, but for the safety concerns of the bearings, the final speed was stopped at 100,000 rpm.

3. Rotor design of ultrahigh-speed SRMs over 1,000,000 rpm

For ultrahigh-speed electric machines that run at speeds over 1 million rpm, one needs to follow a totally different design procedure from "regular" or what is commonly referred to as a "high-speed" machine design. The first thing to be considered is the rotor structure, because the conventional rotor geometries are not suitable for ultrahigh-speed applications over 1 million rpm. In this section a new rotor structure for ultrahigh-speed SRMs is described in detail [36]. The rotor lamination has smooth surfaces on both sides without any shaft bore in the middle. The shaft surrounds the rotor stack on both sides with two clamping arms. Compared to the conventional design, the proposed design has many advantages, such as high strength, high torque density, high efficiency, and high reliability.

3.1 Problem of conventional rotor designs

At over 1 million rpm, the conventional rotor structures cannot be applied due to the following reasons.

3.1.1 Localized stress concentration

Figure 5 shows the finite element stress analysis of the conventional rotor structure with an outer diameter (OD) of 4 mm under the operational speed of 1.2 million rpm. It can be seen that the maximum stress is 606 MPa, which is located at the sharp corners. However, most of the lamination steels only have a yield strength less than 400 MPa. This localized stress concentration is not a big issue when at low speeds. But it would cause failure when the rotor is rotating at ultrahigh speeds.

3.1.2 Too small space for the shaft

Although the rotor size can be reduced to decrease the highest stress to be less than the yield strength of the lamination material, there is another critical problem that prevents the conventional designs from being used at ultrahigh speeds. Usually the rotor OD is limited to a maximal value of 3 to 4 mm when operating at such high speeds. This will result in the rotor shaft OD less than 1 mm (see **Figure 5**). Shafts with such thin OD are extremely difficult to manufacture and are obviously not strong enough at high speeds.

3.1.3 Too much windage loss

Another problem is the high windage loss at ultrahigh speeds due to the fact that the air drag loss is proportional to the third power of the rotational speed. Moreover, the traditional rotor structure is actually not well designed in the perspective of aerodynamics due to the rotor double-salient structure. But the rotor saliency, in

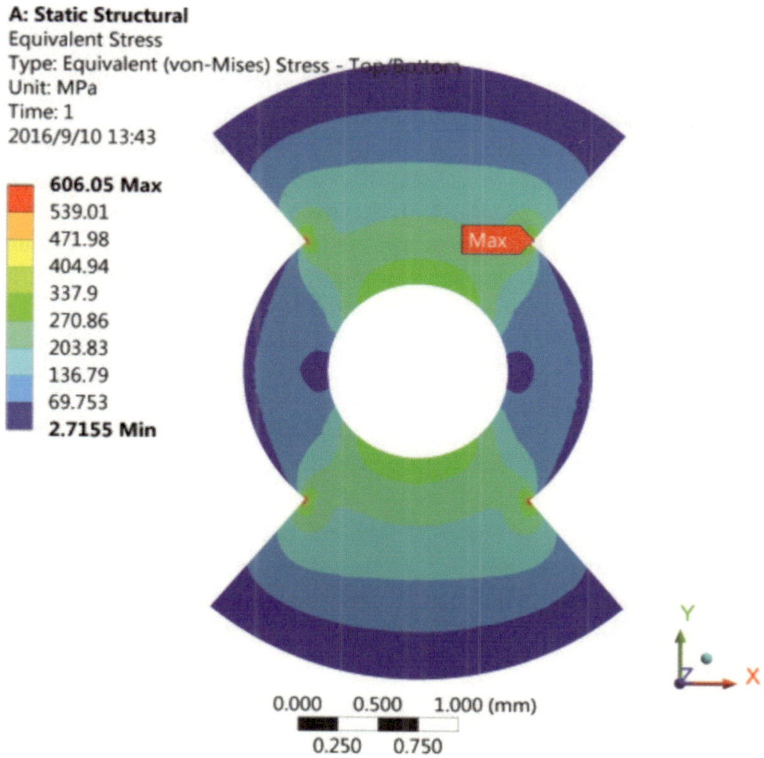

Figure 5.
Stress distribution of the regular rotor geometry.

turn, is the source of the output torque. This implies that the intrinsic properties of the SRM rotors are contrary in terms of aerodynamics and electromagnetics.

3.2 Possible solutions in the literature

3.2.1 Using bolts rather than a shaft

To solve the shaft bore problem, a "shaft-less" rotor design (shown in **Figure 6**) has been proposed in [37]. However, this design still needs bolts fed through the rotor laminations, which increases the assembling difficulty and is also not possible in very tiny scales under ultrahigh-speed cases. On the other hand, the double-salient geometry will still lead to very high windage loss.

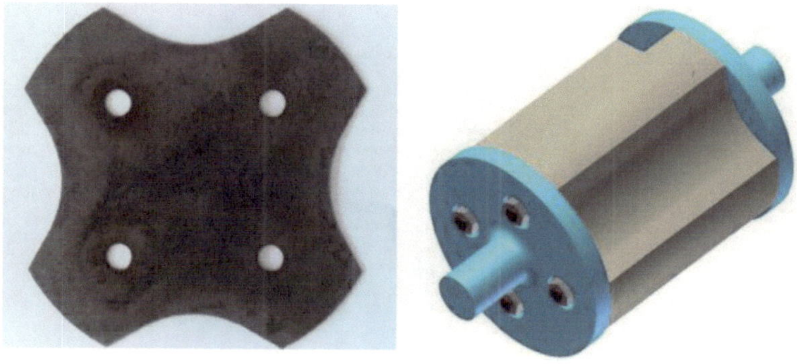

Figure 6.
High-speed rotor lamination with no shaft bore (left) and rotor assembly with end plates (right) [37].

3.2.2 Using rotor sleeves

Some efforts has been made to solve the high windage loss problem in the literature, such as using a rotor sleeve that is made from titanium or carbon fiber, just as in high-speed PM machines [28, 38]. It is not a good design in the perspective of electromagnets, although it is mechanically well designed. A typical value of the air gap of ultrahigh-speed SRMs is 0.1–0.25 mm in order to increase the torque density [26]. These nonmagnetic rotor sleeves are equivalent to an extra air gap in the flux path, which increases the equivalent air gap length by twice or more in the radial direction. This will lead to a low torque density [38]. As shown in **Figure 7(i)**, the equivalent air gap length is 0.6 mm, which is the summation of the 0.35 mm actual air gap length and the 0.25 mm rotor sleeve thickness. This large air gap results in a low maximum flux density of 0.86 T in the stator teeth.

3.2.3 Design with "flux bridges"

Another design to solve the high windage loss problem is using "flux bridges" to connect the salient rotor poles [39, 40]. This unique design has advantages in the perspective of aerodynamics. However, it requires the flux bridge to be thin enough to be magnetically saturated. As can be seen from **Figure 7(ii)**, such thin flux bridge is obviously not mechanically strong enough to sustain the high centrifugal forces at ultrahigh speeds. From the stress finite element analysis, it can be seen that at 1.2 million rpm the highest stress is 1055 MPa, which is located at the connection points of the flux bridge.

3.3 A novel rotor geometry for ultrahigh-speed SRMs over 1,000,000 rpm

From the analysis above, it can be concluded that a new geometry has to be proposed for UHSSRMs over 1 million rpm. It has to be mechanically strong enough to endure the high centrifugal force at ultrahigh speeds. Also it should not have any holes or rotor sleeves. In addition, it should have a good aerodynamic performance. In this subsection, a novel rotor geometry that combines all these advantages is proposed in detail [36, 41].

3.3.1 Design details

To design an SRM, the first step is to select a suitable pole/slot combination. The less poles an SRM has, the less core losses and switching losses it will have because

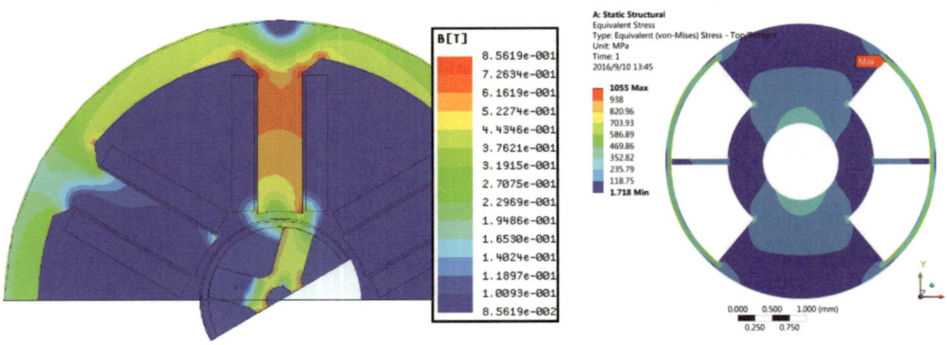

(i) (ii)

Figure 7.
(i) High-speed SRM with rotor sleeves (left). (ii) stress distribution of the rotor with "flux bridges" (right).

of the less fundamental frequency. Eq. (2) shows the relationship between the fundamental frequency and the rotor pole number.

$$f = \frac{N_r \cdot N_{rpm}}{60} \tag{2}$$

where N_r denotes the number of rotor poles and N_{rpm} denotes the rated speed in rpm. So the rotor pole number is chosen to be two in the proposed design.

Figure 8(i) shows a new rotor structure for ultrahigh-speed SRMs. There are two components to form a rotor: a rotor stack (black) and a clamping shaft (gray). The rotor stack is composed of electrical laminations that are made from magnetic materials. The clamping shaft is made from nonmagnetic materials with high mechanical strength such as carbon fiber or titanium alloy. The main idea of the design is to keep the rotor stack as integrated as possible and transfer the stress that is imposed on the rotor laminations to the shaft, which can be made from the nonmagnetic materials having much higher yield strength. Both sides of the rotor laminations are also designed to be smooth curves to avoid any localized stress concentration. Moreover, there is no shaft bore in the middle of the rotor laminations, which significantly reduces the highest stress caused by the large centrifugal force. The rotor stack is mechanically supported by the two contacting curved surfaces between the rotor stack and the two "clamping arms" (yellow shadow in **Figure 8(i)**), which realizes the function of the interference fit between the rotor and the shaft in regular machines or the using of bolts in [37]. It needs to be pointed out that if the rotor is manufactured to be totally symmetrical and balanced, no net force will be produced by the clamping arms when the shaft is rotating, except for the supporting force against the gravity of the rotor stack, because there is no trend of relative movement between the rotor stack and the clamping shaft. The net radial electromagnetic force exerted on the rotor stack is also zero due to the symmetrical geometry. **Figure 8(ii)** shows a prototype of the design using carbon fiber as shaft material.

3.3.2 Advantages

The new design has several advantages for ultrahigh-speed SRMs as follows.

- **High strength**. There is no sharp corner in the rotor laminations as in traditional rotor geometries, and the surfaces of the rotor stack are all smooth. Plus, there is no shaft bore in the center of the rotor laminations. These two features greatly reduce the highest stress in the rotor laminations. Furthermore, the OD of the shaft is the same as the rotor stack, which greatly increases the robustness of the shaft.

- **High torque density**. Unlike using rotor sleeves, which significantly increases the equivalent air gap, there is no increase of the equivalent air gap length at the two rotor pole ends in the radial direction for the new rotor structure. More specifically, when the rotor is aligned with the stator teeth, the equivalent air gap length is just the physical distance from the stator pole to the rotor pole, without adding any additional sleeve thickness. This implies that higher power density is achieved.

- **High efficiency**. Thanks to the cylindrical rotor structure, the windage loss is dramatically reduced to a minimum value.

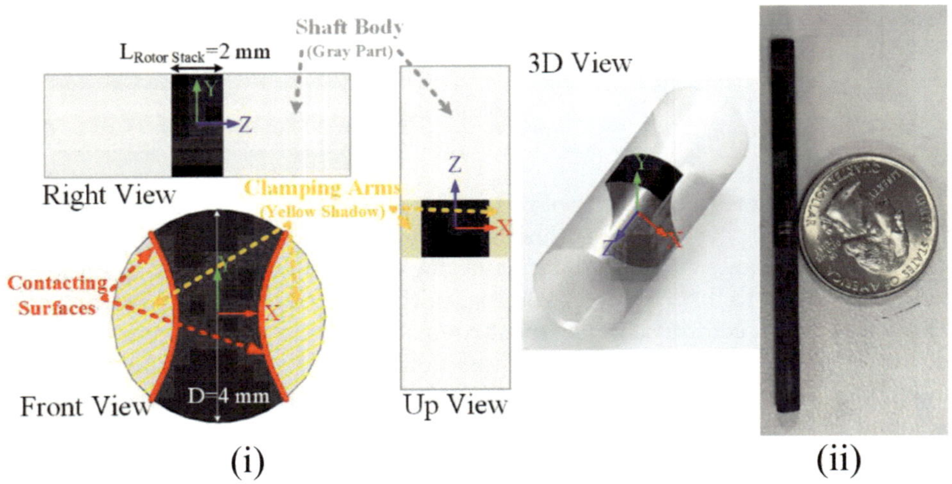

Figure 8.
(i) Three-view and 3D drawing of the proposed design (left). (ii) prototype of the shaft made from carbon fiber (right).

	Regular	Flux bridge [40]	Rotor sleeve [28, 38]	Through bolts [37]	Proposed
Practical to construct at ultrahigh speeds?	No	No	Yes	No	Yes
Rotor lamination strength	Moderate	Low	Moderate	High	High
Torque density	High	Moderate	Low	High	High
Efficiency	Low	Moderate	High	Low	High
Reliability	Low	Low	High	Low	High
Axial length	Long	Long	Long	Long	Short

Table 3.
Comparison of different rotor geometries.

- **High reliability**. The new rotor structure does not need bolts or other kinds of mechanical connections between the rotor and the shaft, which implies high simplicity and reliability.

Table 3 shows a comparison of different rotor geometries in the literature regarding ultrahigh-speed applications. A detailed 3D FEA of the stress distribution of the rotor stack and the clamping shaft can be found in [36, 42]. A rotor dynamics analysis can be found in [43].

4. Electromagnetic design of an ultrahigh-speed SRM over 1,000,000 rpm

After the rotor is designed, the stator and windings should also be designed properly to meet the power and torque requirement. At such high speed, the stator copper loss and core loss will be significant. The former is due to the skin effect of the stator windings, while the latter is because of the high switching frequency. In addition, a proper value of the air gap length should also be determined. It should be as small as possible in order to reduce the electric loading, but it cannot be too small to be manufactured. In this section a complete design of an ultrahigh-speed SRM for applications beyond 1 million rpm is proposed [44]. Pole pair number, dimension calculation, air gap calculation, material selection, winding selection, and windage torque calculation are included in details. The design is validated using finite element analysis. A prototype of the design is built using aerostatic bearing.

4.1 Air gap design

Choosing the right air gap length is of great importance in the design of ultrahigh-speed electric machines. Various types of electric machines require different air gap length. Usually SRMs need smaller air gaps than permanent magnet machines due to the lack of magnets in the rotor. For instance, the high-speed PM motor in [45] has a physical air gap of 0.5 mm. Nevertheless, the equivalent air gap length is larger due to the use of the retaining sleeve. The typical value of the length of the air gap in high-speed SRMs is in the range of 0.1–0.25 mm according to [26]. For example, the SRM proposed in [26] has an air gap of 0.2 mm. The designed value of the air gap in [46] was 0.3 mm and had to be changed to 0.5 mm due to manufacturing limitations. The air gap length of the SRM in [38] is 0.35 mm, plus an additional 0.25 mm of the sleeve thickness, resulting in an overall equivalent air gap of 0.6 mm. This large equivalent air gap leads to a low flux density in the stator and rotor core.

Considering the tiny size, ultrahigh-speed target, as well as the manufacturing limitations, the final air gap length is determined to be 0.3 mm. This value is comparatively large for such a tiny scale motor. But it can give better mechanical robustness when the machine is passing through the natural frequencies before reaching the rated speed of 1 million rpm, which is very important. The trade-off is that the electric loading will be higher to achieve the same torque density.

4.2 Stator winding design

Due to the high fundamental frequency of 33.3 kHz, the skin effect of the stator windings is significant at such high frequency. So Litz wire is used in the stator windings [47]. Different types of Litz wires are recommended for different operational frequencies. Selecting the most suitable type of Litz wire requires a trade-off among different aspects such as the switching frequency, the winding area, the maximum allowed current, and so on. For the fundamental frequency of 33.3 kHz of the proposed design, it can be assumed that the switching frequency of the power switches is about 10 times this frequency. Thus, the Litz wire of strand gauge of 42 is chosen to be used here [48].

Among all different types of Litz wires with the strand gauge of 42, those who have a strand number of 66, 105, and 165 are of particular interest considering the maximum allowed current. The rated RMS current is finally chosen to be 3.5A based on [38, 47, 49], which is a compromise between high current value and low winding area. This choice of maximum current will be verified in the next section.

After choosing the Litz wire, the winding cross section area can be determined. Because of the large air gap, the AC loading or the number of turns per phase has to be greater than usual. In order to produce enough flux to overcome the air gap to generate enough torque, the number of turns per coil is chosen to be 50 by an iterative design procedure. This is also a compromise between high-torque and small winding area. Assuming a filling factor of 0.25, the slot area is finally calculated to be 40 mm^2.

4.3 Stator design

Because of the ultrahigh speeds, laminations that are suitable for high switching frequency have to be applied. Metglas 2605SA1 (an iron-based amorphous metal) is reported to have less than 10th of the losses of the thinnest standard iron laminations [50, 51] at high switching frequencies [52]. However, amorphous alloys are well known to be extremely difficulty to work with, not only due to their necessary thickness but also because of their inherently extreme brittleness. Only very limited shapes are available because it's almost impossible to be stamped and laser cutting has its own problems as well [53]. For this reason, 0.006" Hiperco 50 laminations, which are also used in the rotor stack, will be applied to make the stator stack in order to reduce the manufacturing cost. Another benefit of using Hiperco 50 is to increase the torque density of the motor thanks to the high saturation flux density of the material.

The stator dimension is determined based on [18], particularly the stator and rotor pole arcs. The final OD is calculated to be 30 mm so as to provide enough winding area of 42 mm^2. A CAD model is shown in **Figure 9** to verify that the designed stator has enough space to contain 50 turns of Litz wire inside.

The yoke thickness has a significant effect on the noise, vibration, and harshness (NVH) behavior of the stator. Motors with thicker yokes tend to have higher natural frequencies [54], which are beneficial for high-speed operations. Thus, the yoke thickness is designed to be 2 mm although that 1 mm yoke is thick enough to keep away from saturation. The teeth can be modeled as cantilevers that are attached to the stator back iron in [55]. So an uneven wedge-shaped tooth that is thicker at the connecting point and thinner in the tip is designed to increase the mechanical stability of the stator. **Figure 9** shows the CAD model of the stator.

4.4 Windage torque estimation

At ultrahigh speeds, air friction losses might be very significant due to the fact that the windage losses are proportional to the third power of the angular speed. This phenomenon is particularly prominent for double-salient motors such as SRMs. However, thanks to the cylindrical design of the new rotor geometry, the windage losses are significantly reduced. The ultimate goal is to drive the rotor to as high speed as possible to achieve beyond 1 million rpm. So the motor will be tested at no external load at the first stage, which means the only load for the motor to overcome is the air friction.

The general equation for windage losses of a cylinder is [52]:

$$P_{f,air} = c_f \pi \rho_{air} \omega^3 r^4 l \tag{3}$$

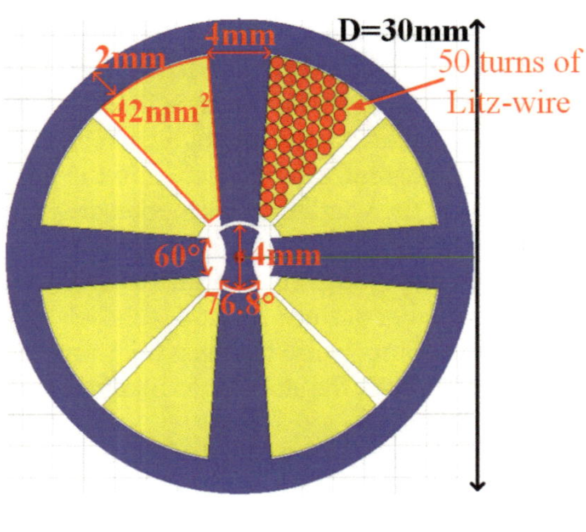

Figure 9.
CAD model of the machine with 50 turns of Litz wire.

where ρ_{air} is the air density, ω is the angular frequency, r is the radius, and l is the length of the cylinder. The proposed question can be simplified to be one cylinder encased in another cylinder, despite the fact that the inner surfaces of the stator tooth tips are actually separated with each other rather than forming a cylinder. This two-cylinder model itself is a very complicated research topic in aerodynamics. Many different methods can be used to calculate the losses. According to the method in [52], the air flow in the air gap can be categorized into three different situations: laminar Couette flow, laminar flow with Taylor vortices, and turbulent flow, which depends on the rotational speed and the Taylor number Ta ($Ta < 41.3$, $41.3 < Ta < 400$, and $Ta > 400$ for the three situations, respectively). The Taylor number Ta is defined as follows [52]:

$$Ta = \frac{r\omega\delta}{\nu}\sqrt{\frac{\delta}{r}} \tag{4}$$

where ν is the kinematic viscosity of the air and δ is the air gap length.

For laminar Couette flow, the friction coefficient c_f can be calculated using [52]:

$$c_f = \frac{1.8}{Re}\left(\frac{\delta}{r}\right)^{-0.25}\frac{(r+\delta)^2}{(r+\delta)^2 - r^2} \tag{5}$$

where Re is the Reynolds number and can be defined as follows:

$$Re = \frac{r^2\omega}{\nu} \tag{6}$$

According to a different model in [38], the c_f can be calculated using:

$$c_f = \begin{cases} 0.515\dfrac{\left(\frac{\delta}{r}\right)^{0.3}}{Re^{0.5}}, & 500 < Re < 10^4 \\ 0.0325\dfrac{\left(\frac{\delta}{r}\right)^{0.3}}{Re^{0.2}}, & Re > 10^4 \end{cases} \tag{7}$$

Using the above equations, the corresponding Ta is calculated to be 1315 at 1 million rpm. This indicates that the machine has already been in the turbulent flow region, where the windage loss increases dramatically. Normally it is extremely difficult to calculate the windage loss in the turbulent flow area from analytical equations. In order to ensure that the machine has the torque capability to overcome the air drag at such high speeds, the windage loss is estimated assuming the same relationship of the friction coefficient c_f as described in [52]. The final friction loss is estimated to be 38w at 1 million rpm and 64w at 1.2 million rpm, which corresponds to a friction torques of 0.36 and 0.51 mN•m, respectively. Assuming a safety factor of two, the machine should be designed to output a maximum torque of 0.7 mN•m and 1 mN•m at 1 million and 1.2 million rpm, respectively. Although this assumption may not necessarily be true, it still gives a good estimation of the windage losses of the proposed machine in the turbulent flow area, especially in the design stage.

4.5 Finite element analysis

The complete design parameters are shown in **Table 4**.

Figures 10 and **11** show the torque and current profiles of the finite element analysis at 1 million rpm and 1.2 million rpm, respectively. The estimated average

Stator		Rotor		Windings		Electrical	
Outer diameter	30 mm	Outer diameter	4 mm	Wire type	Litz	Rated power	100–200 w
Stack length	2 mm	Stack length	2 mm	Strand gauge	42	Rated speed	1,000,000 rpm
Pole number	4	Pole number	2	Equivalent gauge	24	Rated torque	0.7 mN•m
Stator pole arc angle	60°	Rotor pole arc angle	76.8°	Turns per pole	50	Max speed	1,200,000 rpm
Material	Hiperco 50	Material	Hiperco 50	Fill factor	0.25	Max torque	1 mN•m
Back iron thickness	2 mm	Air gap	0.3 mm	Max current	3.5A	Supply voltage	0~150 V

Table 4.
Complete design parameters.

torques are around 0.7 mN•m and 1 mN•m, respectively, which satisfies the design target well. The RMS currents are 3.5 and 4 A, respectively, which is a little higher than the maximum value at 1.2 million rpm. So additional cooling may be needed to keep the motor running for a long operation time.

4.6 Prototype

A prototype of the motor is built based on the electromagnetic design. To fix the stator and reduce the deformation due to resonance, four outer tabs with a hole of 4 mm in diameter each are added at the position of the four teeth. Fifty turns of Litz wire are then wound manually layer by layer around each stator tooth, as shown in

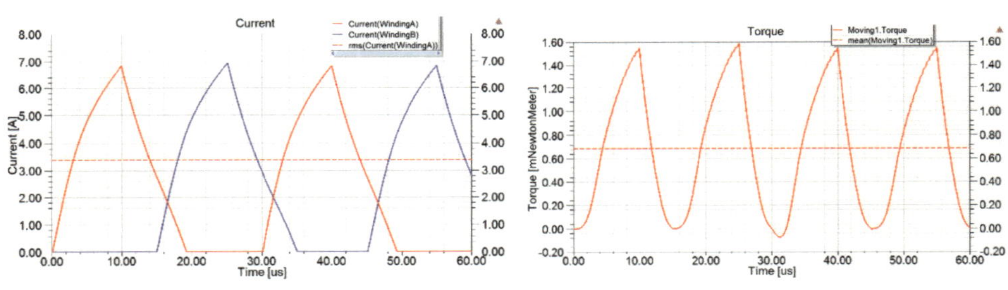

Figure 10.
Current (left) and torque (right) profile at 1 million rpm.

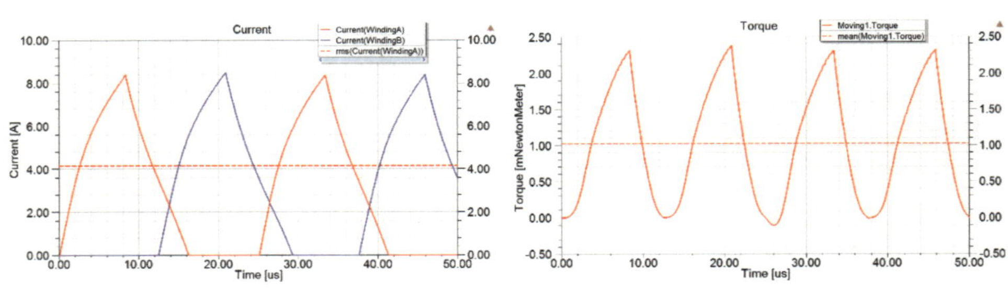

Figure 11.
Current (left) and torque (right) profile at 1.2 million rpm.

Figure 12(i). Two aerostatic bearings are used in the prototype as shown in **Figure 12(ii)**.

In order to integrate the stator with the rest of the motor, two aluminum cases have been designed [56]. The cases also act as spacers to prevent the windings from contacting the air bearings. The final assembly of the complete motor is shown in **Figure 13.** The air compression system for the air bearings is shown in **Figure 14(i)**. The compressed air passes through two air filters and an air regulator to function the two bearings. The working air pressure is about 90 psi. Finally, the completed prototype of the motor is shown in **Figure 14(ii)**.

4.7 Experimental result

Experiments are conducted to verify the FEA models. A video can be found in [57]. First, L_d and L_q are measured using Keysight E4990A impedance analyzer. Due to the significant end effects of the unusually large number of turns of each phase windings, a 3D FEA model is built in addition to the 2D model as shown in **Figure 15(i)**. The inductances of both phases A and B are measured at 0° (unaligned position), 30°, 60°, and 90° (aligned position). **Figure 15(ii)** shows the comparison of the inductance of 2D model, 3D model, and experimental results. It can be seen that, because of the fringing flux from the 3D end effect, the estimated inductance of 3D FEA is larger than 2D FEA. The measurement result is somewhere between the 2D and 3D model. Nevertheless, the slopes of the inductance profile from unaligned position (0°) to aligned position (90°), which determine the output torque, are similar among the three.

Figure 12.
(i) Stator prototype with windings (left). (ii) aerostatic bearings used in the prototype (right).

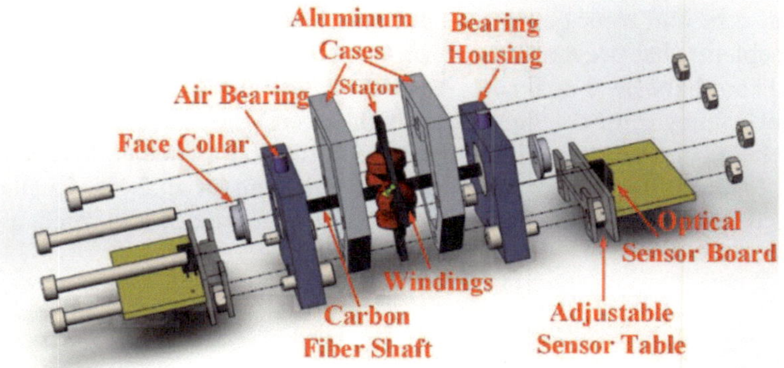

Figure 13.
Final assembly of the proposed ultrahigh-speed SRM.

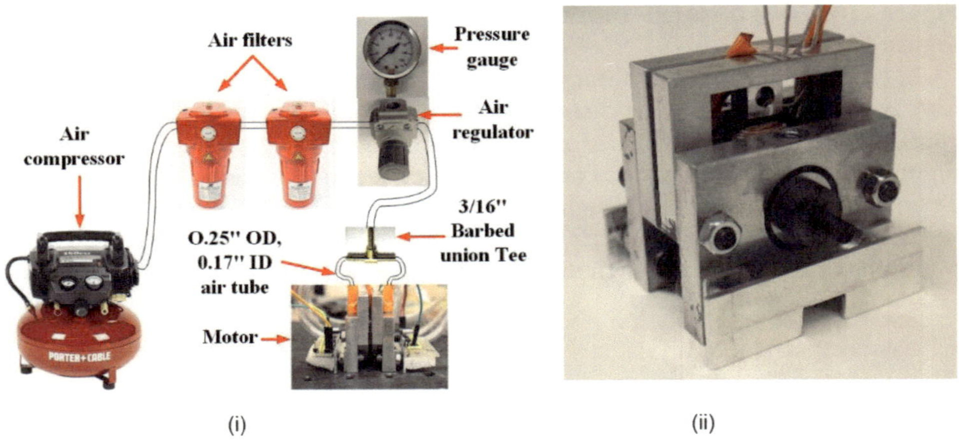

Figure 14.
(i) The air compression system for the aerostatic bearings. (ii) final prototype.

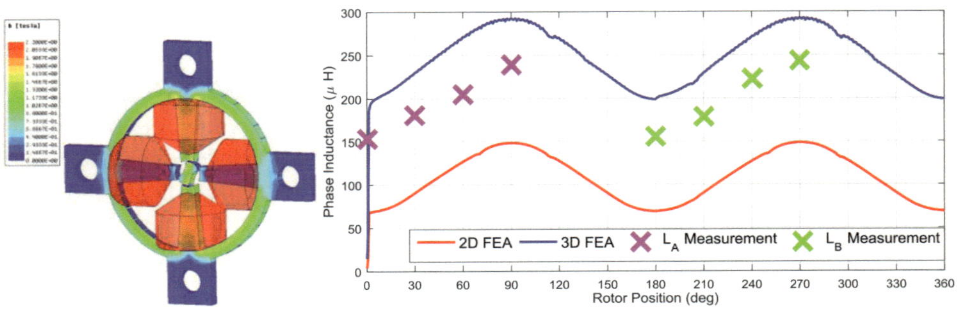

Figure 15.
(i) 3D FEA of the motor. (ii) comparison of inductance of 2D FEA, 3D FEA, and experiment.

5. Conclusion

In this chapter, the state-of-the-art research of ultrahigh-speed switched reluctance machines is introduced and reviewed regarding the design and control. First, different control methods are compared with an emphasis on the introduction of a noncontact direct position control. This technique uses low-cost, noncontact optical sensors to detect the relative rotor positions with respect to each stator pole and achieves one pulse control for each strike, which is validated on a 100,000 rpm 4/2 SRM. Next, different rotor geometries of high-speed SRMs have been compared, and the problems that prevent them from being applied at ultrahigh speeds over 1,000,000 rpm have been analyzed. A novel high-strength, high-power-density, and high-efficiency rotor design for ultrahigh-speed SRMs has been introduced. Then, a detailed electromagnetic design of a 4/2 ultrahigh-speed SRM for applications over 1,000,000 rpm is proposed using finite element analysis. Finally, for the first time in the literature, the proposed design is integrated, prototyped, and tested with aerostatic bearings.

Author details

Cheng Gong* and Thomas Habetler
Georgia Institute of Technology, Atlanta, GA, USA

*Address all correspondence to: cgong30@gatech.edu

References

[1] Gong C, Zhang S, Zhang F, Jiang J, Wang X. An integrated energy-efficient operation methodology for metro systems based on a real case of Shanghai metro line one. Energies. 2014;7: 7305-7329

[2] Yang G, Zhang F, Gong C, Zhang S. Application of a deep deterministic policy gradient algorithm for energy-aimed timetable rescheduling problem. Energies. 2019;**12**:3461

[3] Flankl M, Tüysüz A, Gong C, Stolz T, Kolar JW. Analysis and modeling of Eddy-current couplings for auxiliary power generation on a freight train wagon. IEEE Power and Energy Technology Systems Journal. 2018;**5**: 139-147

[4] Li S, Gong C, Du L, Mayor JR, Harley RG, Habetler TG. Fast calculation of the magnetic field and loss distributions in the stator core end packets and finger plates of large synchronous generators. In: 2018 IEEE Energy Conversion Congress and Exposition (ECCE). 2018. pp. 822-828

[5] Li S, Gong C, Du L, Mayor JR, Harley RG, Habetler TG. Parametric study for the Design of the end Region of large synchronous generators based on three-dimensional transient finite element analysis. In: 2018 IEEE Energy Conversion Congress and Exposition (ECCE). 2018. pp. 7356-7362

[6] Gong C. Design and Control of Ultra-High Speed Switched Reluctance Machines over 1 Million Rpm. Georgia Institute of Technology; 2019

[7] Li S, Gong C, Gallandat NA, Mayor JR, Harley RG. Analyzing the impact of press plate structure on the flux and loss distributions in the end region of large generators by transient 3-dimensional finite-element method with an improved core loss model. In: Electric Machines and Drives Conference (IEMDC), 2017 IEEE International. 2017. pp. 1-8

[8] Zhao L, Ham C, Zheng L, Wu T, Sundaram K, Kapat J, et al. A highly efficient 200 000 RPM permanent magnet motor system. IEEE Transactions on Magnetics. 2007;**43**: 2528-2530

[9] Pfister P-D, Perriard Y. Very-high-speed slotless permanent-magnet motors: Analytical modeling, optimization, design, and torque measurement methods. IEEE Transactions on Industrial Electronics. 2010;**57**:296-303

[10] Zwyssig C, Kolar JW, Round SD. Megaspeed drive systems: Pushing beyond 1 million r/min. IEEE/ASME Transactions on Mechatronics. 2009;**14**: 564-574

[11] Li S, Zhang S, Gong C, Habetler T, Harley R. An enhanced analytical calculation of the phase inductance of switched reluctance machines. IEEE Transactions on Industry Applications. 2019;**55**:1392-1407

[12] Gong C, Tüysüz A, Flankl M, Stolz T, Kolar J, Habetler T. Experimental analysis and optimization of a contactless Eddy-current-based speed sensor for smooth conductive surfaces. IEEE Transactions on Industrial Electronics (Early Access). 2019. Available from: https://ieeexplore.ieee.org/abstract/document/8871175

[13] Xu L, Wang C. Accurate rotor position detection and sensorless control of SRM for super-high speed operation. IEEE Transactions on Power Electronics. 2002;**17**:757-763

[14] MacMinn SR, Jones WD. A very high speed switched-reluctance starter-generator for aircraft engine

applications. In: Proceedings of the IEEE 1989 National Aerospace and Electronics Conference (NAECON). 1989. pp. 1758-1764

[15] Lee D-H, Ahn J-W. A novel four-level converter and instantaneous switching angle detector for high speed SRM drive. IEEE Transactions on Power Electronics. 2007;**22**:2034-2041

[16] Calverley SD, Jewell G, Saunders R. Design of a High Speed Switched Reluctance Machine for Automotive Turbo-Generator Applications. SAE Technical Paper 0148-7191; 1999

[17] Radun AV. High-power density switched reluctance motor drive for aerospace applications. IEEE Transactions on Industry Applications. 1992;**28**:113-119

[18] Dang J, Haghbin S, Du Y, Bednar C, Liles H, Restrepo J, et al. Electromagnetic design considerations for a 50,000 rpm 1kW switched reluctance machine using a flux bridge. In: 2013 IEEE International Electric Machines & Drives Conference (IEMDC). 2013. pp. 325-331

[19] Wichert T. Design and construction modifications of switched reluctance machines. [Ph.D. thesis]. Poland: Institute of Electrical Machines, Warsaw University of Technology; 2008

[20] Ferreira CA, Jones SR, Heglund WS, Jones WD. Detailed design of a 30-kW switched reluctance starter/generator system for a gas turbine engine application. IEEE Transactions on Industry Applications. 1995;**31**:553-561

[21] Lee D-H, Ahn J-W. Performance of high-speed 4/2 switched reluctance motor. Journal of Electrical Engineering and Technology. 2011;**6**:640-646

[22] Brauer HJ, De Doncker RW. Thermal modeling of a high-speed switched reluctance machine with axial air-gap flow for vacuum cleaners. In: Proceedings of the 2011-14th European Conference on Power Electronics and Applications (EPE 2011). 2011. pp. 1-10

[23] Won SH, Choi J, Lee J. Windage loss reduction of high-speed SRM using rotor magnetic saturation. IEEE Transactions on Magnetics. 2008;**44**: 4147-4150

[24] Bateman CJ, Mecrow BC, Clothier AC, Acarnley PP, Tuftnell ND. Sensorless operation of an ultra-high-speed switched reluctance machine. IEEE Transactions on Industry Applications. 2010;**46**:2329-2337

[25] Kim J, Krishnan R. High efficiency single-pulse controlled switched reluctance motor drive for high speed (48k rpm) application: Analysis, design, and experimental verification. In: 2008 IEEE Industry Applications Society Annual Meeting. 2008. pp. 1-8

[26] Calverley SD. Design of a high-speed switched reluctance machine for automotive turbo-generator applications. [Ph.D. dissertation]. Department of Electronic and Electrical Engineering, University of Sheffield; 2002

[27] Bui MD. Maximum torque control of a high speed switched reluctance starter/generator used in more/all electric aircraft [Ph.D thesis]. Germany: Technical University of Berlin; 2014

[28] Morel L, Fayard H, Fos HV, Galindo A, Abba G. Study of ultra high speed switched reluctance motor drive. In: 2000 IEEE Industry Applications Conference. 2000. pp. 87-92

[29] Kachapornkul S, Somsiri P, Pupadubsin R, Nulek N, Chayopitak N. Low cost high speed switched reluctance motor drive for supercharger applications. In: 2012 15th International

Conference on Electrical Machines and Systems (ICEMS). 2012. pp. 1-6

[30] Kozuka S, Tanabe N, Asama J, Chiba A. Basic characteristics of 150,000 r/min switched reluctance motor drive. In: 2008 IEEE Power and Energy Society General Meeting-Conversion and Delivery of Electrical Energy in the 21st Century. 2008. pp. 1-4

[31] Gong C, Habetler T, Restrepo J, Soderholm B. Direct position control for ultra-high speed switched reluctance machines based on non-contact optical sensors. In: 2017 IEEE International Electric Machines and Drives Conference (IEMDC). 2017. pp. 1-6

[32] Gong C, Li S, Habetler T, Restrepo JA, Soderholm B. Direct position control for ultrahigh-speed switched-reluctance machines based on low-cost nonintrusive reflective sensors. IEEE Transactions on Industry Applications. 2019;**55**:480-489

[33] Gong C, Li S, Habetler T, Zhou P. "Acoustic Modeling and Prediction of Ultra-High Speed Switched Reluctance Machines Based on Finite Element Analysis," in 2019 IEEE International Electric Machines & Drives Conference (IEMDC), San Diego, CA2019. pp. 336-342

[34] Gong C, Habetler T. Constant volts per hertz control of ultra-high speed switched reluctance machines. In: 2017-43rd Annual Conference of the IEEE Industrial Electronics Society (IECON). 2017. pp. 1868-1873

[35] 100,000 rpm switched reluctance motor [Online]. Available: https://youtu.be/yd-5vlgyPdY

[36] Gong C, Habetler T. A novel rotor design for ultra-high speed switched reluctance machines over 1 million rpm. In: 2017 IEEE International Electric Machines and Drives Conference (IEMDC). 2017. pp. 1-6

[37] Besharati M, Widmer J, Atkinson G, Pickert V, Washington J. Super-high-speed switched reluctance motor for automotive traction. In: 2015 IEEE Energy Conversion Congress and Exposition (ECCE). 2015. pp. 5241-5248

[38] Kunz J, Cheng S, Duan Y, Mayor JR, Harley R, Habetler T. Design of a 750,000 rpm switched reluctance motor for micro machining. In: 2010 IEEE Energy Conversion Congress and Exposition. 2010. pp. 3986-3992

[39] Won SH, Choi J, Lee J. Windage loss reduction of high-speed SRM using rotor magnetic saturation. IEEE Transactions on Magnetics. 2008;**44**: 4147-4150

[40] Dang J, Mayor JR, Semidey SA, Harley R, Habetler T, Restrepo J. Practical considerations for the design and construction of a high-speed SRM with a flux-bridge rotor. IEEE Transactions on Industry Applications. 2015;**51**:4515-4520

[41] A rotor structure for ultra-high speed switched reluctance motors. [Online]. Available: https://industry.gatech.edu/technology/high-strength-rotor-structure

[42] Gong C, Li S, Habetler TG. Analysis of rotor robustness of ultra-high speed switched reluctance machines over 1 million rpm using cohesive zone model. In: 2018 IEEE Energy Conversion Congress and Exposition (ECCE). 2018. pp. 2401-2406

[43] Gong C, Li S, Habetler TG. Rotor dynamic analysis of ultra-high speed switched reluctance machines over 1 million rpm. In: 2018 IEEE Energy Conversion Congress and Exposition (ECCE). 2018. pp. 1704-1709

[44] Gong C, Habetler T. Electromagnetic design of an ultra-high speed switched reluctance machine over 1 million rpm. In: 2017 IEEE Energy

Conversion Congress and Exposition (ECCE). 2017. pp. 2368-2373

[45] Luomi J, Zwyssig C, Looser A, Kolar JW. Efficiency optimization of a 100-W 500 000-r/min permanent-magnet machine including air-friction losses. IEEE Transactions on Industry Applications. 2009;**45**:1368-1377

[46] Dang J. Switched Reluctance Machine Electromagnetic Design and Optimization. Atlanta, GA, USA: Georgia Institute of Technology; 2015

[47] Zwyssig C, Kolar J, Thaler W, Vohrer M. Design of a 100 W, 500000 rpm permanent-magnet generator for mesoscale gas turbines. In: Fourtieth IAS Annual Meeting. Conference Record of the 2005 Industry Applications Conference, 2005. 2005. pp. 253-260

[48] Round Litz Wire Data Sheet. Available at: https://www.newenglandwire.com/products/litz-wire-and-formed-cables/round

[49] Zwyssig C, Duerr M, Hassler D, Kolar J. An ultra-high-speed, 500000 rpm, 1 kW electrical drive system. In: Power Conversion Conference-Nagoya, 2007. PCC'07. 2007. pp. 1577-1583

[50] Li S, Gong C, Mayor JR, Harley RG, Habetler TG. Efficient calculation of the Strand Eddy current loss distributions in the end stepped-stator region of large synchronous generators. In: 2018 IEEE Energy Conversion Congress and Exposition (ECCE). 2018. pp. 1783-1789

[51] Li S, Gong C, Gallandat NA, Mayor JR, Harley RG. Implementation of surface impedance boundary conditions in the quasi three-dimensional finite-difference simulations of generator end regions. In: Electric Machines and Drives Conference (IEMDC), 2017 IEEE International. 2017. pp. 1-7

[52] Zwyssig C, Round S, Kolar J. Analytical and experimental investigation of a low torque, ultra-high speed drive system. In: Conference Record of the 2006 IEEE Industry Applications Conference Forty-First IAS Annual Meeting. 2006. pp. 1507-1513

[53] PES Tutorial on Magnetic Materials. Available at: https://www.pes.ee.ethz.ch/uploads/tx_ethpublications/APEC2012_MagneticTutorial.pdf

[54] Long S, Zhu Z, Howe D. Vibration behaviour of stators of switched reluctance motors. IEE Proceedings-Electric Power Applications. 2001;**148**: 257-264

[55] Watanabe S, Kenjo S, Ide K, Sato F, Yamamoto M. Natural frequencies and vibration behaviour of motor stators. IEEE Transactions on Power Apparatus and Systems. 1983:949-956

[56] Gong C, Li S, Habetler T. Practical considerations in the design and manufacture of ultra-high speed switched reluctance machines over 1 million rpm. In: 2019 IEEE International Electric Machines & Drives Conference (IEMDC), San Diego, CA. 2019. pp. 25-30

[57] Ultra-high Speed SRM with Aerostatic Bearings [Online]. Available at: https://youtu.be/7Lq9LI3K5KE

Section 2

Control

Chapter 8

Switched Reluctance Motor Drives: Fundamental Control Methods

Manuel Fernando Sequeira Pereira, Ana Mamede and Rui Esteves Araújo

Abstract

The switched reluctance motor has been gaining interest both at the academic and industrial levels. Its most appreciated characteristics are the nonuse of rare elements and the low construction price. The control of the switched reluctance motor is not as easy as other traditional machines. In the control, it is necessary to reduce the torque ripple as much as possible, which has been the most studied problem in the field. This chapter aims to give the reader a concise but focused knowledge on how the switched reluctance motor can be controlled. Fundamental control methods and their impact in the motor performance are described. A simulation section is presented, where the simulation files are provided for the reader. The simulations are based on the theory described and have the advantage of using a non-linear switched reluctance motor model from a real motor.

Keywords: switched reluctance motor, electric drive, modelling, linear control

1. Introduction

The switched reluctance motor (SRM) is an electrical machine that converts electric power into mechanical power. This electric motor, similarly to conventional motors (induction motor, synchronous motor and others), can operate in the four quadrants: In other words, it can accelerate and decelerate for clockwise and anticlockwise rotor rotations.

SRM control cannot be compared either with the alternated current motor, which is fed with sinusoidal current waveforms or with the direct current motor. In the simplest control, it needs each motor phase (winding) to be magnetized and demagnetized at the right moments of rotor position, so square current waveforms can be used in the simplest control. But as the motor is highly non-linear, the electromagnetic torque does not depend only on the instantaneous current value but also on a non-linear inductance profile, which means that using only square current waveforms will create torque with a high ripple component.

Our aim in this chapter is to describe a variety of fundamental methods to operate the SRM in the four quadrants. Indeed, there are many works published on this subject, [1–5], and due to this reason, we propose a chapter for beginners who wants to really understand the foundations of SRM drives. For a more extensive and systematic exposition of SRM, the reader is referred to the many good books on the subject. For example, in [6] there is a complete description on every subject

IntechOpen

related to the SRM. In [7] a description about the SRM and a brief description about the controllers are given. Although these works are very complete, they do not give a clear path for the reader to create a SRM drive. So, in this chapter, simple controllers and all fundamental methods to develop a SRM controller and thus contribute to a very steep learning curve in SRM drives are described. And while it is aimed at beginners, we hope that even experts will have something to learn from this chapter.

Linear controllers are presented: voltage impulse, current and torque controllers, respectively. Depending on the required performance and cost, a choice can be made from the cheapest with the lowest performance of impulse control to the most expensive with the best performance of torque control.

As already stated, the motor is highly non-linear, so linearization needs to be performed for control design. The small-signal linearization method which although being very limited is widely used in real systems is described. It will be seen that an ideal current waveform can be calculated to produce constant torque, instead of using a square waveform current. About the power electronic device, a description is given of how the traditional SRM converter works and how it can be controlled to maximize performance.

A simulation environment is also developed in the numerical tool *Simulink/ Matlab®*. It is based on the controllers and methods described in the chapter, to which the reader can have access and make his own simulations. This way, one student who works through this text will get an excellent grounding in SRM drives. He or she will actually learn what SRM control is about and how it can be used and will be in an excellent position to go on to more advanced controllers using this material.

In the description of the theory and in the simulation results, the SRM model parameters described in [8] were used.

The chapter is divided into a description of the SRM in *Section 2*, the fundamental linear control methods in *Sections 3*, a study of the SRM model and its linearization in *Section 4*, the current waveform for constant torque in *Section 5*, a description of the traditional SRM converter and its control in *Section 6*, a presentation of numerical results using the tool *Simulink/Matlab®* in *Section 7* and to end, the conclusions and bibliography.

2. Switched reluctance motors

The SRM is constituted by a stator with concentrated windings disposed around polar cores and by a rotor composed by salient poles free of windings or magnets. Normally, each stator phase is composed by a couple of windings diametrically opposed. In **Figure 1** a SRM configuration of 6 stator poles and 4 rotor poles is shown. That is the configuration of reference for this chapter.

When a stator phase is magnetized, a closed magnetic field is generated between the stator, the air gap and the rotor. This magnetic field tends to minimize the reluctance by reducing the air gap which creates a rotor movement. When a stator pole is aligned with a rotor pole, it is said that they are in the position of minimum reluctance, and when they are completely unaligned, it is said they are in the position of maximum reluctance.

This characteristic of the motor makes it possible to create a rotational movement of the rotor by magnetizing and demagnetizing each phase in the right position of the rotor.

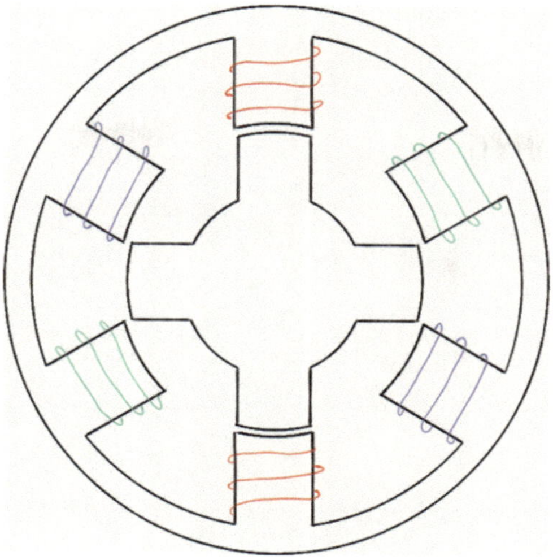

Figure 1.
Switched reluctance motor structure.

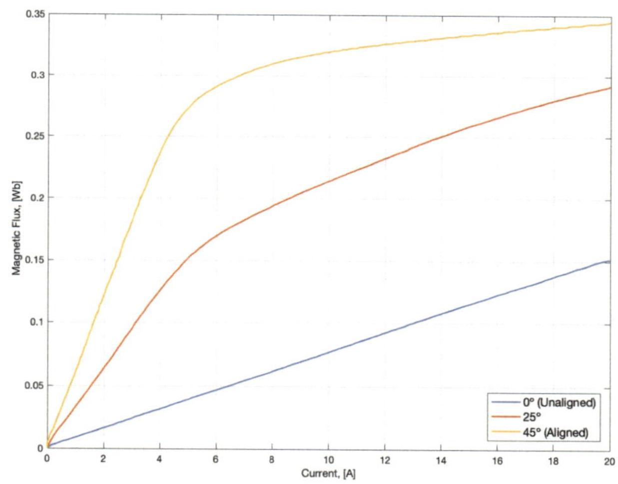

Figure 2.
SRM flux characteristic.

2.1 Torque characteristic

The SRM suffers from high saturation. The inductance profile is non-linear, making the flux characteristic also non-linear. This can be seen in **Figure 2**, where the blue line is the unaligned rotor position and the red line the aligned rotor position. Saturation starts to influence the motor characteristic from about 5 A. It is around this point where the torque ampere ration significantly increases.

The SRM torque characteristic graphic is shown in **Figure 3**. It represents one phase for 90° of the rotor, as the rest of the 360° are just a repetition for the other 3 rotor poles. The blue line represents a low current profile of 5 A, and the red line represents a high current profile of 20 A, which is the maximum motor current. It can be seen that from 0 to 45°, positive torque is produced and from 45 to 90°

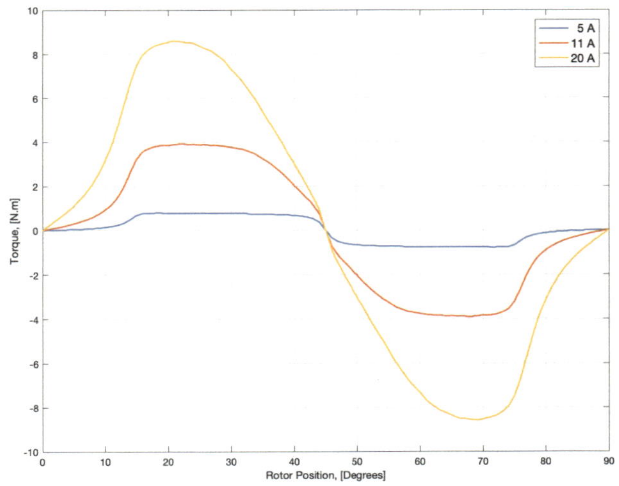

Figure 3.
SRM torque characteristic.

negative torque is produced. It can also be seen that the produced torque is not constant for constant current.

3. Control techniques

Various control schemes have been investigated essentially for torque ripple reduction and the increase of efficiency, for example, in [2, 9]. These can be essentially divided into voltage impulse control, by keeping a constant current or by keeping a constant torque.

3.1 Voltage impulse control

The impulse control is the simplest controller for the SRM. It is used, essentially, for high speed control and for systems that do not require high performance of the drive. Its typical implementation scheme, represented in **Figure 4**, is composed by a speed controller, a phase control method, a block that generates the transistors signals, and an overcurrent protection.

For the speed controller, a linear PI controller is usually used, as in other electric drives with other electrical machines. It receives the speed error and gives a reference voltage signal.

The phase control has the task of triggering each phase and decide how the reference voltage must be applied. For this part, it is essential to have good accuracy

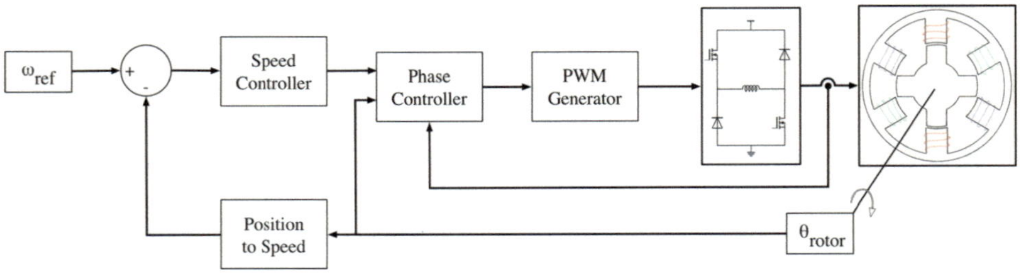

Figure 4.
Speed control loop based on impulse control.

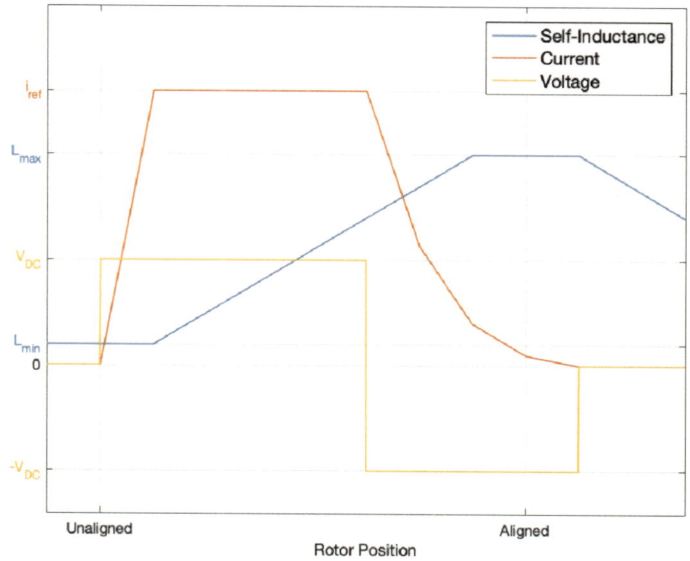

Figure 5.
Impulse control method.

in rotor position measure in order to obtain an effective control. There are various methods for this block, the most complete one being the switching angle, which is demonstrated in **Figure 5**.

This method can be divided into three parts. The first part happens when the inductance is minimum, so the current can grow fast. A voltage is applied for the current to reach the reference value. In the second part, the self-inductance increases, and the current is expected to be kept constant. And in the third part, the negative source voltage is applied, for the current to decrease to zero before it starts producing negative torque.

For very high speeds, advancing the start and end angles of commutation may be necessary, [10]. In this case the magnetizing and demagnetizing times are much shorter, and the back-EMF makes the current dynamic slower.

This controller also needs an overcurrent protection, so some current estimation or current sensor is necessary.

A description of variants of the impulse control can be found in [11] where the effect of advance angle, switching off angle, and the time of the dwell angle are addressed.

3.2 Current control

The voltage is kept constant in the last controller, expecting the current to be constant as well. Due to back-EMF and different inductance values for different rotor positions, this does not happen. So, in order to solve this limitation, a current controller is necessary for each phase.

The configuration of this controller, as shown in **Figure 6**, is in cascade form, where the current is controlled in the inner loop and the speed in the outer loop. There is also a decision block which dictates when each phase must be conducting current or not according to the rotor position.

As seen in **Figure 3**, the torque characteristic does not depend only on the current, which means that keeping the current constant does not make the produced torque constant. In this line of thought, an improvement for this control is

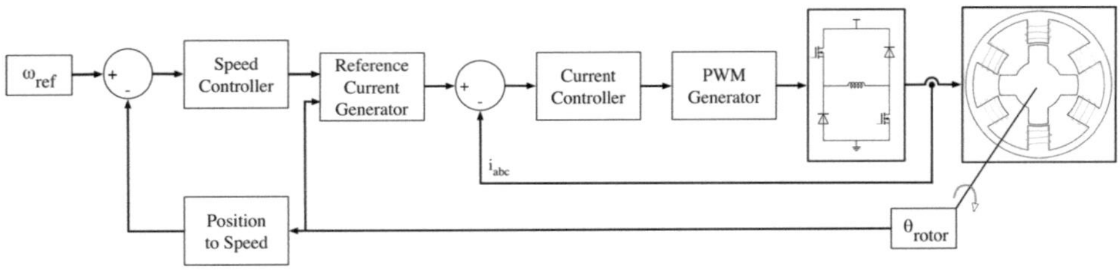

Figure 6.
Block diagram of speed control based on current control.

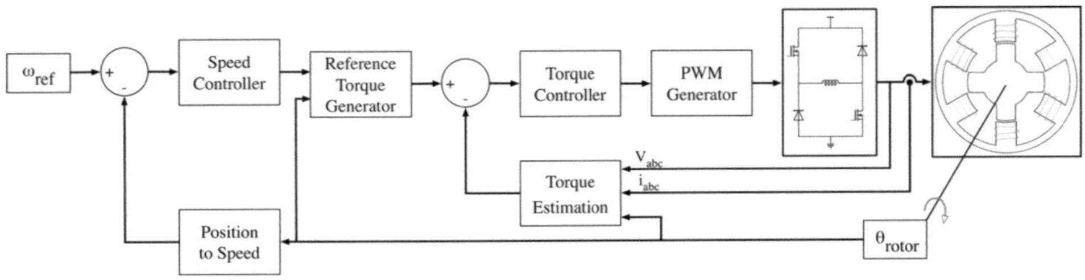

Figure 7.
Torque control diagram.

the use of functions like cosine and exponential to dictate not only when each phase must be conducting but also to give a weight for each rotor position in order to decrease the torque ripple. This subject will be discussed in detail in *Section 5*.

3.3 Torque control

One of the main objectives in a speed controller is to have constant torque with the lowest possible ripple. Measuring the instantaneous torque with a sensor can be too expensive for most applications. The typical solution is to measure currents and voltages in order to estimate the flux and using look-up tables with the motor model, as, for example, in [12, 13]. The torque control diagram is seen in **Figure 7**.

Although the feedback is the torque, a reference generator with the task of dividing the required torque in the different phases is still necessary. If this block is not well designed, and the instants of conduction are wrong or, it does not concern current constraints, the system can become unstable and impossible to control.

4. Linear controller design

Design of linear controllers can be very tricky for a complicated and non-linear model. This is the case of the SRM which is highly non-linear. To overcome this problem, the small-signal method can be used. Although it is a limited method, this solution is frequently applied in practice.

4.1 Model equations

The SRM model can be represented by the electrical equation:

$$V = R_s.i + L(\theta)\frac{di}{dt} + i\frac{dL(\theta)}{d\theta}\omega \tag{1}$$

where V, R_s, i, L, ω are, respectively, phase voltage, stator resistance, current, self-inductance, and rotor speed and by the mechanical equation:

$$\frac{1}{2}i^2\frac{dL(\theta)}{d\theta} - T_L = J\frac{d\omega}{dt} + B.\omega \tag{2}$$

where T_L, J, B are, respectively, mechanical load, motor inertia, and viscous friction coefficient.

In these equations the assumptions of non-saturating magnetic structure and steady-state in phase current in order to obtain an equivalent self-inductance that only depends on the rotor position, $L(i, \theta) \to L(\theta)$, are made. Even so, the model has a continuous dependency with the rotor position and is still highly non-linear.

To resolve the rotor dependency, $L(\theta)$ and $\frac{dL(\theta)}{d\theta}$ can be approximated, with some loss of information, by the self-inductance mean value and by the derivative of the maximum and minimum self-inductance points, respectively.

To resolve the rest of the non-linearities, it is necessary to linearize the model. A common solution is the linearization around an operation point—small-signal linearization. For this, each system state has to be represented by a steady-state value and by a small-signal value as:

$$i = i_o + \delta i \tag{3}$$

$$\omega_r = \omega_{ro} + \delta\omega_r \tag{4}$$

$$V = V_o + \delta V \tag{5}$$

$$T_L = T_{Lo} + \delta T_L \tag{6}$$

where the subscript o is used to represent the steady-state operation point and the δ is used to represent incremental deviations of the state in the vicinity of the steady-state point.

Using the new states defined in Eqs. (3)–(6) in the SRM model equations (1)–(2), the model for small signals is obtained as:

$$\frac{d\delta i}{dt} = -\frac{R_{eq}}{L}\delta i - \frac{\delta_e}{L} + \frac{\delta V}{L} \tag{7}$$

$$\frac{d\delta\omega_r}{dt} = \frac{K_b}{J}\delta i - \frac{B}{J}\delta\omega_r - \frac{\delta T_L}{J} \tag{8}$$

with the constants as:

$$R_{eq} = R_s + \frac{dL}{d\theta}\omega_r \tag{9}$$

$$K_b = \frac{dL}{d\theta}i_o \tag{10}$$

$$\delta_e = \frac{dL}{d\theta}i_o.\delta\omega_r \tag{11}$$

4.2 Laplace model

Applying the Laplace transform to Eqs. (7)–(8), Eqs. (12)–(13) are obtained in the frequency domain and the block diagram shown in **Figure 8**.

$$\delta I(s) = \frac{\delta V(s) - \delta_e}{Ls + R_{eq}} \tag{12}$$

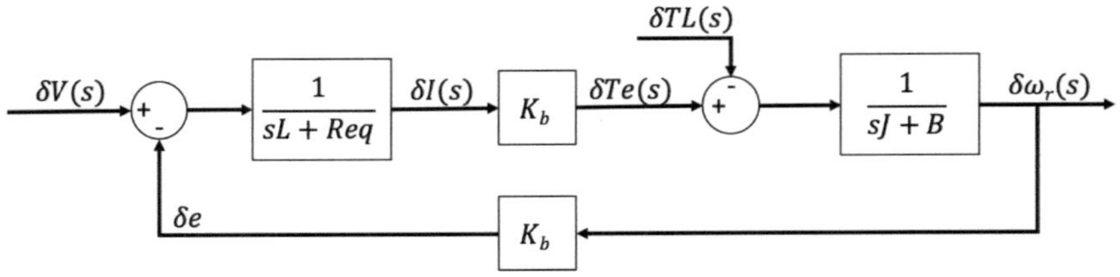

Figure 8.
Linearized SRM model.

$$\delta\omega_r(s) = \frac{K_b.\delta I(s) - \delta T_L(s)}{Js + B} \tag{13}$$

The inputs of the open loop plant diagram are the voltage and load torque variation, and the output is the speed variation.

With the linearized model, it is now easy to design the controllers. Two design methods, using the same linearized model, can be found in [6, 14].

As already stated, the model is a linearization for an operating point, so if the controllers are designed to have fixed gains, a study of the system for other operating points needs to be carried out. Also, it is possible to design adaptive controllers where the gains are a function of the inductance and the operating point. It is expected that more detail will bring a better dynamic for the system.

5. Reference generator

In this section the signal processing block with the function of dividing the output of speed controller between each phase to serve as a reference for the inner controller, whenever there is a current or torque controller, is discussed. The cases are presented for constant current and for constant torque.

As already known, the SRM is not an alternated current motor; the negative current has the same effect as the positive, so the motor is designed to work only with positive currents. Also, the currents have to be commuted in the function of the rotor position for that the rotor to have a constant movement.

The overlapping between phases is explained by the angular measure of a rotor pole, which has $360°/4\,poles = 90°$ for each pole. This means that during a rotor movement of 90°, each phase has to act sequentially. As the motor studied has three phases, this means that each phase is spaced at $90°/3\,phases = 30°$ from the start of the last one. It also means that two phases can be conducting at the same time during 15°.

5.1 Constant current

As already seen, the torque is the result of the squared current times the inductance derivative according to the rotor position, for constant current. This second term is non-linear, so the torque is not a direct calculation of the current. When using constant current, and assuming that current has no dynamic, it will generate the electromagnetic torque represented in **Figures 9** and **10**. In these figures, SRM torque characteristics for the reference motor are shown. In **Figure 9** it is shown the calculations for a medium value of current, specifically 11 A, and in **Figure 10** for a high value of current, specifically 20 A.

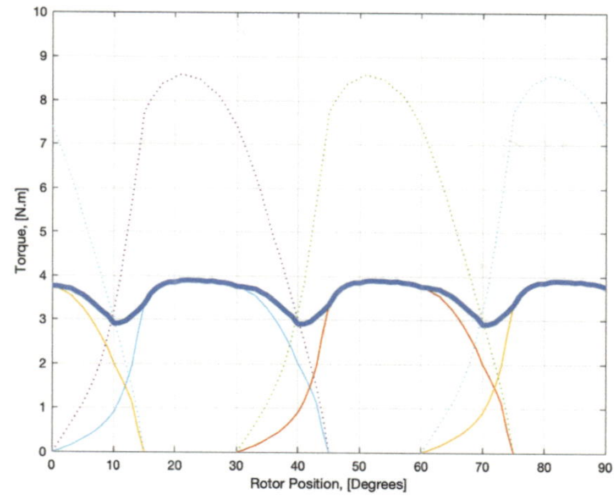

Figure 9.
Torque for medium current value.

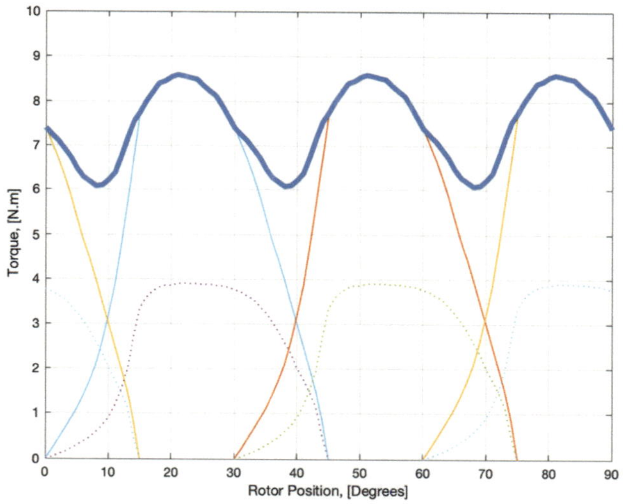

Figure 10.
Torque for maximum current value.

Although the motor has twice the current in the phase transitions, because it has two phases conducting with the same current value, for example, between 30 and 45°, it does not have enough torque as it does when only one phase is conducting, for example, between 15 and 30°.

The SRMs are not all similar; in some of them, this example can be the opposite, where in the phase transitions more electromagnetic torque is produced than when only one phase is conducting, so this study has to be done for each SRM.

5.2 Constant torque

For the SRM characteristic used as the example, it is then necessary to have different values of current for each rotor position to have constant torque.

In this section, the current waveforms which are necessary to have constant torque at medium and high values of electromagnetic torque are generated, as shown in **Figures 11** and **12**.

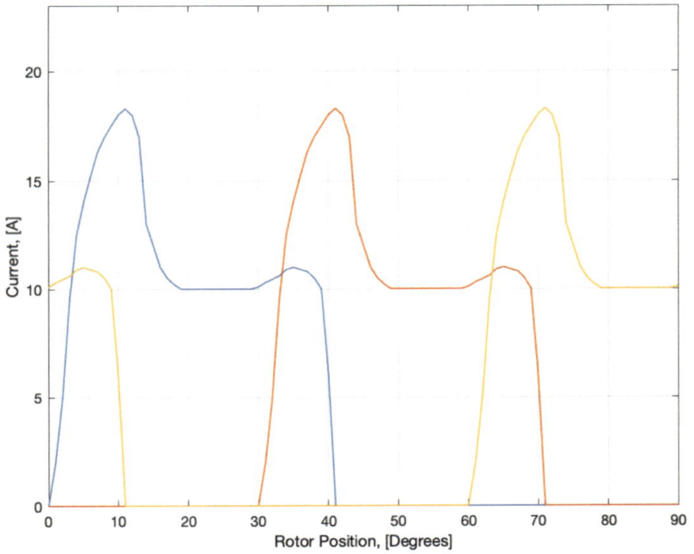

Figure 11.
Current for medium torque value.

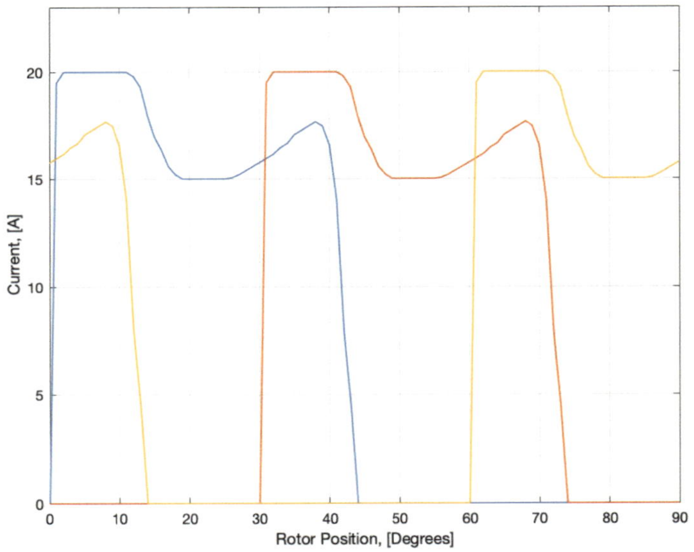

Figure 12.
Current for high torque value.

These waveforms were generated in order to produce constant torque but also to have the minimum necessary current when the phase is demagnetizing.

In both figures the waveforms have similarities. When the phase is magnetizing, it requires much more current than when only one phase is conducting. In **Figure 12** it is seen that current hits its maximum value, so the phase that is demagnetizing has to compensate the torque demand.

The current waveform profile has the name of torque sharing functions in literature, and in most cases it is generated by the use of sine or exponential waves with the rotor position as input, as seen, for example, in [11, 15, 16].

Another solution is the use of look-up tables which are much easier to change and give the right values in the right vector position.

5.3 Current dynamic constraints

In the last two sections, it was assumed that the current had no dynamic and the current controllers were ideal. That, of course, is not true, and from the inductance profile (**Figure 13**), it can be seen that its dynamic changes with the rotor position and with the current itself. The minimum inductance, when the current has a faster dynamic, happens when the phase is magnetized. And the maximum inductance, when the dynamic is slower, happens when the phase is demagnetized.

In **Figures 11** and **12**, it can be seen that the transitions can be very fast, and the current may not be able to follow the reference precisely. The fast-current transitions happen in the phase magnetization and in the phase demagnetization. In the magnetization, the current can start earlier, because as seen in the torque characteristic (**Figure 3**) around rotor position 0°, no significant torque is produced, and this is where the inductance is minimum. The problem is in the demagnetization, where the produced torque is significant, so the current must be of zero amperes or very low in the transition or it will produce a significant negative torque.

Another constraint that influences the current dynamic is the back-EMF. The back-EMF can be represented by a voltage that increases with increase in speed. This means that at high speeds the equivalent voltage applied to the phase is less than the source voltage, decreasing the current dynamic. Solutions can be applied by anticipating the trigger moments or by diminishing the range of the trigging moments.

5.4 Deaccelerating and speed inverting

The last sections were only described for one quadrant, which corresponds to positive speed and accelerating. The case of deaccelerating is a little different. It is necessary to produce torque with the opposite direction of the speed. So instead of conducting current from 0 to 45°, it must conduct current from 45 to 90°.

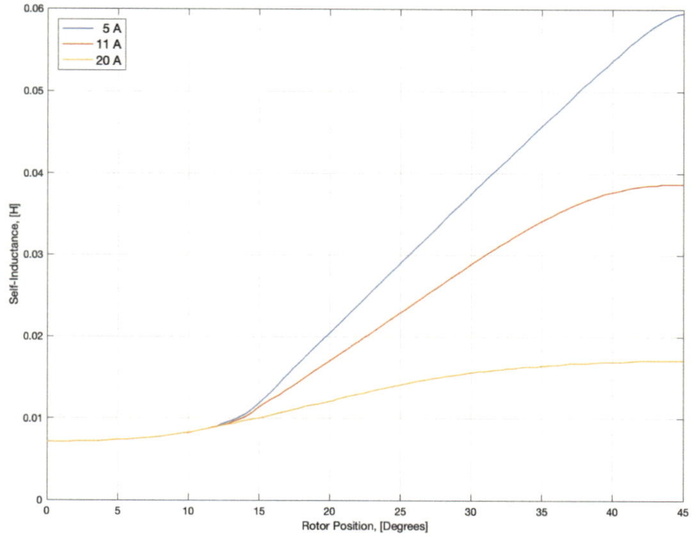

Figure 13.
Self-inductance profile.

In the case of deaccelerating, the magnetization happens in the zone of maximum inductance, making the task of designing waveforms to keep constant torque much harder.

The case for negative speeds is very similar to positive speeds. In positive speeds the rotor angle increases from 0 to 90°, while for negative speeds, it is the opposite, and the rotor angle decreases from 90 to 0°. Thus, to accelerate negatively, the current must be applied from 90 to 45°, and to deaccelerate negatively, the current must be applied from 45 to 0°.

6. Asymmetric bridge converter

One advantage of common electric motors, like the induction motor, is the possibility to plug it to the AC electric socket, and it can run at a certain power. In the SRM this is not possible; it requires a controller and a converter.

The most used converter for the SRM is the asymmetric bridge converter (**Figure 14**). It is composed by two diodes and two transistors for each phase. Its most appreciated advantages are the phase independent control, the possibility of generating three levels of voltage, and the fact that it regenerates current for the voltage source. Other converter configurations are found in [6, 17, 18].

6.1 Converter states

As said before, this converter can apply three voltage levels to the motor, $V_{\{DC\}}$, $0V$, and $-V_{\{DC\}}$. These converter states are seen in **Figure 14**. The $V_{\{DC\}}$ voltage level is applied when both transistors are conducting. The $-V_{\{DC\}}$ voltage level is applied when both transistors are not conducting, and still there is current in the winding, so the current goes by the diodes. And the $0V$ voltage level is applied when only one transistor is conducting, where it is represented by a freewheeling of the top or bottom mesh.

As the SRM only needs to work with positive currents, the converter cannot allow a negative current. The asymmetric bridge meets this objective because the current in the winding always has the same direction.

6.2 Converter control

For the control, both transistors can have the same control signal or different ones. When using the same signal, it is only possible to apply two voltage levels

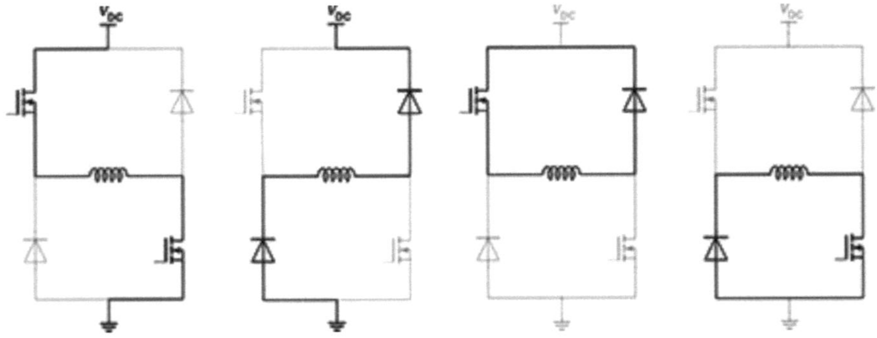

Figure 14.
Asymmetric converter states.

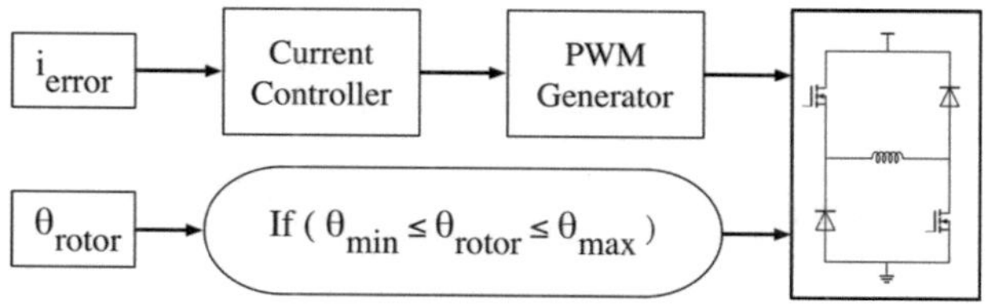

Figure 15.
Technique for generation of two signal of converter command.

when conducting, $V_{\{DC\}}$ and $-V_{\{DC\}}$. In this case, the converter's full capability cannot be used.

When using two signals, it is possible to use the three voltage levels at any time. Various methods can be used. A simple one is represented in **Figure 15** where the converter receives two signals for each transistor.

In this method the first signal is the result of the current controller and the PWM generator. The second signal is the trigger of defined values of rotor position. For example, if applying current between 0 and 45° is desired, the second signal will be active during 0 and a little less than the 45°, specifically when it is necessary for the current to decrease, for example, around 40°. So, in this method, the converter applies $V_{\{DC\}}$ and $0V$ from the beginning until the desired moment of demagnetization where $-V_{\{DC\}}$ is applied and $0V$ until the current returns to zero amperes. Although it limits the number of states in different zones, the method is simple and effective.

7. Numerical simulations

To demonstrate the fundamental control of the switched reluctance motor, a numeric simulation environment was developed in *Simulink/Matlab*® (version R2018a), available in the link: https://bit.ly/2KfPUsG . To open the application, the user has to type "Hello" in the *Matlab* command window. A window appears, as in **Figure 16**, and it is then possible to choose to see the SRM model parameters and flux characteristic or to open the simulation. The SRM is a non-linear model based on a real motor described in [8]; the main parameters are presented in **Table 1**.

The simulation is in a closed-loop speed control with a direct control of the current. The current waveform method for constant torque was applied, and the asymmetric bridge converter was used with the control described in *Section 6.2*.

Some simulations are shown. First the steady-state operation of the motor is analyzed, and then the transient response showing the four quadrants of operation. Controller gains are presented in **Table 2**.

7.1 Steady-state condition

In the first set, a torque load of 3 N m is applied, and two reference speeds of 750 rpm and 1250 rpm, respectively, are shown in **Figure 17**.

The currents have the same waveform as described in *Section 5.2*, where in the magnetization higher current values are required than the rest of the phase. Looking for results at 750 rpm speed reference, this method keeps the torque in a ripple less than 2 N m. For the 1250 rpm speed reference, the speed is higher so more current is necessary, and the time for magnetizing and demagnetizing is less. Because of that, the current cannot be as fast as the current reference, which means that during the

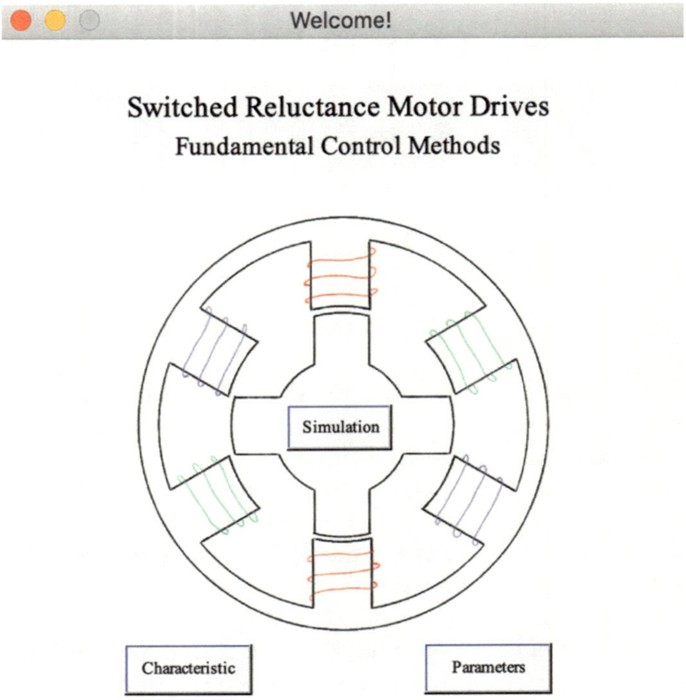

Figure 16.
Initial window for the simulation environment.

Stator poles	6
Rotor poles	4
Voltage	150 V
Stator resistance	1.3 Ω
Inertial value	0.0013 $Kg\ m^2$
Friction coefficient	0.0183 $N\ \ m/(rad/s)$
Minimum inductance	8e-3 H
Maximum inductance	60e-3 H

Table 1.
SMR model parameters.

Current proportional gain
20
Current integrative gain
1100
Speed proportional gain
0.22
Speed integrative gain
5.50

Table 2.
Controllers gains.

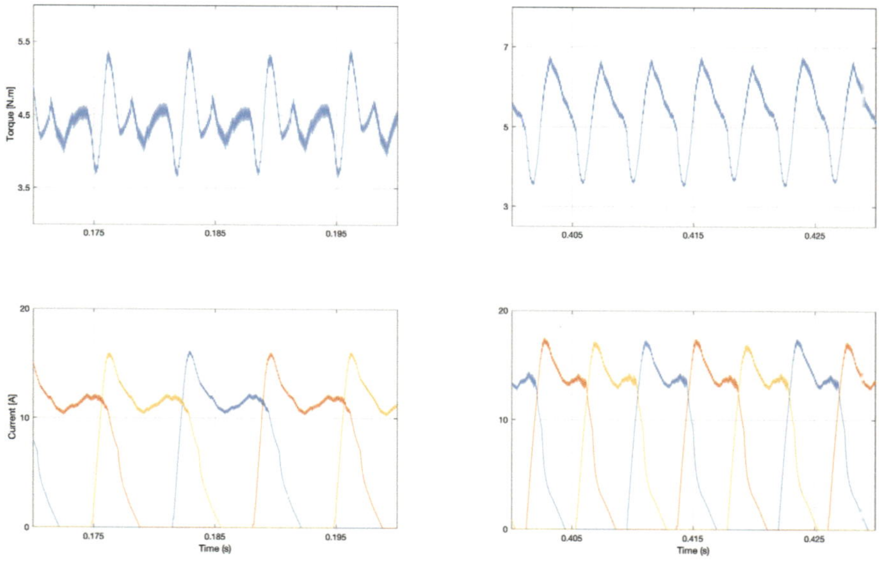

Figure 17.
Steady-state condition for same load and different speeds.

phase transition it is not possible to have enough current to keep the torque as constant as in the operation at lower speed.

In a second set of simulations, shown in **Figure 18**, the speed reference is kept constant at 1000 rpm, and a load torque is applied of 2 and 4 N m, respectively.

For the 2 N m load torque simulation, the response torque has a ripple of less than 1.5 N m. For the 4 N m load, it is necessary to have more current in the phase transition due to the SRM torque characteristic, so during this period, the torque has a low value.

7.2 Transient condition

For the transient response of the motor, all four quadrants of operation are simulated, accelerating, and deaccelerating for positive and negative speed. A reference speed is given as 1000 rpm from 0 to 0.5 s, and −1000 rpm from 0.5 to 1 s. The load torque is of 2 N m from 0 to 0.25 s and 0.75 to 1 s and of −2 N m from 0.25 to 0.75 s. Results are shown in **Figure 19**.

Firstly, the controller applies the maximum current so the motor can quickly reach 1000 rpm. The torque has some ripple during the acceleration phase, but when the speed reaches the set point, the ripple decreases. The speed response to the step has a minimum of overshoot.

At 0.25 s the load torque signal is changed, so the load is helping the rotor to move. The speed controller sees this as a perturbation and tries to compensate, which takes no longer than 0.05 s to recover the reference speed. This situation is similar at 0.75 s where the difference is in the signals of speed and load.

At time 0.5 s the reference speed is changed from 1000 to −1000 rpm. The motor has to break, stop, and accelerate negatively. This process happens with no problem. It is possible to see that the sequence of current phases is changed from orange, yellow, and blue to blue, yellow, and orange.

7.3 Discussion

In general, with the developed controller and with the use of the described fundamental control methods, the results of the numerical simulations are very

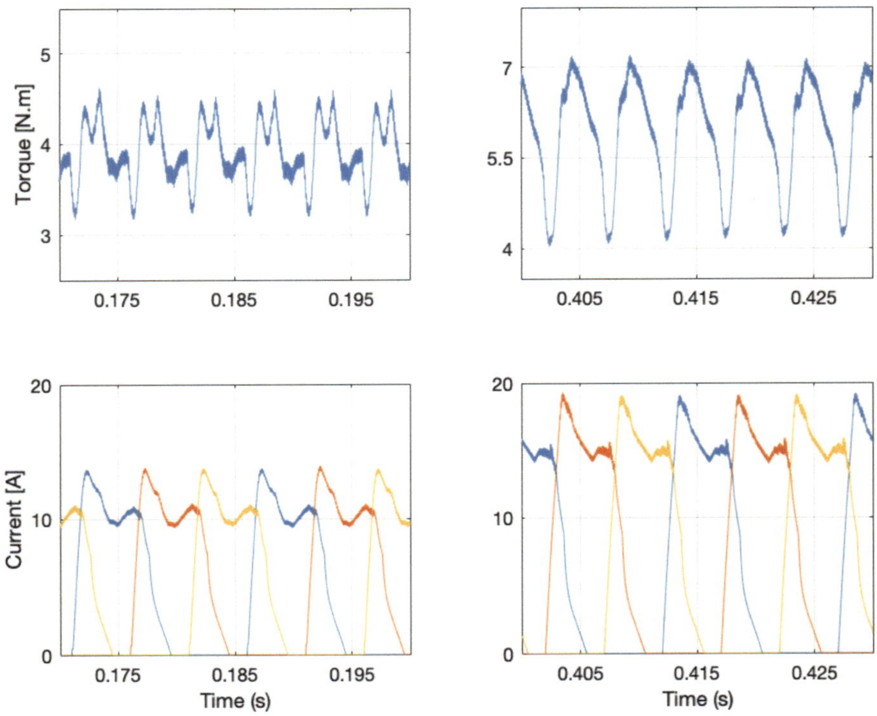

Figure 18.
Steady-state condition for same speed and different loads.

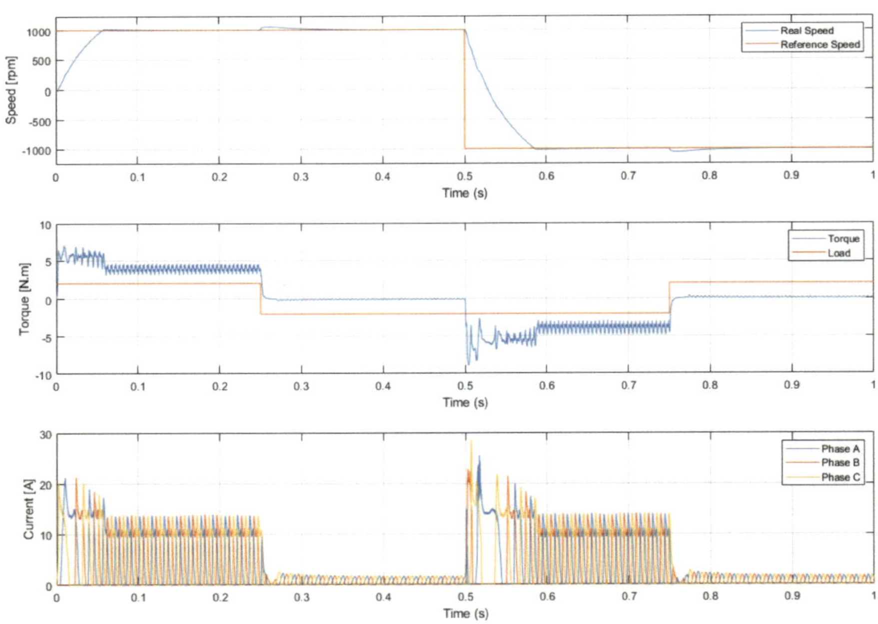

Figure 19.
Operation in the four quadrants.

satisfactory. However, there is a margin for improvement and design of a more robust controllers and methods.

As seen in *Section 7.1*, the required current waveform for always having the best performance changes essentially with speed and load. For high speeds, it is necessary to anticipate the current magnetization and demagnetization, and with high loads, it is necessary to have more current in the phase transitions. For improvement, a series of optimal control problems could be solved with a posterior extrapolation of results in order to have the ideal waveform for any operation point.

Current controller performance can be optimized too. As seen the inductance value is non-linear, changing with rotor position and current itself. An improvement could be the use of adaptive controllers that change its gains, at least, with the rotor position.

Other works, such as [13], are recommended to readers with further interest in direct torque control. For this type of electric drives, the feedback variable is the torque, so a torque estimation is necessary. There's also additional reading material if the reader wants to go more in depth. For example, in [19] the sliding mode technique to control the current is used, and in [20] an adaptive control methodology for the speed control is used.

8. Conclusions

The control of the SRM may not be as easy as conventional electric motors. Torque ripple, which is one of the main problems, may and must be reduced by the controller.

Different types of controllers can be used—the fundamental ones are voltage impulse, current, and torque controllers. Impulse control is essentially used for high speeds, while current and torque controllers are used for high performance.

As the produced torque does not depend only on the current but also on the non-linear derivative of the inductance, a constant current control is not sufficient for high performance. Reference current waveforms, or torque waveforms, have to be calculated for the response to have the minimum ripple torque component possible. Also, current dynamic has to be considered in the controller design. The inductance changes along the rotor position and so does the current dynamic. The back-EMF also has an effect in the current dynamic, especially at high speeds.

Numerical simulations of a real SRM model were performed. Steady-state results of the reference operating point and on a speed and load change were analyzed. Moving away from the used operating point to design the controller changes the response, as expected.

The dynamic behaviour was also simulated. It contemplates all four quadrants of operation. The speed response follows the reference speed with good accuracy. The load changes provoke perturbations on speed, which are compensated very quickly by the controller.

Finally, we hope that seeing the simulation files will help the reader to feel a concrete experimentation.

Author details

Manuel Fernando Sequeira Pereira[1*], Ana Mamede[2] and Rui Esteves Araújo[3]

1 Faculty of Engineering, University of Porto, Porto, Portugal

2 Department of Electrical Engineering, Universidade Federal de Uberlândia (UFU), Uberlândia, Brazil

3 INESC TEC, Faculty of Engineering, University of Porto, Porto, Portugal

*Address all correspondence to: ee12314@fe.up.pt

References

[1] Xue XD, Cheng KWE, Ho SL. Optimization and evaluation of torque-sharing functions for torque ripple minimization in switched reluctance motor drives. IEEE Transactions on Power Electronics. 2009;**24**(9): 2076-2090

[2] Vujičić VP. Minimization of torque ripple and copper losses in switched reluctance drive. IEEE Transactions on Power Electronics. 2012;**27**(1):388-399

[3] Ye J, Bilgin B, Emadi A. An extended-speed low-ripple torque control of switched reluctance motor drives. IEEE Transactions on Power Electronics. 2015;**30**(3):1457-1470

[4] Cheng H, Chen H, Yang Z. Average torque control of switched reluctance machine drives for electric vehicles. IET Electric Power Applications. 2015;**9**(7): 459-468

[5] Mikail R, Husain I, Sozer Y, Islam MS, Sebastian T. Torque-ripple minimization of switched reluctance machines through current profiling. IEEE Transactions on Industry Applications. 2013;**49**(3):1258-1267

[6] Krishnan R. Switched Reluctance Motor Drives: Modeling, Simulation, Analysis, Design, and Applications. Boca Raton: CRC Press; 2001. p. 398

[7] Boldea I, Nasar SA. Electric Drives. Boca Raton: CRC Press; 2005. p. 544

[8] Soares F, Branco PJC. Simulation of a 6/4 switched reluctance motor based on Matlab/Simulink environment. IEEE Transactions on Aerospace and Electronic Systems. 2001;**37**(3): 989-1009

[9] de Doncker R, Pulle DWJ, Veltman A. Advanced electrical drives: Analysis, modeling, control. Power System. 2011;**52**:1-443

[10] Kim J, Krishnan R. High efficiency single-pulse controlled switched reluctance motor drive for high speed (48k RPM) application: Analysis, design, and experimental verification. In: Conference Record – IAS Annual Meeting (IEEE Industry Applications Society). 2008

[11] Ahn J-W. In: Lamchich MT, editor. Switched Reluctance Motor, Torque Control. IntechOpen; 2011. Available from: https://www.intechopen.com/books/torque-control/switched-reluctance-motor

[12] Inderka RB, De Doncker RW, Krehenbrink M. On-line estimation of instantaneous torque in switched reluctance machine control. In: IEEE International Symposium on Industrial Electronics. 2000. pp. 385-389

[13] Nakao N, Akatsu K. Torque control of a switched reluctance motor by using torque estimation and excitation angle control. In: 2012 IEEE Energy Conversion Congress and Exposition, ECCE 2012. 2012. pp. 4314-4321

[14] Pereira M, Araújo RE. Analysis and Design of a Speed Controller for switched reluctance motor drive. U. Porto Journal of Engineering. 2019;**5**(1): 46-58

[15] Ye J, Bilgin B, Emadi A. An offline torque sharing function for torque ripple reduction in switched reluctance motor drives. IEEE Transactions on Energy Conversion. 2015;**30**(2):726-735

[16] Dowlatshahi M, Nejad SMS, Ahn JW. Torque ripple minimization of switched reluctance motor using modified torque sharing function. In: 2013 21st Iranian Conference on Electrical Engineering, ICEE 2013. 2013

[17] Skvarenina TL. The Power Electronics Handbook. Boca Raton: CRC Press; 2002

[18] Pittermann M, Fort J, Diesl J, Pavlicek V. Converters for switched reluctance motor - topology comparison. In: Proceedings of the 2018 18th International Conference on Mechatronics - Mechatronika, ME 2018. 2019. pp. 1-8

[19] Ye J, Malysz P, Emadi A. A fixed-switching-frequency integral sliding mode current controller for switched reluctance motor drives. IEEE Journal of Emerging and Selected Topics in Power Electronics. 2015;**3**(2):381-394

[20] Namazi MM, Rashidi A, Koofigar H, Saghaiannejad SM, Ahn JW. Adaptive control of switched reluctance motor drives under variable torque applications. Journal of Electrical Engineering and Technology. 2017; **12**(1):134-144

Chapter 9

Mathematical Modeling of Switched Reluctance Machines: Development and Application

Marcelo Vinícius de Paula, Thiago de Almada Lopes, Tárcio André dos Santos Barros, Paulo Sergio Nascimento Filho and Ernesto Ruppert Filho

Abstract

The usage of switched reluctance machines (SRMs) grows following the power electronics development. For this reason, a precise mathematical model is crucial for the development of SRM automatic control projects due to the nonlinearities caused by the machine topology and working principle. This chapter focuses on SRM characterization procedure, enlightening the nonlinear characteristics and the importance of the magnetization curves to accomplish precise automatic control of SRM. Different methods found in the literature are commented. The blocked rotor test is detailed, and an automatic acquirement system to obtain the SRM magnetization curves is reasoned. Magnetization curves are processed to create the mathematical model of the SRM. The computational algorithm used to process the acquired data is presented with the purpose of clarifying the production of the lookup tables used in the mathematical model. The developed mathematical model is implemented in Matlab/Simulink® environment. The system simulates the SRM operating both in motoring and generating mode. The mathematical simulation results are compared to experimental results. The developed model is accurate and may be used to study SRM behavior and control systems for SRM applications.

Keywords: switched reluctance machine, nonlinear characteristics, magnetization curves, mathematical modeling, computational simulation

1. Introduction

Switched reluctance machines (SRMs) have a doubly salient structure, which means that both the stator and the rotor have salient poles. Note that only the stator poles have field coils. The operation of SRM as motor is based on the principle of minimum reluctance, that is, when the winding on a pair of stator poles is energized, the closest poles of the rotor are attracted to a position of minimum reluctance, which generates rotating torque in the rotor. If a coordinated drive is performed, rotational movement occurs [1–3]. Usually, SRMs operate in the saturation region. Moreover, the presence of doubly salient structure and the working principle of SRM imply that SRMs have highly nonlinear characteristics. The nonlinear operation makes it difficult to represent the SRM by an efficient linear model as it is done for induction machines and synchronous machines [4]. To study SRMs, it is imperative to use models that accurately incorporate the nonlinear characteristics and quantity behavior of the system to be simulated. The most precise models for SRMs are based on experimentally acquired data. Several nonlinear models have been studied and tested in the literature [4–6].

In order to model the electrical and mechanical behavior of SRMs, it is necessary to obtain the magnetization curves $\phi(I, \theta)$ of the machine. The curves express the nonlinear behavior of the SRM and can be used to relate the input voltage (V_{DC}), electric current (I), rotor position (θ), and magnetic flux (ϕ) to the electromagnetic torque (T_e) [6, 7]. The methods for acquisition of the magnetization curves are currently divided into direct and indirect methods. Indirect methods consist of determining the magnetization curves from the SRM static torque characteristics. However, these methods are highly complicated to perform with good accuracy, as any mechanical deviation leads to large errors in the magnetization curves. On the other hand, the direct methods consist of applying voltage to the SRM phases and determining the magnetic flux [8–11].

Once the magnetization curves are obtained, the SRM model is generally built through lookup tables or neural network [4, 12]. For both methods, a large amount of data is required. The number of curves used to create the model is a determinant factor for the accuracy of the model. Diversely, the number of curves obtained experimentally is limited by the precision and execution time of the test. Therefore, a mathematical processing procedure is commonly used to achieve a larger number of magnetization curves. In some cases, analytical equations are developed to obtain intermediary magnetization curves [12–14]. However, experimental process errors, such as sensor errors, noise, residual currents, and numerical integration disparities, are encountered in the data obtained experimentally in magnetization tests [12, 15]. These errors may result in inconsistent characteristics of SRM operation. Thus, the mathematical processing to develop the lookup tables for current $I(\phi, \theta)$ and electromagnetic torque $T_e(I, \theta)$ must be robust and precise.

This chapter focuses on the characterization of SRM. Initially, the nonlinear characteristics of SRMs are presented. An automatic system to obtain the magnetization curves of SRM is uncovered. The procedure to process the experimental data is detailed, and the algorithms used in the Matlab/Simulink environment are made available. Mathematical simulations are performed, and the results are compared to those of experimental application to identify the accuracy of the developed model. Finally, a brief statistical analysis is employed to evaluate the precision of the mathematical model established.

2. Mathematical modeling of switched reluctance machines

2.1 Nonlinear characteristics of switched reluctance machines

To simplify the analysis of the nonlinear characteristics intrinsic to SRMs, consider some simplifications: (1) the magnetic coupling between phases is negligible, (2) there are no Eddy current and hysteresis losses, (3) the phase resistance is constant, and (4) the phase inductance depends on the rotor position and on the phase current.

With the assumptions made, an elementary circuit of one SRM phase may be derived as in **Figure 1**. The sum of the resistance voltage drop and the change rate of flux linkage must be equal to the applied voltage to a motor phase; thus Eq. (1) can be derived, where V is the voltage applied to the phase, i is the phase current, R is

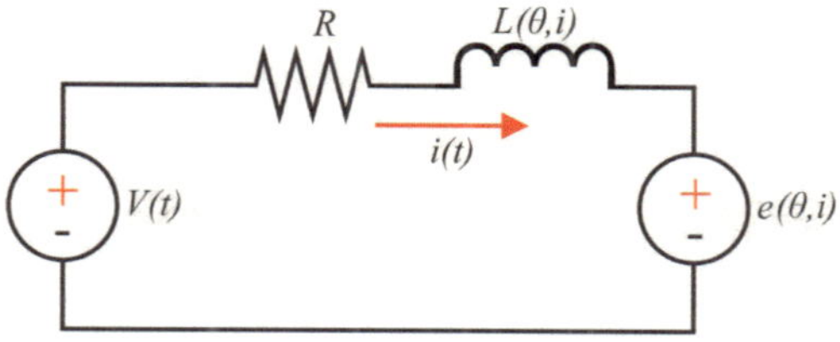

Figure 1.
Equivalent circuit for one SRM phase.

the phase resistance, λ is the phase flux linkage, θ is the rotor position, and e is the back electromotive force.

$$V = Ri + \frac{d\lambda(\theta, i)}{dt} \tag{1}$$

The phase flux linkage can be expressed as

$$\lambda(\theta, i) = L(\theta, i)i \tag{2}$$

Note that the phase inductance varies according to the rotor position, because of the reluctance variation, and according to the phase current due to the magnetic saturation. Furthermore, both the rotor position and the phase current may vary with time.

Substituting Eq. (2) into Eq. (1) and expanding, the phase voltage equation can be extended into Eq. (3):

$$\begin{aligned} V = Ri + \frac{d\lambda(\theta, i)}{dt} &= Ri + L(\theta, i)\frac{di}{dt} + i\frac{dL(\theta, i)}{dt} \\ &= Ri + L(\theta, i)\frac{di}{dt} + i\frac{d\theta}{dt}\frac{dL(\theta, i)}{d\theta} \\ &= Ri + L(\theta, i)\frac{di}{dt} + i\omega\frac{dL(\theta, i)}{d\theta} \end{aligned} \tag{3}$$

In Eq. (3), the mechanical speed is given by the variation of the rotor position in accordance with time: $\omega = d\theta/dt$. Thus, note that the phase voltage is composed of three components. The first term is the voltage drop in the phase resistance, the second term is the voltage drop in the phase inductance, and the third term is the induced back electromotive force, or back-emf (e). The back-emf is highlighted in Eq. (4) to focus on the perception of the direct relation between the rotor speed and the back-emf:

$$e = i\omega\frac{dL(\theta, i)}{d\theta} \tag{4}$$

During the phase energization, a certain amount of magnetic flux linkage is generated. As the magnetic circuit depends on the rotor position, the amount of flux linkage generated depends on the electric current and rotor position. The magnetic flux characteristics according to the electric current are known as magnetization curves and can be used to model the SRM dynamics. The modeling procedure will be explained in detail further in this chapter. Magnetization curves of a generic SRM are presented in **Figure 2(a)**. Note that the magnetization curves are obtained as the magnetic flux, versus the DC electric current in one phase, for a fixed angle (θ).

When the phase current is low, the SRM works out of the saturated region. However, recall that usually SRMs work in the saturated region. Furthermore, the alignment and misalignment angles are marked in **Figure 2(a)** along with two intermediate angles (θ_1 and θ_2).

Figure 2(b) shows the magnetization curve for the intermediate angle θ_1. For a given point P_1 in the curve, there is an electric current value I_1 and a corresponding flux linkage value λ_1. In any case, the electric energy applied to the winding is stored in the form of magnetic energy (hatched region $O\lambda_1P_1$). The stored magnetic energy (W_f) is calculated as [16]

$$W_f = \int_0^t Vi\,dt = \int_0^t \frac{d\lambda}{dt} i\,dt = \int_0^{\lambda_1} i\,d\lambda \tag{5}$$

The hatched area limited by the points OP_1I_1 represents the coenergy (W'_f). The coenergy has no physical meaning. On the other hand, the mechanical work realized by the machine in a given period of time is equal to the variation in coenergy during the same period. The coenergy can be calculated as

$$W'_f = \int_0^{I_1} \lambda\,di \tag{6}$$

Now, assume that the rotor position is driven to the second intermediate angle θ_2 and the current is maintained I_1, as in **Figure 2(c)**. The new magnetic energy will be calculated as the region limited by the points $O\lambda_2P_2$ and the coenergy as the region limited by the points OP_2I_1. Then, one can calculate the coenergy variation,

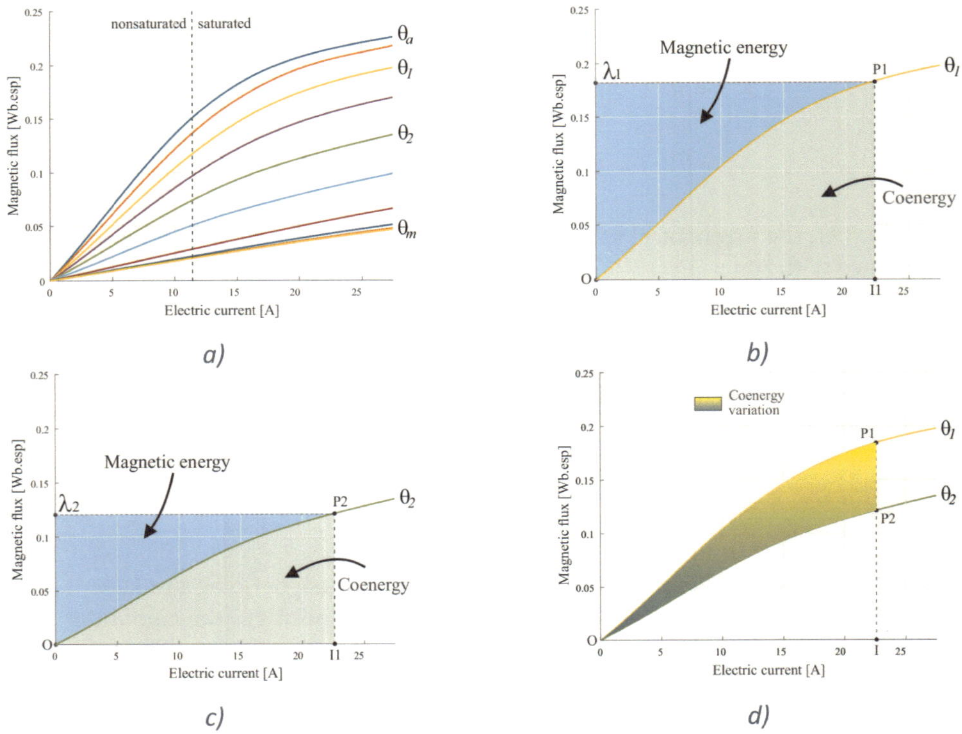

Figure 2.
Graphic representation of coenergy variation: (a) magnetic flux linkage according to rotor position and electric current intensity; (b) magnetic energy and coenergy for θ_1 position; (c) magnetic energy and coenergy for θ_2 position; and (d) coenergy variation between angles θ_1 and θ_2.

that is, the mechanical work realized to take the rotor from position θ_1 to position θ_2. **Figure 2(d)** shows the coenergy variation as the hatched area limited by the points OP_1P_2. The variation of the electric energy (ΔW_e) applied is calculated as in Eq. (7), where L_1 and L_2 are the phase inductance in positions θ_1 and θ_2 for the current I_1, respectively.

$$\Delta W_e = \int Vi\, dt = \int \frac{d\lambda}{dt} i\, dt = I_1(\lambda_2 - \lambda_1) = (L_2 - L_1)I_1^2 \tag{7}$$

Considering the variation in the linear portion of the magnetization curve, the coenergy variation is half the variation of electric energy applied to the phase winding, as in Eq. (8):

$$\Delta W_f' = \frac{1}{2}(L_2 - L_1)I_1^2 \tag{8}$$

The average electromagnetic torque ($\overline{T}$) produced when an electric current I_1 flows through a phase winding is given by Eq. (9):

$$\overline{T} = \frac{\Delta W_f'}{\Delta \theta} = \frac{1}{2} I_1^2 \frac{\Delta L}{\Delta \theta} \tag{9}$$

Considering infinitesimal variations, the instantaneous electromagnetic torque (T_e) can be calculated as in Eq. (10), where i is the instantaneous DC current in the phase:

$$T_e = \frac{\partial W_f'(i,\theta)}{\partial \theta} = \frac{1}{2} i^2 \frac{\partial L(i,\theta)}{\partial \theta} \tag{10}$$

Observing Eq. (10), some information may be inferred. First, torque production does not depend on the electric current direction but has a quadratic relation with its amplitude. Moreover, the produced torque direction depends strictly on the inductance derivative at the energizing moment, while the inductance is a function of rotor position and electric current.

The nonlinearities caused by the saturation and reluctance variation during the rotor excursion are carried to the magnetization curves, electric current, and electromagnetic torque characteristic. For this reason, a well-founded model is of utmost importance for studies covering SRM dynamics and control.

2.2 Magnetization curves

The acquisition of the magnetization curves $\phi(I,\theta)$ is crucial to model the electrical and mechanical behaviors of switched reluctance machines (SRM) as these machines operate essentially in the saturation region [1]. As aforementioned, more precise magnetization curves can be obtained through experimental tests. These tests are implemented to obtain the curves for specific rotor positions. The indirect methods consist of using the static torque characteristics to acquire the magnetization curves and may lead to substantial variation if mechanical deviations occur. The direct methods, conversely, consist of applying voltage to the SRM winding and determining the magnetic flux.

One simple way to obtain the magnetization curves is to apply sinusoidal AC voltage to the machine winding in different positions [17]. Once the root mean square current (I_{rms}) and the lag angle between voltage and electric current (θ_1) are known, one can calculate the flux linkage as

$$\phi = LI = \frac{V_{rms}\sin(\theta_l)}{I_{rms}2\pi f} I_{rms} \tag{11}$$

This method returns errors if magnetic saturation occurs, which generally happens when working with SRMs. To work around this issue, the most implemented method is the blocked rotor test, as discussed in previous literature [8]. Thus, in this chapter, the blocked rotor test will be used. This test consists of locking the rotor in place while applying voltage steps for each position. The electric current and voltage signals must be stored in order to calculate the magnetic flux. Once the phase resistance is known, the magnetic flux is given by

$$\phi(t) = \int_0^t (V - Ri)dt \tag{12}$$

The model accuracy is directly related to the number of curves and points in each curve. However, setting up the experiment for each angle and voltage is cumbersome. Hence, an automatic system can be developed to obtain the magnetization curves for as many positions as desired.

2.3 Automatic characterization system

The experimental characterization test is set up according to **Figure 3**. The test consists of positioning the SRM rotor at known positions and applying a voltage step. When the voltage step is applied, the current dynamics is stored. A stepper motor and a microstep driver are used to control the position. The stepper driver must have high resolution so that the position is precisely controlled. The maximum resolution used in this book chapter is 40,000 steps per mechanical cycle.

A Digital Signal Processor (DSP) model TMS320F28335 from Texas Instruments® is responsible for managing the rotor position control. An absolute encoder model RE36SC06112B Resnishaw® with 12-bit resolution (0.087°) is connected to the DSP through SPI communication and used to measure the rotor position. The DSP adjusts the position by controlling the stepper motor driver. When the rotor reaches the desired position, the electromagnetic brake locks the SRM rotor. Then, the DSP enables the energizing of the desired phase. Phase voltage and current waveforms are retrieved through an oscilloscope (Lecroy® 24MXs-B) using an

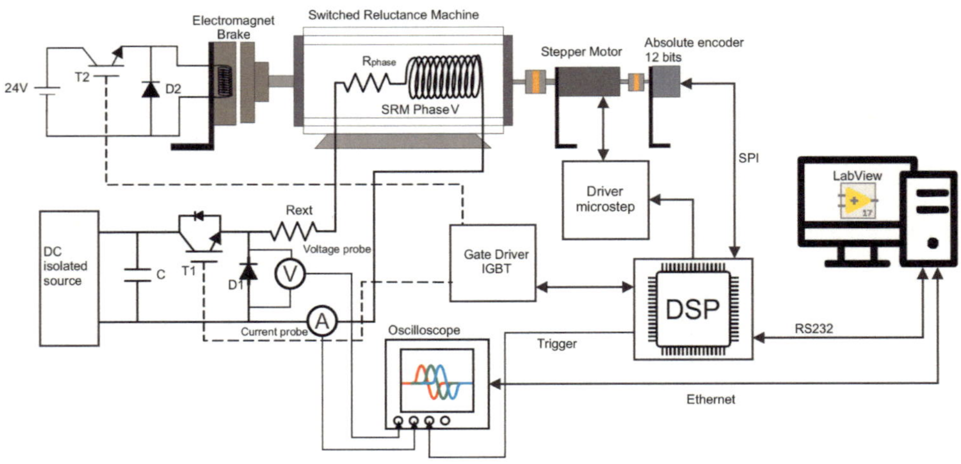

Figure 3.
General diagram of the automatic characterization system to obtain the magnetization curves [18].

isolated voltage differential probe and an isolated current probe, both with high resolution.

For better control and management of the experiment, a human machine interface (HMI) was developed in LabView®. The communication between the HMI and the DSP board is made through serial communication protocol (RS-232), while the communication between the HMI and the oscilloscope is made through the Ethernet using the IP protocol. **Figure 4** shows the flowchart for the software developed in LabView®.

Initially, some parameters, such as motor topology, test angles, phase resistance, and external resistance, are passed to the software. Next, the DSP is commanded to calibrate the encoder position according to the rotor position. In this step the electromagnetic brake is released, and the tested phase is energized until the rotor pole is aligned to the stator pole. This position is set as the initial position for the test. Afterward, the new desired position is informed to the DSP. The DSP sets the rotor position and sends a confirmation signal to the HMI.

When the rotor position is confirmed, the HMI configures the oscilloscope (through the Ethernet cable) to await an external trigger signal in order to acquire the data. The DSP sends a trigger pulse to the oscilloscope and then energizes an SRM phase. The acquired data are read in the oscilloscope and sent to the HMI via the Ethernet cable. After, the magnetic flux is computed through Eq. (12), and the $\Phi(I,\theta)|_{\theta\ tested}$ curve is created. The commanded position is then incremented, and the process is repeated until the final position is reached. At this stage, all $\Phi(I,\vartheta)$ collection has been obtained.

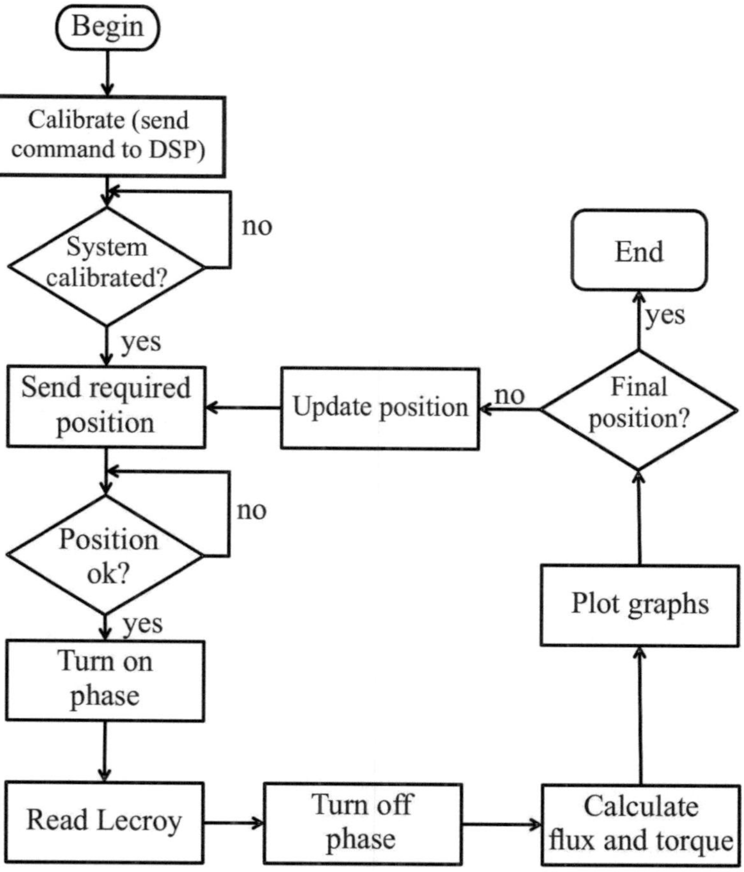

Figure 4.
Flowchart of the test management scheme [18].

Figure 5 shows the assembled system for automatic characterization, wherein an industrial SRM was subjected to the test. The tested machine is a 12/8 three-phase SRM with 1.5 kW nominal power, 1500 rpm nominal speed, 180–280 V supply voltage (DC voltage), 1.05 Ω phase resistance, 0.45 Wb.esp maximum magnetic flux, and 0.002 kg m^2 inertia moment. The machine parameters are gathered in Appendix C. The datasheet information provided by the manufacturer is used to stipulate the DC current source sizing, considering the SRM maximum current. In order to extrapolate the model, higher voltage must be applied to the machine terminal; thus, an external resistor must be added in series to the phase winding under test. The phase resistance, if not provided by the manufacturer, can be measured with a precision multimeter by performing a four-probe test. The resistance value precision is of utmost importance to the test accuracy.

The magnetization curve acquisition experiment must be performed varying the rotor position from the aligned to the misaligned position, which for the 12/8 SRM corresponds to 22.5°. A fixed step of 2.5° was used for each iteration, totalizing 10 magnetization curves. The HMI and the magnetization curves for the tested SRM are presented in **Figure 6**. On the left side, there is a gauge and a text box indicating the actual rotor position. Finally, there is a waveform graph with the phase current and voltage applied to the phase terminals. The algorithm that generates the curves derived from the acquired data is presented in the Appendix A. The curves generated will be further processed to obtain the lookup tables.

2.4 Switched reluctance machine mathematical model

The mathematical model of SRM is divided in electrical and mechanical portions. The electrical portion is mathematically represented by Eq. (1), while the mechanical portion is represented by Eq. (13), where T_{mec} is the mechanical torque, B is the viscous friction coefficient, and J is the moment of inertia:

$$T_{mec} = T_e - B\omega - J\frac{d\omega}{dt} \tag{13}$$

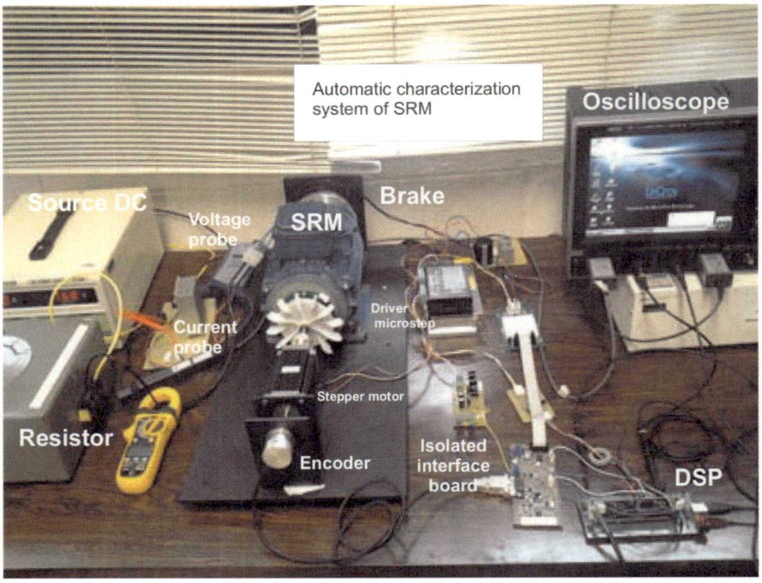

Figure 5.
Experimental setup [18].

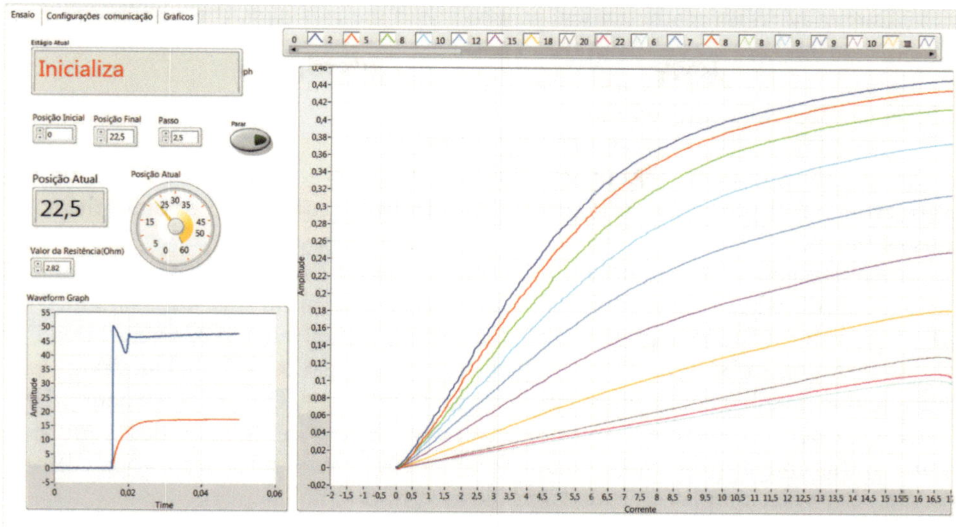

Figure 6.
Human machine interface for the management software developed in Labview® magnetization test results.

A block diagram representing the mathematical model is presented in **Figure 7**. To implement the model, the $I(\Phi, \theta)$ and $T(I, \theta)$ points must be known. The data obtained experimentally can be processed to obtain two lookup tables as will be discussed in the following sections.

For a precise model, the relation between magnetic flux, electric current, and phase excitation voltage must be well established, as in the magnetization curves. In some cases, fuzzy logic is used to calculate the magnetic characteristics and to model the SRM [19, 20]. Alternatively, $I(\Phi, \theta)$ analytical equations may be created [21]. Fourier series expansion is used to represent the necessary equations to simulate the SRM in this case.

The usage of $I(\Phi, \theta)$ and $T(I, \theta)$ lookup tables (LUT) allows faster simulation than the aforementioned methods. However, to obtain an accurate model, the number of magnetization curves required is unfeasibly high to be performed

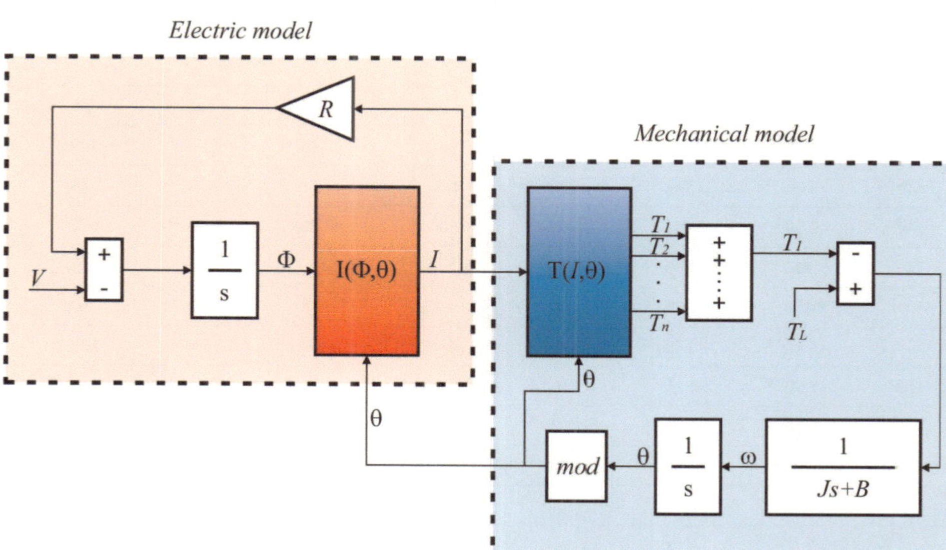

Figure 7.
Complete model for mathematical simulation of an SRM.

experimentally. With mathematical processing of the test data, it is possible to obtain intermediate curves from the experimental ones.

2.5 Processing the acquired data

During the automatic characterization test, 10 different magnetization curves were obtained as presented in **Figure 6**. The acquired data is used to build the lookup tables.

The initial objective is to obtain $I(\Phi, \theta)$ LUT with N columns (as position) and N lines (as flux), where the flux varies from 0 to the maximum flux Φ_{max} obtained when the rotor is in the aligned position and the position must vary from 0 to 45°. The flux step must be Φ_{max}/N and the position step must be 45°/N. The value of N must enable detailed comprehension regarding the model. On the other hand, the value must not be too large to minimize calculation time in mathematical simulations. In this chapter, the chosen value for N is 200, which returned a good compromise between simulation speed and likeness of the model.

2.5.1 Polynomial magnetization curves

With experimental curves $\Phi(I, \theta)$ for each known position, polynomial fits are used on these data to obtain functions that describe the curve. The functions have the format of Eq. (14). The coefficients for the obtained data are shown in **Table 1**. The results of the polynomial fits are very precise ones, for each position, there are 10,000 acquired points (I, ϕ). **Figure 8** shows the curves obtained from the polynomial regression:

$$\Phi(i)_{\theta=n|_{0^\circ}^{22.5^\circ}} = p_0^n + p_1^n i + p_2^n i^2 + p_3^n i^3 + p_4^n i^4 + p_5^n i^5 + p_6^n i^6 \tag{14}$$

Now, the magnetization curves are expressed as functions of current for fixed positions. However, there are only 10 curves (one for each tested position) as depicted in the first table of **Figure 9**. The next step is to set the number of current points we desire in the LUT. Consider that N points are desired. Then, using Eq. (13), the current value can be varied from 0 to I_{max} in steps of I_{max}/N in order to obtain Table b in **Figure 9**.

θ	$p0$	$p1$	$p2$	$p3$	$p4$	$p5$	$p6$
0.0°	$-6.51e^{-9}$	$3.05e^{-7}$	$-5.78e^{-6}$	$5.37e^{-5}$	$-2.49e^{-4}$	$7.28e^{-3}$	$-3.29e^{-4}$
2.5°	$-5.06e^{-9}$	$1.94e^{-7}$	$-2.87e^{-6}$	$2.02e^{-5}$	$-7.97e^{-5}$	$7.23e^{-3}$	$-1.89e^{-4}$
5.0°	$-7.56e^{-9}$	$3.04e^{-7}$	$-4.56e^{-6}$	$3.00e^{-5}$	$-7.91e^{-5}$	$8.01e^{-3}$	$-2.92e^{-4}$
7.5°	$6.81e^{-9}$	$-4.70e^{-7}$	$1.18e^{-5}$	$-1.40e^{-4}$	$6.61e^{-4}$	$1.10e^{-2}$	$-1.83e^{-4}$
10.0°	$8.89e^{-9}$	$-1.03e^{-6}$	$3.64e^{-5}$	$-5.39e^{-4}$	$2.89e^{-3}$	$1.54e^{-2}$	$-2.63e^{-4}$
12.5°	$-8.81e^{-9}$	$-3.85e^{-7}$	$3.29e^{-5}$	$-6.46e^{-4}$	$3.89e^{-3}$	$2.13e^{-2}$	$-1.30e^{-3}$
15.0°	$-1.52e^{-8}$	$-1.23e^{-7}$	$3.16e^{-5}$	$-7.03e^{-4}$	$4.22e^{-3}$	$2.94e^{-2}$	$-2.04e^{-3}$
17.5°	$-6.87e^{-9}$	$-6.83e^{-7}$	$4.72e^{-5}$	$-9.12e^{-4}$	$5.11e^{-3}$	$3.58e^{-2}$	$-2.69e^{-3}$
20.0°	$2.72e^{-8}$	$-3.01e^{-6}$	$1.06e^{-4}$	$-1.57e^{-3}$	$7.77e^{-3}$	$4.00e^{-2}$	$-3.91e^{-3}$
22.5°	$9.38e^{-8}$	$-6.95e^{-6}$	$1.94e^{-4}$	$-2.45e^{-3}$	$1.10e^{-2}$	$4.06e^{-2}$	$-3.01e^{-3}$

Table 1.
Polynomial fits coefficients.

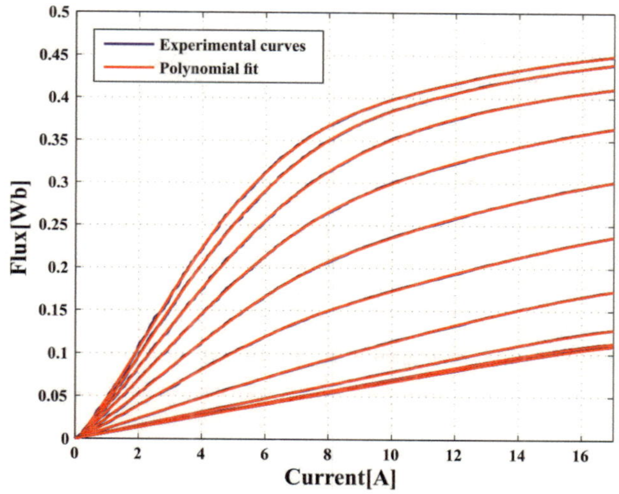

Figure 8.
Polynomial fit and experimental acquired magnetization curves [18].

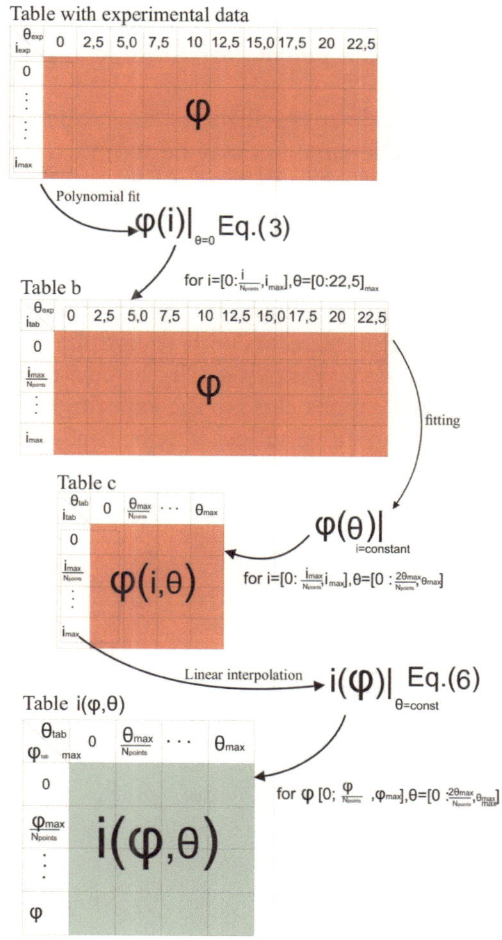

Figure 9.
Procedure to obtain the $I(\Phi, \theta)$ table [18].

With these data, it is possible to obtain the points $(\theta, \Phi)|_{I=constant}$, that is, how the flux behaves when the rotor position varies for a constant current. For each current value, there are 10 (θ, Φ) points; however, more points are required to achieve an accurate model. The next sessions deal with further data processing.

2.5.2 Procedure to obtain $I(\phi, \theta)$ and $T(i, \theta)$ lookup tables

In an SRM, for a constant current applied to the phase winding when the rotor is in the aligned position (θ = 0°), the reluctance of the magnetic circuit is minimal, and, therefore, the magnetic flux is maximum. As the rotor poles align with the stator poles, the reluctance increases, and the magnetic flux decreases until the rotor reaches the misaligned position in θ = 22.5° for an 12/8 pole SRM. Note that if the rotor and stator pole widths are different, during the next misalignment position, the flux may remain constant. Errors in experimental tests may lead to differences in magnetic flux when the rotor is close to the aligned and misaligned positions [22]. These errors cause inaccuracies even in models obtained with experimental data.

Again, for each value of current, there are 10 (θ, Φ) points that must be turned into a larger quantity of points. In order to do that, a robust regression must be applied. The attainment of the functions may be compromised by the use of less robust types of regressions, which may lead to errors. Hence, the generation of the $(\theta, \Phi)|_{I=constant}$ curves is tested with polynomial, linear, and cubic spline and smoothing spline regressions.

Polynomial fit may present abrupt oscillations, especially close to the regression interval [23]. This undesired phenomenon may be accentuated due to the magnetic flux variation for a given current when the rotor is close to the aligned and misaligned positions. As depicted in **Figure 10**, the polynomial regression represents improper magnetic flux behavior. The flux should decrease smoothly between aligned and misaligned positions.

Piecewise linear interpolation is a simple and easy technique to be implemented. The main idea is to connect the data points with straight lines as addressed in [2]. Though, when the number of data points is small, the curves do not present a smooth profile, notably in the regions close to the experimental data points.

To attain smoother behavior, cubic splines may be used [22, 24]. In this case, cubic polynomials are determined to connect the experimental data points. The accuracy of this method depends on the amount of experimental data. This method

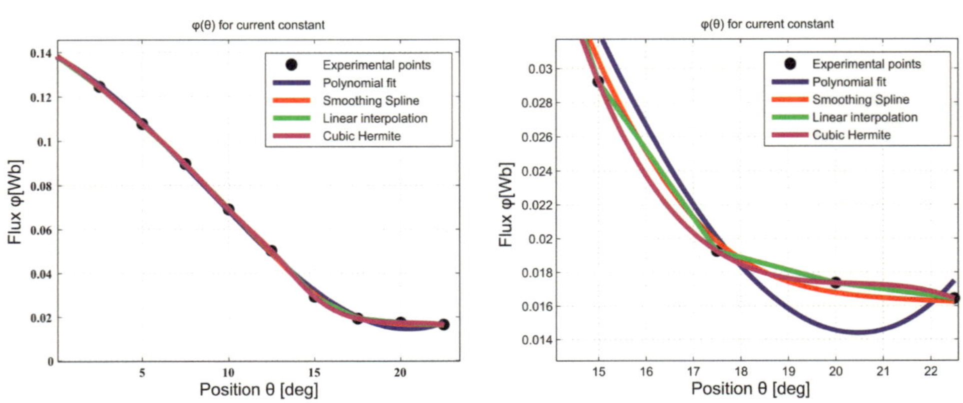

a) *b)*

Figure 10.
(a) Magnetic flux according to rotor position; (b) zoom at critical position [18].

ensures that the obtained curve passes through all experimental data points, which means that all experimental data acquirement errors are carried to the model, especially in positions close to alignment and misalignment.

Thus, to overcome these particularities, a more robust technique must be used. Smoothing spline technique minimizes Eq. (15), which represents the sum of square errors (first term) and the roughness of smoothness factor (second term) [25]. The parameter p defines the smoothing spline and can be calculated as Eq. (16), where h is the average spacing between points [26]. The smoothness obtained using the smoothing spline technique can be observed in **Figure 11**.

$$p\sum_{i}(y_i - s(x_i))^2 + (1-p)\int\left(\frac{d^2s}{dx^2}\right)^2 dx \tag{15}$$

$$p = \frac{1}{1+\frac{h^3}{6}} \tag{16}$$

Now we have N functions, each for a fixed current and for the current ranging from 0 to I_{max} in steps of I_{max}/N. With these curves we can find the $\Phi(\theta, I)$ points for varying the position from 0 to $\theta_{max} = 22.5°$ in steps of θ_{max}/N and then, create Table c in **Figure 9**.

As previously discussed, for the SRM mathematical model, we need the $I(\Phi, \theta)$ LUT. Thus, as observed in Table c in **Figure 9**, we get the data points (Φ, θ) through Eq. (17). With these points, the $I(\Phi)$ functions are obtained through a linear regression. In this case, the linear regression is applied because it enables extrapolating the current value of the test, if necessary:

$$I(\Phi)|_{\theta=constant} = i_0 + (i_1 - i_0)\frac{\Phi - \Phi_0}{\Phi_1 - \Phi_0} \tag{17}$$

Table $I(\Phi, \theta)$ is obtained through Eq. (16) by varying the magnetic flux from 0 to Φ_{max} in steps of Φ_{max}/N. This table represents the current behavior in respect to the magnetic flux varying from 0 to Φ_{max} and in respect to the rotor position varying from 0 to 22.5°. The 12/8 SRM has a polar angle of 45°; thus, the table must be reflected and added to the end of the original table. The final $I(\Phi, \theta)$ LUT is presented in **Figure 11**.

Henceforth, the data from Table c in **Figure 9** is used to estimate the coenergy through the trapezoidal method as in Eq. (18). The estimated magnetization curves

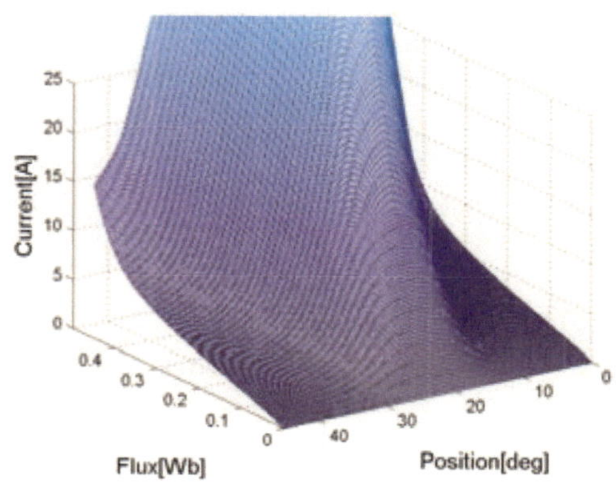

Figure 11.
$I(\Phi, \theta)$ lookup table [18].

and coenergy as a function of the electric current are presented in **Figure 12(a)** and **(b)**, respectively.

$$W'_n(i)|_{\theta=constant} = trapz(i[1:n], \phi[1:n])|_2^N \tag{18}$$

For a constant current value, the pairs (W', θ) can be calculated using Eq. (18). Then, the electromagnetic torque may be calculated for a given current as a function of rotor position as observed in Eq. (19). If the same procedure is implemented for all current values in the $I(\Phi, \theta)$ LUT, one can create the $T(I, \theta)$ LUT by interpolating the points obtained with Eq. (19):

$$T = \frac{180}{\pi} \frac{\left(W'_n - W'_{n-1}\right)}{\left(\theta_n - \theta_{n-1}\right)}\bigg|_{n=2}^{N} \tag{19}$$

Different interpolation methods lead to slight disparities in the inductance profile that may cause major discrepancies between the model and the real system behavior. Given the relation between the produced torque and the inductance derivative, the inductance profile generated may deviate from the actual profile and cause divergences in the torque lookup table. For example, if linear interpolation is used, the inductance profile will be composed as the junction of lines with different slopes, specifically in the region bounding the experimentally acquired points. That will cause the torque LUT to present subtle variations in the returned torque according to position. Another common problem is the unexpected change of slope caused by the polynomial fit and Hermite cubic interpolations. Observed in **Figure 10(b)**, when the Hermite cubic interpolation is used, the slope of inductance changes from negative to positive and back to negative, which causes the torque profile to have abrupt variation at the transition from negative to positive torque produced. When the polynomial fit is applied, the slope becomes positive at the end of the inductance profile, causing a change from generating to motoring torque before the real transition.

The usage of polynomial regression, linear interpolation, and Hermite cubic interpolation to obtain the $\Phi(\theta)_{I=constant}$ functions implies that variations on the torque and inductance profiles are not suitable to the SRM real behavior [18]. These variations may cause problems in projects that require accurate models, such as torque ripple minimization and position estimation. Therefore, the most suitable technique to model the performance of the SRM is the smoothing splines provided that no disturbances or unexpected behavior appeared in the inductance and torque

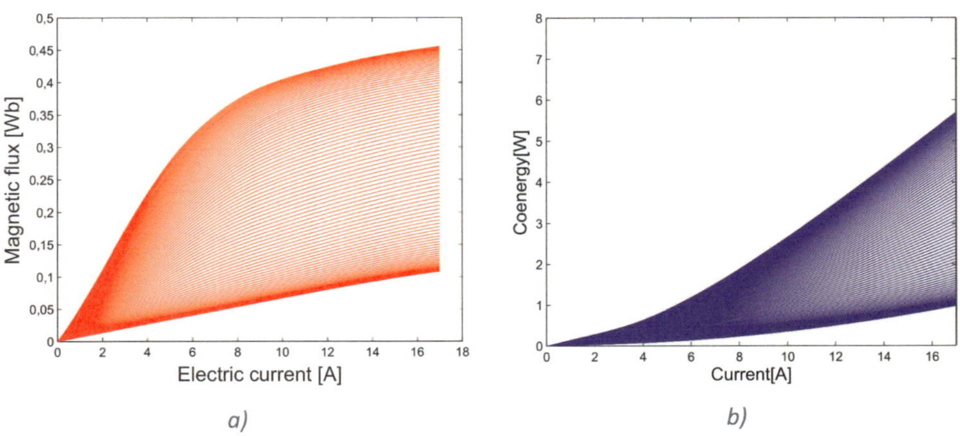

Figure 12.
(a) Magnetization curves estimated through the acquired data; (b) estimated coenergy.

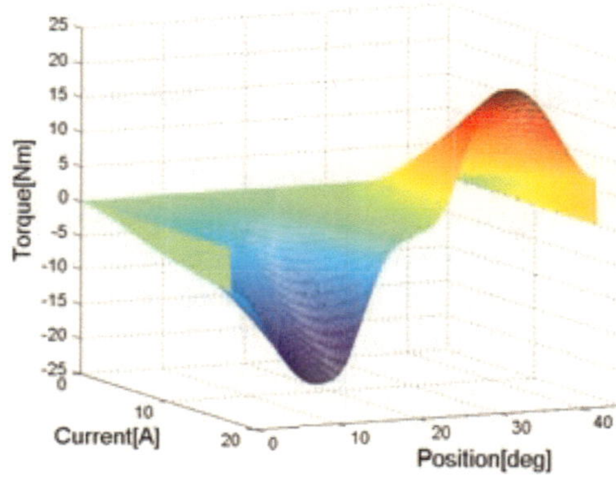

Figure 13.
$T(I, \theta)$ lookup table [18].

profiles. The $T(I, \theta)$ LUT surface obtained using the smoothing spline interpolation is presented in **Figure 13**; note the smooth transition between negative and positive torque in the table.

Note that the experimentally acquired data was directly used to develop the tables from the initial position ($\theta = 0°$) to half the dwell angle ($\theta = 22.5°$). The other hemisphere of the table was synthesized by mirroring the developed part.

The algorithm developed in Matlab/Simulink® environment for realization of the procedure described in this section is provided at the Appendix B. These algorithms include the data processing procedure to obtain both the $I(\Phi, \theta)$ and $T(I, \theta)$ lookup tables. With the supplied code and the information presented, the reader should be able to replicate the results achieved in this book chapter.

3. Model implementation

With the obtained LUTs $I(\Phi, \theta)$ and $T(I, \theta)$, the SRM model is implemented using the Matlab/Simulink® software, as presented in **Figure 14**. The electric

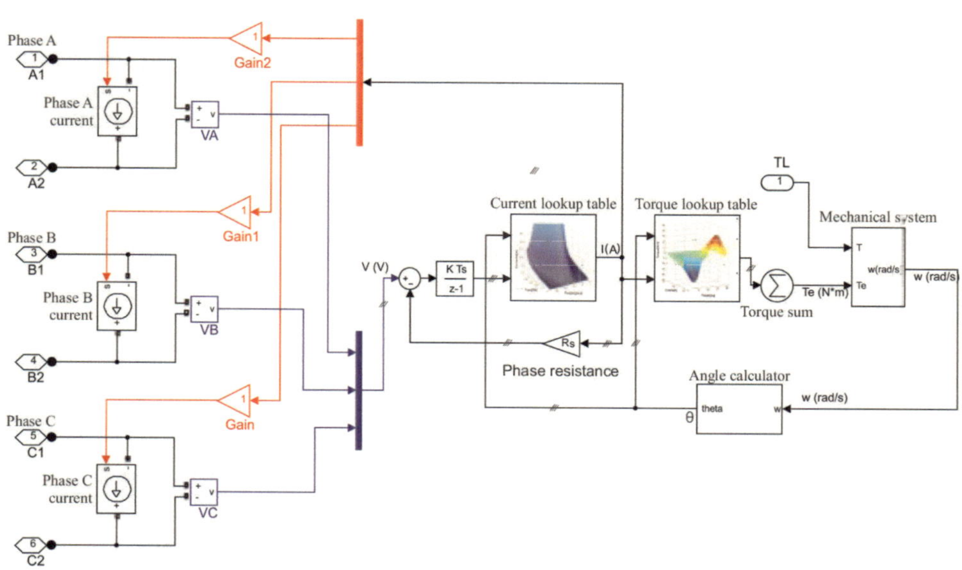

Figure 14.
SRM model diagram developed in Simulink®.

voltage in each phase is measured, and the magnetic flux is calculated through a discrete integral block, as in Eq. (12). The rotor position is also obtained by integrating the SRM rotational speed so that the angles from each phase may be calculated.

The magnetic flux and the phase positions are used as input to the $I(\Phi, \theta)$ LUT. The output current is then used along with the phase position as input for the $T(I, \theta)$ LUT, which outputs the electromagnetic torque. In both cases, for nonexisting input values, the output is calculated through cubic interpolation. The total electromagnetic torque is obtained as the sum of the torque produced by each phase. The output currents from the $I(\Phi, \theta)$ LUT are generated in the model terminals using current-controlled sources.

4. Model validation

To validate the obtained SRM model, two control systems were implemented in Matlab/Simulink® environment and also in an experimental bench. In the first system, the SRM is operating as a motor and speed control is implemented. In the second implemented system, the SRM is operating as a generator and the DC link voltage is controlled. Next sessions address the implementation and obtained results for both systems.

4.1 Motor operation-speed control

The block diagram for the speed control system simulated is presented in **Figure 15**. The SRM operates as a motor at a constant load of 0.5 N.m. The error between reference and instantaneous speed is processed through a PI controller to create the current reference. The current regulator block generates the driving signals for the AHB converter according to the turn-off (θ_{off}) and turn-on (θ_{on}) angles.

The experimental setup is assembled in the same conditions as the simulation. A DSP TMS320F28335 is used to control the speed. The simulations must be discretized to accurately represent the experimental behavior, and the discretized control algorithm is embedded in the DSP.

Figure 16 depicts the experimental setup. There are labels indicating the components in the setup. The AHB converter and gate drives are indicated by the "a" label and the oscilloscope and the sensors board are indicated by the "b" and "c" labels, respectively. The signal conditioning and DSP board are indicated by the "d" label. The induction machine (IM) used as load is indicated by the "e" label. The

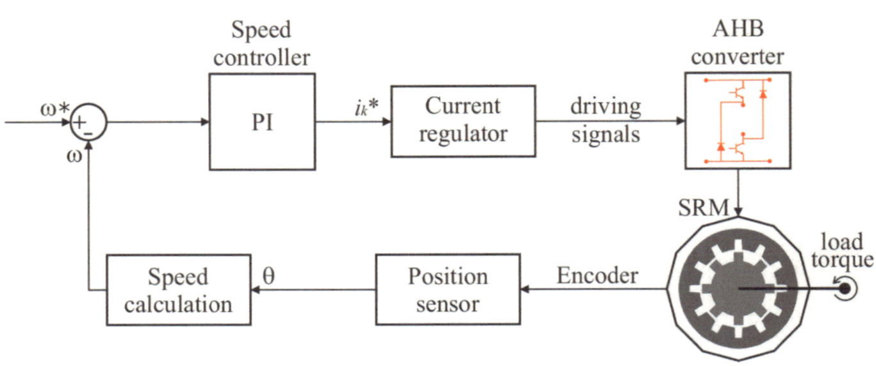

Figure 15.
Block diagram for the SRM speed control system.

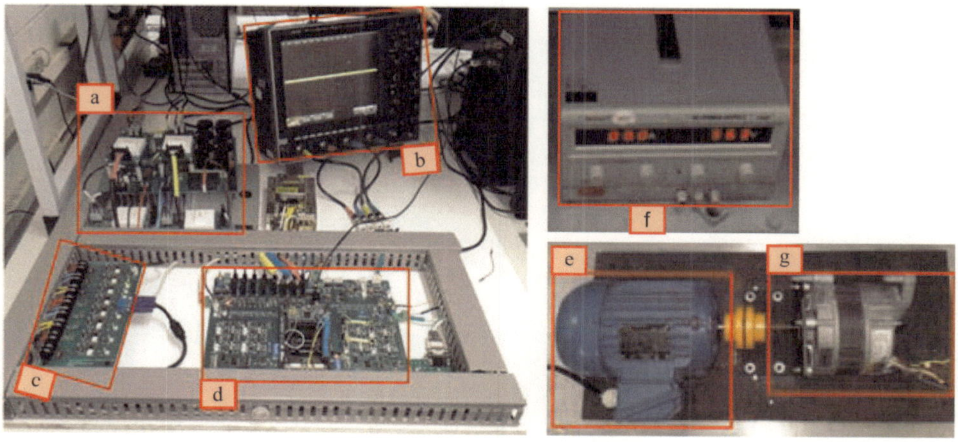

Figure 16.
SRM experimental setup.

voltage source and the SRM are indicated by the "f" and "g" labels, respectively. For both the experiment and the simulation, an asymmetric half bridge (AHB) converter is used to drive the SRM in motor mode. In the experimental setup, the IM is used as load.

Two simulations are conducted to study the behavior of the model at below base speeds and above base speeds. For the first simulation setup, named S_1, the speed reference is set to 30 rad/s, to simulate below base speeds, and the current and turn-off angle are controlled. The simulation results are presented in **Figures 17** and **18**. **Figure 17** shows the electric current of the three phases of the SRM, while **Figure 18** presents the phase A voltage and current.

For the second simulation conducted, named S_2, the reference speed was set to 130 rad/s, to simulate above base speeds. For this simulation, only the turn-off angle is controlled because voltage single-pulse operation takes place. **Figure 19** depicts the electric currents of all the three phases. **Figure 20** shows the voltage single pulse applied to the phase and the electric current produced.

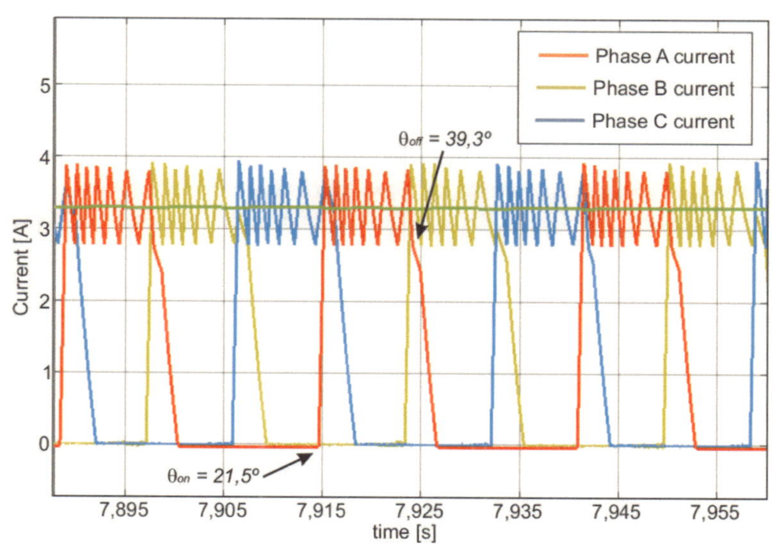

Figure 17.
Phase currents for S_1 setup.

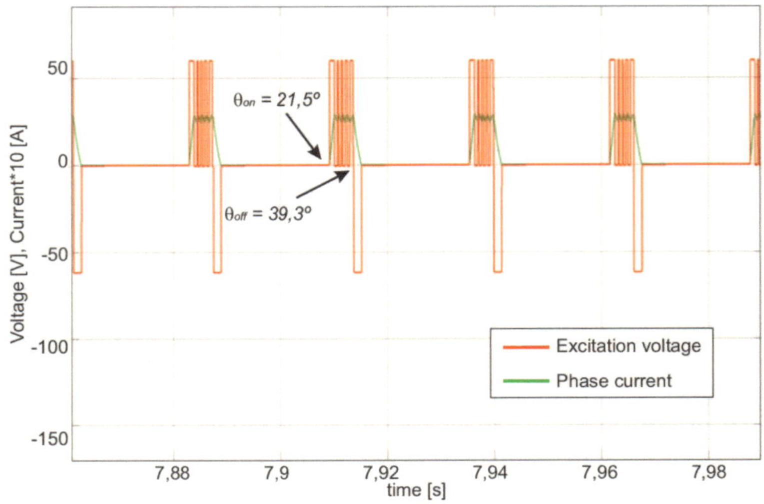

Figure 18.
Phase A current and voltage for S_1 setup.

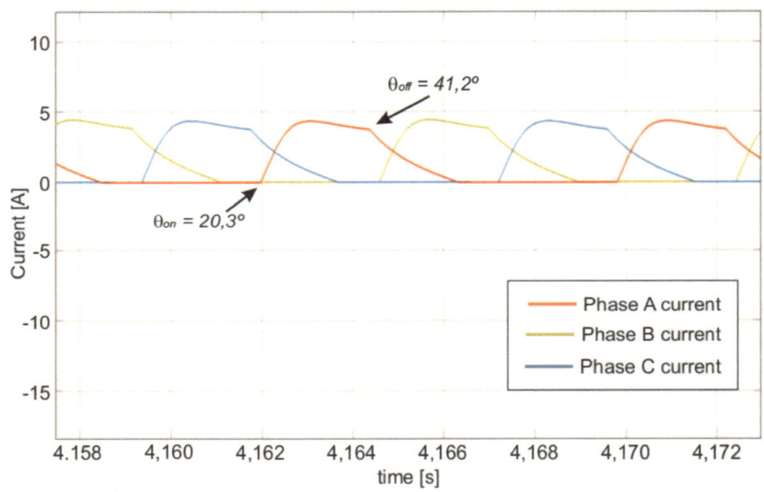

Figure 19.
Phase currents for S_2 setup.

The simulation setups S_1 and S_2 are then replicated experimentally as the E_1 and E_2 experiments, respectively. The same waveforms from the simulations are collected for comparison. **Figure 21** depicts the electric currents in the SRM phases and the reference current, while **Figure 22** shows the phase A current and applied voltage (VA), both for the E_1 setup.

Figures 23 and **24** are collected using the E_2 setup. **Figure 23** presents the phase electric currents of the SRM and the maximum permitted current, while **Figure 24** depicts the excitation voltage (VA) and the electric current in phase A.

Comparing the waveforms from S_1 and E_1, one can observe that the model is consistent for below base speeds. Some minor differences occur due to other parameters of the systems, such as the power switches and data sampling period. Nonetheless, the accuracy of the model is evidentiated.

For above base speeds, the model proves to be reliable as well. Comparing the S_2 and E_2 results, we ascertain that minor differences occur in the current waveform. Still, the model precisely predicts the real behavior of the SRM.

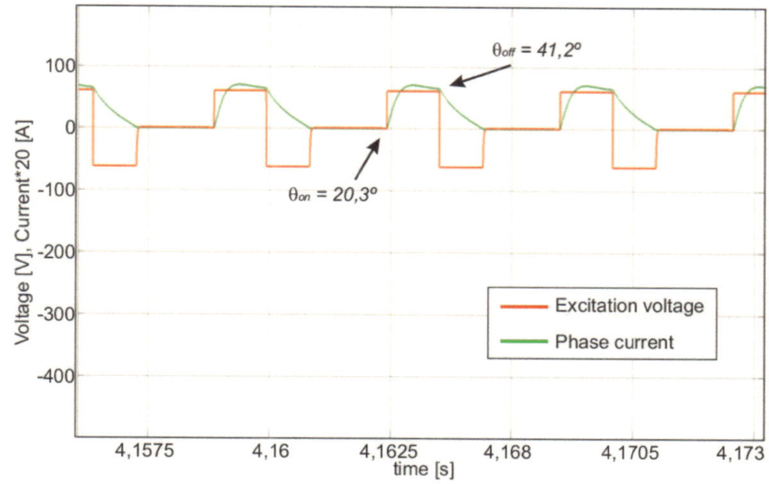

Figure 20.
Phase A current and voltage for S_2 setup.

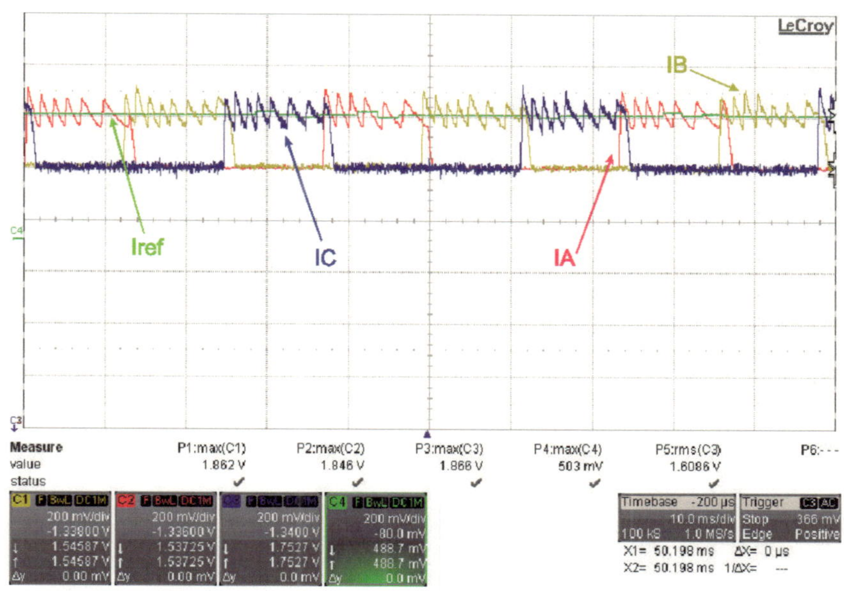

Figure 21.
Phase currents for E_1 setup.

4.2 Generator operation-voltage control

Figure 25 shows a simplified block diagram of the simulated system. The switched reluctance generator (SRG) operates at a constant speed of 1200 rpm. The SRG feeds a resistive load, and the control is responsible for maintaining the bus voltage on the reference value V_{ref}. A PI controller processes the voltage error and regulates the turn-off angle (θ_{off}), thus controlling the SRG level of magnetization to maintain a constant voltage at the DC link.

The experimental setup must be assembled with the same conditions as the simulation. A DSP TMS320F28335 can be used to control the DC voltage. The controller in the simulations must be discretized (sample time of 40 kHz) and embedded in the DSP.

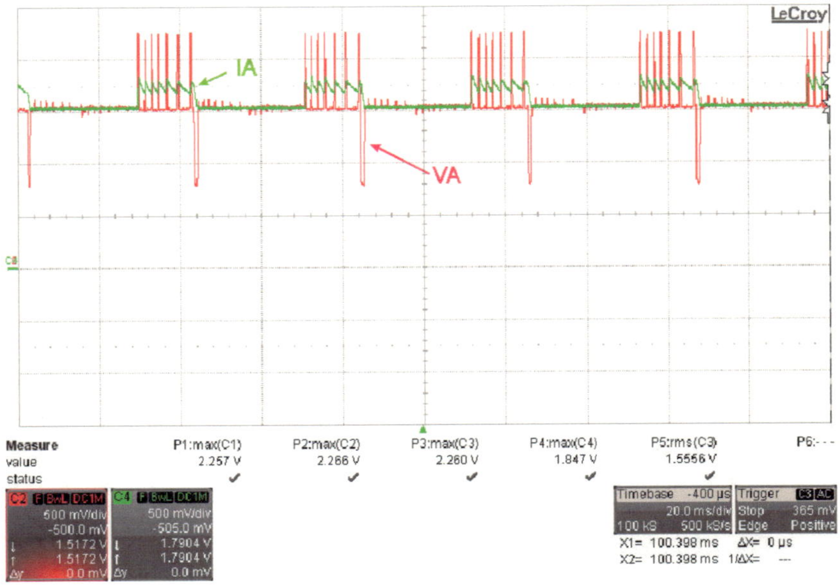

Figure 22.
Phase A current and voltage for E_1 setup.

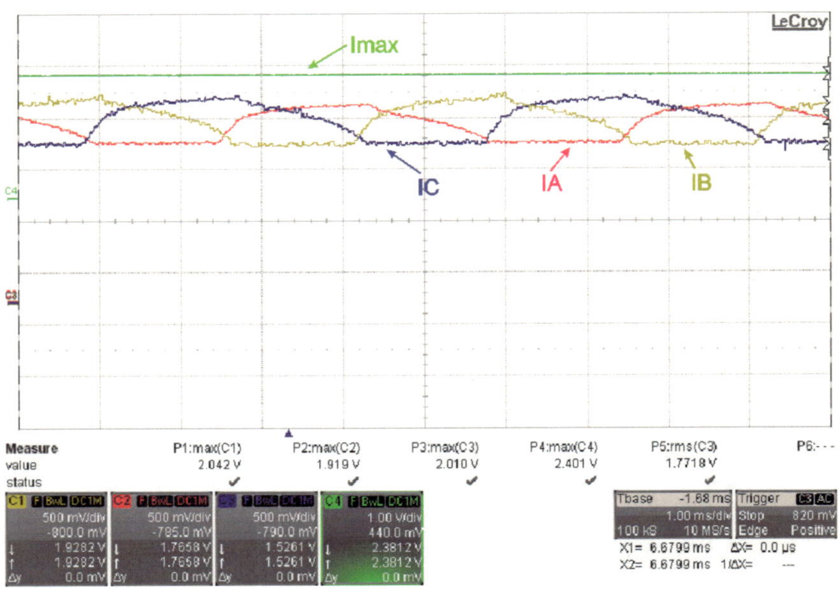

Figure 23.
Phase currents for E_2 setup.

An AHB converter is used to drive the phases of the SRG. An IM driven by an inverter may be used as the prime mover. An absolute encoder is used to detect the rotor position. A pulse generator drives the machine phase according to the driving angles (θ_{on} and θ_{off}). **Figure 26** shows the experimental setup.

The simulation was set to start with a 180 Ω load and $V_{ref} = 100$ V. At 5 s of simulation, V_{ref} is set to 160 V. The DC bus voltage, turn-off angle (θ_{off}) and phase A current are shown in **Figure 27** at the load change moment. To operate the SRG in self-excited mode, an initial magnetization is necessary, which is provided by a 2250 μF capacitor at the DC bus. Note that the control acts on the turn-off angle and increases the SRG excitation voltage to supply the energy to be generated, thus

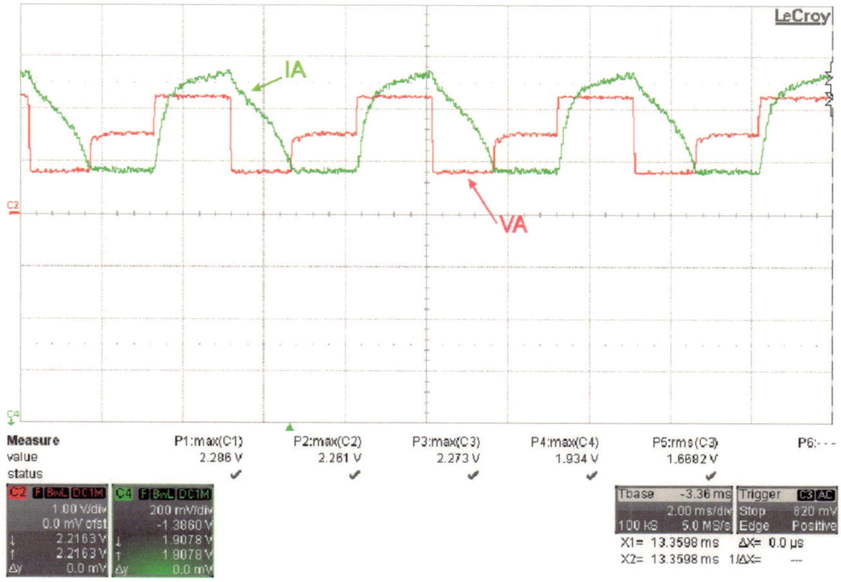

Figure 24.
Phase A current and voltage for E_2 setup.

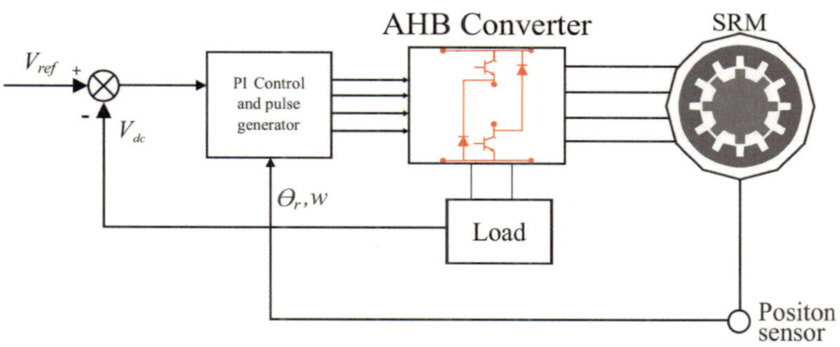

Figure 25.
Simulated control system for the SRG.

Figure 26.
SRG experimental setup [18].

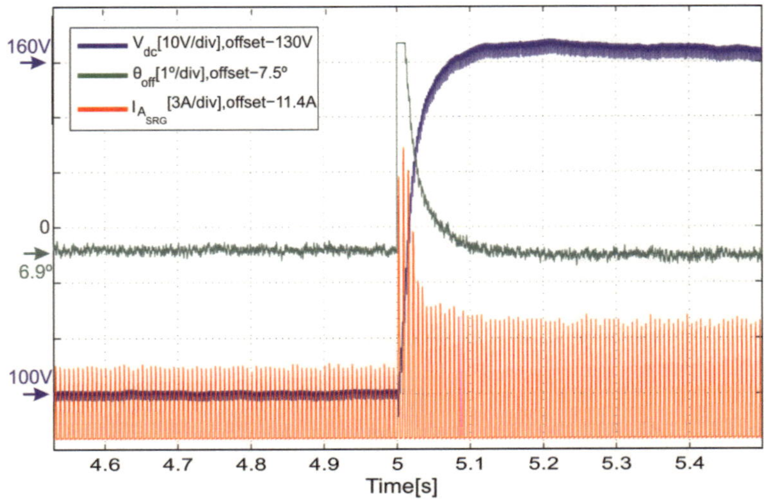

Figure 27.
Simulation results [18].

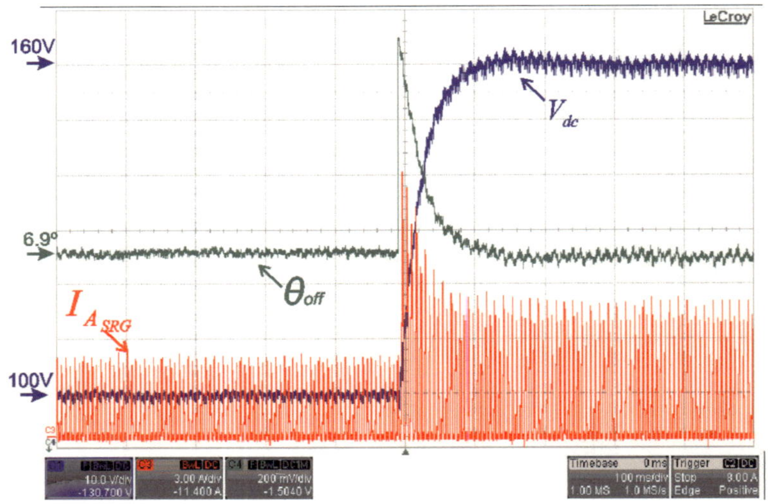

Figure 28.
Experimental results [18].

keeping the DC bus at the voltage reference value. Several techniques may be applied to control the DC bus voltage [27]. The DC bus capacitor is also responsible for filtering the voltage ripple. **Figure 28** depicts the experimental results at the same conditions as in the simulation.

The electric currents of the experiment and simulation test are compared in **Figure 29**. To evaluate the accuracy of the obtained model, statistical parameters were calculated, such as root mean square error (RMSE), mean absolute error (MAE), sum of square errors (SSE), and R-square (R^2). RMSE and MAE relative errors for the SRG electric current are defined as Eqs. (19)–(23), where I_{exp} and I_{sim} are the measured and simulated SRG phase currents, n is the number of data points, and e_j and e_{avg} are relative errors and their average value, respectively:

$$RMSE = \sqrt{\frac{1}{n}\sum_{j=1}^{n}\left(e_{avg} - e_j\right)^2} \tag{20}$$

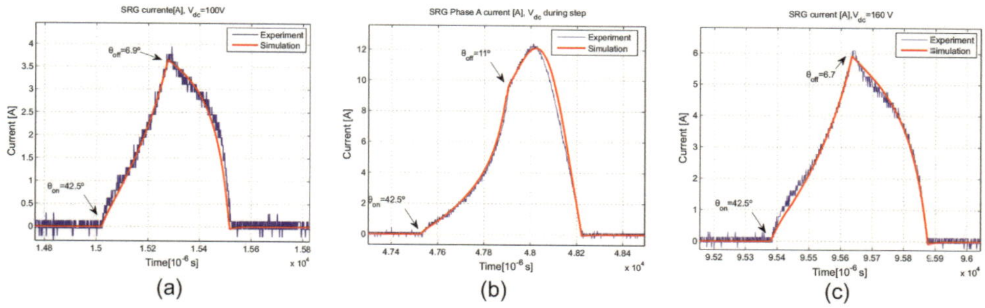

Figure 29.
Simulated and experimental currents: (a) $V_{DC} = 100$ V*; (b)* V_{DC} *step; (c)* $V_{DC} = 160$ V *[18].*

T_{srm}	Error	MAE	R^2	RMSE	SSE
100 V	0.21 A	6.80%	0.974	0.1446	8.94
160 V	0.017 A	4.99%	0.9888	0.1572	10.57
Step	0.28 A	5.73%	0.9877	0.4023	85.45

Table 2.
Statistic goodness-of-fit parameters.

$$e_j = \frac{|I_{exp} - I_{sim}|}{I_{exp}} 100\% \tag{21}$$

$$e_{avg} = \frac{1}{n} \sum_{j=1}^{n} e_j \tag{22}$$

$$MAE = mean(e_j) \tag{23}$$

Table 2 shows the results of the statistical analysis comparing the simulated and experimental electric current. Observe that the model obtained with the developed instrumentation and using the smoothing splines describes accurately the nonlinear magnetic characteristics of the SRM.

5. Conclusions

In this chapter, an automatic characterization system for SRM was introduced and developed. The system is able to obtain automatically and accurately the magnetization curves. With the obtained curves, it is possible to synthesize a nonlinear model consistent with the real SRM behavior. The model allows precise simulations, which support the development of many automatic control systems. The application of smoothing spline technique in data processing granted accurate models. Simulations and experimental implementation in motoring and generating modes were performed. For the motoring mode, a speed control with variable angles was developed, while for generating mode, voltage control was performed. The results endorse the similarity between the model and the real behavior of the SRM in both, motoring and generating modes. The accuracy of the model proves that the mirroring method used to complete the table is efficient for the creation of the tables. That is only possible because the interpolation method used, for instance smoothing splines, returns a smooth curve at the borders.

Acknowledgements

This chapter was a result of researches supported by the São Paulo Research Foundation, FAPESP, under Grants 2017/21640-9, 16/08645-9, and 15/03248-9.

A. Appendix

A.1 Procedure to obtain the magnetization curves from the blocked rotor test data

```
close all;
%% Blocked rotor test
% Positions used in the test
ang=[ 0 2.5 5 7.5 10 12.5 15 17.5 20 22.5] ;
% Experimentally acquired flux values for these positions
fluxExp=[ 0 0 0 0.000402 0.000632 0.001532 0.001532 0.001035 0 0.001035] ;
% Experimental current values for the position and their respective flux
iExp=[ 0 0 0 0.1635 0.1829 0.1829 0.1829 0.192 0 0.192] ;
ang = ang(:); % Transposition of angles vector
% Variables initialization
i=0;
d=0;
current=0;
for q = 1:1:10 %goes from 1 to 10 because there are 10 initial positions
  d=1;
i=0;
% This loop consider the maximum extrapolated current and count how
% many different i values may be used
while i<27.5
% since we have a 20 columns table and want to take only 10, we use the 2*q factor
i=Aazul1k525ok(d, (2*q));
      d=d+1;
end
i=0;
  y=1;
% Retrieve the current values
i=Aazul1k525ok(y, (2*q));
% Fills the i space for below 0.15 A currents
while i<0.15
i=Aazul1k525ok(y, (2*q));
      y=y+1;
end
   current(1,q) = 0;
   flux(1,q) = 0;
   figure(1);
   hold off;
   a=iExp(q);
   b=fluxExp(q);
% populate current and flux according to positions
for g = 1:(d-y)
      h = Aazul1k525ok(y+g, (2*q-1))-fluxExp(q);
if h > -1
            current(g+1,q)=Aazul1k525ok(y+g, (2*q))+iExp(q);
            flux(g+1,q)=Aazul1k525ok(y+g, (2*q-1))+fluxExp(q);
else
            current(g+1,q)=0;
            flux(g+1,q)=0;
end
end
```

```
if q <= 19
     hold all;
     plot(current(1:d-y,q),flux(1:d-y,q),'LineWidth',1.5);
end
end
% Figure configuration
xlabel('Current[ A] ','fontsize',14);
ylabel('Flux[ Wb.esp] ','fontsize',14);
box on
title('\phi(I,\theta) Curves');
hold all
text(3.3,0.11,'Aligned\rightarrow ','fontsize',12,'HorizontalAlignment','left')
text(15,0.02,'\uparrow Misaligned','fontsize',12,'HorizontalAlignment','left')
legend off
grid on;
axis([ 0 27 0 0.25] )
```

A.2 Data processing to obtain the lookup tables

```
%% Look Up Table creation Algorithm
% Author: Tárcio André dos Santos Barros
% Coauthor: Marcelo Vinícius de Paula
% University of Campinas - Campinas, SP - Brazil
%% Description:
% This algorithm receive the data from magnetization curves data and calculate the I (theta,flux), T(I,Theta) e I(T,Theta) lookup tables (LUTs)
clc;
% Load the experimentally obtained magnetization curves data
figure(1);
     h = findobj(gca,'Type','line');
iExpData=get(h,'Xdata');
fExpData=get(h,'Ydata');
% These functions carry the magnetization curves data according to position obtained through the blocked rotor test
%% Parameter initialization
Ncurves = 10;       % Number of experimental test curves
Npoints = 201;      % Desired number of data points in the LUTs
IMax = 27.5;        % Maximum current in the test
IMax2 = 27.5;       % Maximum desired current in the model
angMax = 45;        % Maximum angle: 45 for a 12/8 SRM and 60 for a 8/6 SRM
ang = 22.5;         % Maximum angle of SRM curves: 22.5 for 12/8 SRM and 27.5 for 8/6 SRM
fMax = 0.2225;      % Maximum flux obtained in the blocked rotor test
TMax = 15.3020;     % Maximum torque of TTBLD table
TMax2 = 20; % Maximum desired torque for the model
%% Initialize vectors that will contain the curves
flux_i_thetaCte = zeros(Ncurves,Npoints);
fluxCalc = zeros(101,Npoints);
% Vector with the positions used in the blocked rotor test
posExp =[ 22.5 20 17.5 15 12.5 10 7.5 5 2.5 0] ;
posExp = posExp(:);                       % Transposition of the vector
i_tab = 0:(IMax/(Npoints-1)):IMax;        % Create vector for experimental currents
i_tab = i_tab(:);                         % Transpose vector
i_tab2 = 0:(IMax2/(Npoints-1)):IMax2;     % Create vector for model currents
i_tab2 = i_tab2(:);                       % Transpose vector
flux_tab = 0:(fMax/(Npoints-1)):fMax;     % Create flux vector for model
flux_tab = flux_tab(:);                   % Transpose vector
ang_tab = 0:(angMax/200):angMax;          % Create position vector for the model (0-AngMax)
```

```
ang_tab = ang_tab(:);                        % Transpose vector
ang_tab2 = 0:(ang/100):ang;                  % Create position vector for the model (0-AngMax/2)
ang_tab2 = ang_tab2(:);                      % Transpose vector
T_tab = -TMax:2*TMax/(Npoints-1):TMax; % Create torque vector for the model (0-Tmax)
T_tab = T_tab(:); % Transpose vector
T_tab2 = -TMax2:2*TMax2/(Npoints-1):TMax2; % Create torque vector for the model (0-Tmax/2)
T_tab2 = T_tab2(:); % Transpose vector
%% Create figure with polynomial fit of experimental data
set(0,'DefaultFigureColor','w');
%% flux(theta,current) curves obtention
% Realization of fit for the beginning values
for curve = Ncurves:-1:4 % Polynomial fit of experimental data
Xaxle = iExpData{ curve,1} ;
Yaxle = fExpData{ curve,1} ;
Xaxle = Xaxle(:);
Yaxle = Yaxle(:);
% phi(i) for constant theta curves theta = (0:2.5:22.5)
    flux_i_thetaCte10 = fit(Xaxle,Yaxle,'poly6');
% Get the coefficients for each curve
coef_poly(curve,1:7) = coeffvalues(flux_i_thetaCte10);
for curvePoints = 0:(Npoints-1)
% Flux curves from Ncurves to 4
flux_i_thetaCte(curve,curvePoints+1) = flux_i_thetaCte10(curvePoints*IMax/(Npoints-1))-flux_i_thetaCte10(0);
end
end
for curve=3:-1:1 % Linear fit from experimental data
Xaxle = iExpData{ curve,1} ;
Yaxle = fExpData{ curve,1} ;
Xaxle = Xaxle(:);
Yaxle = Yaxle(:);
% Draw curve for specific position
    flux_i_thetaCte10 = fit(Xaxle,Yaxle,'poly6');
% Get the coefficients for each curve
coef_poly(curve,1:7) = coeffvalues(flux_i_thetaCte10);
for curvePoints = 0:(Npoints-1)
% Curves from 3 to 1
flux_i_thetaCte(curve,curvePoints+1) = flux_i_thetaCte10(curvePoints*IMax/(Npoints-1))-flux_i_thetaCte10(0);
end
end
% Diminishment of flux values according to position
flux_i_thetaCte(1,1:4) = flux_i_thetaCte(1,1:4)-0.00002;
flux_i_thetaCte(1,4:25) = flux_i_thetaCte(1,4:25)-0.0001;
% At this point we have 10 phi(i) curves, one for each position (0:2.5:22.5)
%% Table b calculation
% Calculate the phi(theta) functions for each current in the table and determine, through polynomial fit, the flux. Also obtain the flux for intermediate positions
% 201 flux values for each of the 10 positions
for Ncorrente = 1:201
% ang,flux
flux_theta_iCte = fit(posExp,flux_i_thetaCte(:,Ncorrente),'smoothingspline');
for curvePoints = 0:100
%flux(I,theta)
fluxCalc(curvePoints+1,Ncorrente) = flux_theta_iCte(curvePoints*ang/100);
end
end
%% Extra: L(theta) for i constant for each table position
ind=[ ];
for IndCurvePoints=0:200
```

```
ind(:,IndCurvePoints+1)=1000*((fluxCalc(:,IndCurvePoints+1))./(IndCurvePoints*IMax/200));
end
for IndCurvePoints = 101:201
ind(IndCurvePoints,:)=ind((202-IndCurvePoints),:);
end
ind(:,1) = ind(:,2);
indCalc = [];
for IndCurvePoints = 1:201
bufferInd = fit(ang_tab,ind(:,IndCurvePoints),'smoothingspline');
for curvePoints = 0:200
indCalc(curvePoints+1,IndCurvePoints) = bufferInd(curvePoints*angMax/200);
end
end
% At this point we have the phi(I,theta) curves up to Imax
%% Calculation of Table c
% Obtaining new phi(I,theta) curves and extends to maximum current
for Nfluxes=1:101
flux_I_posCte = fluxCalc(Nfluxes,:); % get flux for each position
flux_I_posCte = flux_I_posCte(:); %transpose vector
flux_theta_iCte = fit(i_tab,flux_I_posCte,'linear');% fit phi(I)
for curvePoints = 0:200
flux_theta_I(Nfluxes,curvePoints+1) = flux_theta_iCte(curvePoints*IMax2/200); %flux(theta,Current)
end
end
% At this point we have Table c - phi(I,Theta)
%% I(theta,phi) table creation
% For the computational model, we need the I(theta,phi) table. Calculate the I (theta,phi) and fit the I(phi,theta) and I(theta,phi) tables Calculates the I(phi) curves for all positions, going through half the total N to get to position 22.5 degrees
for Nang = 1:101
% phi(theta,I)
fluxVector = flux_theta_I(Nang,:); % Flux vector for fixed angle through the entire current variation
fluxVector = fluxVector(:);
% I(phi) curves for a given angle
flux_regression = fit(fluxVector,i_tab2,'linear');
for curvePoints=0:(Npoints-1)
IftLUT(Nang,curvePoints+1) = flux_regression(flux_tab(curvePoints+1));
end
end
%% Mirror the IftLUT
IftLUT(:,1)=0;
for curvePoints=102:201
% Adds the mirrored vectors to the table
IftLUT(curvePoints,:) = IftLUT((202-curvePoints),:);
end
%% TitLUT obtention
bufferTable_Phi_Theta_I = flux_theta_I;
bufferTable_Phi_Theta_I(:,1) = 0;
% Mirroring of buffer table for the entire driving period
for curvePoints=101:201
bufferTable_Phi_Theta_I(curvePoints,:)=bufferTable_Phi_Theta_I((202-curvePoints),:); %Mirroring
end
%% TitLUT calculation process
for Mpositions = 1:201
for Nfluxes = 2:201
      flux_vector2 = bufferTable_Phi_Theta_I(Mpositions,1:Nfluxes);
      flux_vector2 = flux_vector2(:);
      coenergy(Nfluxes) = trapz(i_tab2(1:Nfluxes),flux_vector2);
end
```

```
CoenergyLines(Mpositions,:) = coenergy;
end
%% TitLUT final definition
for Mpositions=1:201
%T=dW/dTheta
TTBLDcalc(:,Mpositions) = (180/pi)*diff(CoenergyLines(:,Mpositions))./diff
(ang_tab);
hold on;
end
for Mpositions=1:200
TitLUT(Mpositions+1,:)=TTBLDcalc(Mpositions,:);
end
% Adding table contour
TitLUT(201,:)=0;
TitLUT(1,:)=0;
```

A.3 Switched reluctance motor parameters

This appendix presents the parameters of the SRM used in this chapter:

- Nominal power: 1.5 kW
- Nominal speed: 1500 rpm
- Number of poles: 12/8 (stator/rotor)
- Operation voltage: 180–280 V (DC voltage)
- Winding resistance: 1.05 Ω
- Maximum flux linkage: 0.45 Wb.esp
- Inertia moment (obtained through deceleration test): 0.002 kg m^2

Author details

Marcelo Vinícius de Paula[1], Thiago de Almada Lopes[1], Tárcio André dos Santos Barros[1*], Paulo Sergio Nascimento Filho[2] and Ernesto Ruppert Filho[2]

1 School of Mechanical Engineering, University of Campinas, Campinas, Brazil

2 School of Electrical and Computer Engineering, University of Campinas, Campinas, Brazil

*Address all correspondence to: tarcioandre@hotmail.com

References

[1] Krishnan R. Switched Reluctance Motor Drives: Modeling, Simulation, Analysis, Design, and Applications. Boca Raton, FL, USA: CRC Press; 2001

[2] Miller TJE. Electronic Control of Switched Reluctance Machines (Newnes Power Engineering Series). New York, NY, USA: Elsevier; 2001

[3] Sezen S, Karakas E, Yilmaz K, Ayaz M. Finite element modeling and control of a high-power SRM for hybrid electric vehicle. Simulation Modelling Practice and Theory. 2016;**62**:49-67

[4] Kiyota K, Kakishima T, Chiba A. Comparison of test result and design stage prediction of switched reluctance motor competitive with 60-kW rare-earth PM motor. IEEE Transactions on Industrial Electronics. 2014;**61**(10): 5712-5721

[5] Cardenas R, Pena R, Perez M, Asher JCG, Wheeler P. Control of a switched reluctance generator for variable-speed wind energy applications. IEEE Transactions on Energy Conversion. 2005;**20**(4):691-703

[6] Chen H, Gu JJ. Implementation of the three-phase switched reluctance machine system for motors and generators. IEEE/ASME Transactions on Mechatronics. 2010;**15**(3):421-432

[7] Rafiq M, Rehman S-U, Rehman F-U, Butt QR, Awan I. A second order sliding mode control design of a switched reluctance motor using super twisting algorithm. Simulation Modelling Practice and Theory. 2012;**25**:106-117. Available from: http://www.sciencedirect.com/science/article/pii/S1569190X12000342

[8] Ding W, Liang D. A fast analytical model for an integrated switched reluctance starter/generator. IEEE Transactions on Energy Conversion. 2010;**25**(4):948-956

[9] Cai J, Deng Z, Hu R. Position signal faults diagnosis and control for switched reluctance motor. IEEE Transactions on Magnetics. 2014;**50**(9):8201811

[10] Corda J, Jamil S. Experimental determination of equivalent-circuit parameters of a tubular switched reluctance machine with solid-steel magnetic core. IEEE Transactions on Industrial Electronics. 2010;**57**(1): 304-310

[11] Parreira B, Rafael S, Pires AJ, Branco PJC. Obtaining the magnetic characteristics of an 8/6 switched reluctance machine: From FEM analysis to the experimental tests. IEEE Transactions on Industrial Electronics. 2005;**52**(6):1635-1643

[12] Nasirian V, Kaboli S, Davoudi A, Moayedi S. High-delity magnetic characterization and analytical model development for switched reluctance machines. IEEE Transactions on Magnetics. 2013;**49**(4):1505-1515

[13] Song S, Zhang M, Ge L. A new fast method for obtaining flux linkage characteristics of srm. IEEE Transactions on Industrial Electronics. 2015;**62**(7):4105-4117

[14] Shen L, Wu J, Yang S, Huang X. Fast flux linkage measurement for switched reluctance motors excluding rotor clamping devices and position sensors. IEEE Transactions on Instrumentation and Measurement. 2013;**62**(1):185-191

[15] Zhang P, Cassani P, Williamson S. An accurate inductance profile measurement technique for switched reluctance machines. IEEE Transactions on Industrial Electronics. 2010;**57**(9): 2972-2979

[16] Barros TA d S, Neto PJ d S, Filho PSN, Moreira AB, Filho ER. An

approach for switched reluctance generator in a wind generation system with a wide range of operation speed. IEEE Transactions on Power Electronics. 2017;**32**(11):8277-8292

[17] de Oliveira ESL. Projeto, Construção, Simulação, Implementação e Testes de um Gerador a Relutância Chaveada Monofásico [thesis]. São Carlos, São Paulo: University of São Paulo; 2011

[18] Dos Santos Barros TA, Dos Santos Neto PJ, De Paula MV, Moreira AB, Nascimento Filho PS, Ruppert Filho E. Automatic characterization system of switched reluctance machines and nonlinear modeling by interpolation using smoothing splines. IEEE Access. 2018;**6**:26011-26021

[19] Torrey D, Niu X-M, Unkauf E. Analytical modelling of variable reluctance machine magnetization characteristics. IEE Proceedings - Electric Power Applications. 1995; **142**(1):14-22

[20] Sun J, Zhan Q1.2pt, Guo Y, Zhu J. Back propagation neural network applied to modeling of switched reluctance motor. In: Proceedings of the 12th Biennial IEEE Conference on Electromagnetic Field Computation; 2006. p. 151

[21] Wang S-C. An fully-automated measurement system for identifying magnetization characteristics of switched reluctance motors. Measurement. 2012;**45**(5):1226-1238

[22] Xue XD, Cheng KWE, Ho SL. Simulation of switched reluctance motor drives using two-dimensional bicubic spline. IEEE Transactions on Energy Conversion. 2002;**17**(4):471-477

[23] Kahaner D, Moler C, Nash S. Numerical Methods and Software. Prentice-Hall, NJ, USA: Upper Saddle River Inc.; 1989

[24] Moreira JC. Torque ripple minimization in switched reluctance motors via bi-cubic spline interpolation. In: Proceedings of the 23rd Annual IEEE Power Electronics Specialists Conference (PESC); Vol. 2; 1992. pp. 851-856

[25] Mao S-H, Tsai M-C. An analysis of the optimum operating point for a switched reluctance motor. Journal of Magnetism and Magnetic Materials. 2004;**282**:53-56

[26] Xue XD, Cheng KWE, Ho SL, Kwok KF. Trigonometry based numerical method to compute nonlinear magnetic characteristics in switched reluctance motors. IEEE Transactions on Magnetics. 2007;**43**(4):1845-1848

[27] Silveira AWFV, Andrade D, Fleury A, Gomes L, Bissochi C, Dias R. Voltage control in starter/generator SRM based systems. In: Proceedings of the 2009 IEEE Energy Conversion Congress and Exposition (ECCE); 2009. pp. 2460-2465

Chapter 10

Some Basic and Key Issues of Switched-Reluctance Machine Systems

Chang-Ming Liaw, Min-Ze Lu, Ping-Hong Jhou and Kuan-Yu Chou

Abstract

Although switched-reluctance machine (SRM) possesses many structural advantages and application potential, it is rather difficult to successfully control with high performance being comparable to other machines. Many critical affairs must be properly treated to obtain the improved operating characteristics. This chapter presents the basic and key technologies of switched-reluctance machine in motor and generator operations. The contents in this chapter include: (1) structures and governing equations of SRM; (2) some commonly used SRM converters; (3) estimation of key parameters and performance evaluation of SRM drive; (4) commutation scheme, current control scheme, and speed control scheme of SRM drive; (5) some commonly used front-end converters and their operation controls for SRM drive; (6) reversible and regenerative braking operation controls for SRM drive; (7) some tuning issues for SRM drive; (8) operation control and some tuning issues of switched-reluctance generators; and (9) experimental application exploration for SRM systems—(a) wind generator and microgrid and (b) EV SRM drive.

Keywords: switched-reluctance machine, motor drive, generator system, modeling, current control, speed control, commutation shift, voltage boosting

1. Introduction

Although switched-reluctance machine (SRM) [1, 2] is not the mainstream among the existing machines, it still possesses high potentials both in motor and generator applications [3–6]. Basically, it has the following key features: (i) doubly salient and singly excited with concentrated windings and (ii) without permanent magnets and conductors on its rotor. Hence, it has rigid structure and is suited for high-speed driving applications. In driving control, it possesses highly developed torque and acceleration capabilities. The employed converter is simple topology and has fault tolerance. However, the doubly salient structure of SRM makes it have many inherent drawbacks, such as higher torque ripple, vibration, and acoustic noise. In addition, the nonlinear winding inductance and non-ideal winding current waveform render its dynamic modeling and high-performance control more difficult to achieve. Thus, many key affairs must be treated. The typical ones include (i) motor design [7], (ii) power circuit design and switching control, (iii) proper rotor position sensing and commutation setting, (iv) dynamic modeling, (v) current control [8, 9], (vi) speed control [10], (vii) commutation shift [11–13], (viii) voltage boosting [14–17], (ix) field weakening via commutation shifting [11–13], and (x) torque ripple and vibration reductions. Under high speed and/or heavy load, commutation shift with equivalent field-weakening effects is the effective way to improve the SRM winding current waveforms and thus the torque generating capability. As the speed is further increased, the voltage boosting approach must be applied for enhancing the current tracking response. Some possible solutions will be discussed in this chapter.

SRM can be operated as a generator by properly allocating winding current at negative winding inductance slope region [18–20]. Owing to the rigid mechanical structure and ease of starting for cogging torque-free nature, SRM is suited to be a wind generator. However, the generation behaviors of SRG are also highly influenced by many critical features, especially the effects of back electromotive force (EMF), which is negative in generator mode to assist the current during demagnetizing period. The key issues affecting the performance improvement of SRG include (i) equipment of excitation source, (ii) commutation setting and shifting [19], (iii) current switching control, (iv) voltage controls [19], and (v) power maximization control [18], etc. To effectively counteract the negative effects of back-EMF, the hysteresis current control PWM with hard freewheeling is normally adopted. From the exploration one can be aware that commutation instant tuning is the most effective way in improving the developed power and ripple characteristics of a SRG.

A suited converter for switched-reluctance machine generating quasi-square winding current waveform is needed. The surveys for SRM converters can be found in [21–24]. The asymmetric bridge converter with 2N (N = phase number) switches possesses the most flexible winding current PWM switching control capability. Therefore, it is normally adopted, especially the SRM drive with regenerative braking capability and the SRG.

2. Basics of switched-reluctance machines

2.1 Machine structures

The major features of an SRM are shown in **Figure 1**. Similar to variable-reluctance stepping motor, SRM possesses doubly salient and singly excited structure. It has toothed stator and rotor. The rotor is not equipped with neither windings nor permanent magnets. Due to its rugged structure, highly developed torque and acceleration capabilities, low cost, etc., it possesses high application potential, especially driving at high speed in harsh environments.

For speed drive applications, the stroke angle of an SRM is generally larger than a stepping motor. Generally, the four-phase 8/6, three-phase 6/4, and three-phase 12/8 SRMs are the most popularly adopted. The stroke angle θ_s and stroke rate R_s of SRM can be derived to be.

$$\theta_s = \left| \frac{360°}{n_s} - \frac{360°}{n_r} \right| = |P_s - P_r|, R_s = \frac{360°}{\theta_s} \text{ (strokes/rev)} \tag{1}$$

where n_s is the stator tooth number, n_r is the rotor tooth number, P_s is the stator tooth pitch, and P_r is the rotor tooth pitch.

The stroke frequency f_{st} of an SRM with θ_s at the speed n (rpm) can be derived as

$$f_{st} = \frac{nR}{60} = \frac{6n}{\theta_s} \tag{2}$$

2.2 Governing equations

Figure 2 shows the configuration and phase equivalent circuit of a four-phase 8/6 SRM. A simplified model of the SRM is obtained on the basis of the following assumptions: (a) no mutual flux linkage between phase windings and (b) the magnetic circuit is linear.

2.2.1 Voltage equation

The per-phase voltage equation of an SRM can be expressed as.

$$\begin{aligned} v &= iR + \frac{d\lambda(i,\theta_r)}{dt} = iR + \frac{dL(i,\theta_r)}{dt} = iR + L(i,\theta_r)\frac{di}{dt} + i\omega_r\frac{\partial L(i,\theta_r)}{\partial\theta_r} \\ &= iR + L(i,\theta_r)\frac{di}{dt} + e(i,\omega_r,\theta_r) \end{aligned} \tag{3}$$

where v is the winding terminal voltage, i is the winding current, R is the winding resistance, $L(i,\theta_r) \equiv L(i,\theta_r) \triangleq \partial\lambda(i,\theta_r)/\partial i$ is the incremental winding inductance, θ_r is the rotor angle, ω_r is the rotor speed, and $e(i,\omega_r,\theta_r) \triangleq [\partial L(i,\theta_r)/\partial\theta_r]i\omega_r$ is the back electromagnetic force (EMF).

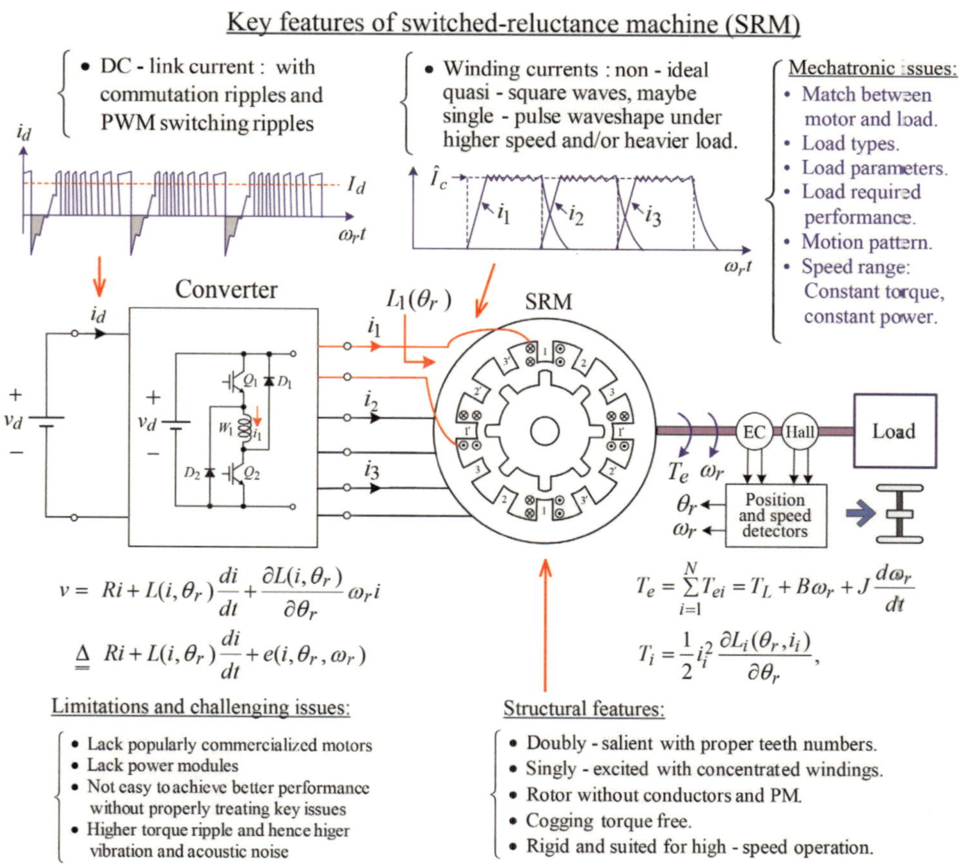

Figure 1.
Key features of switched-reluctance machines.

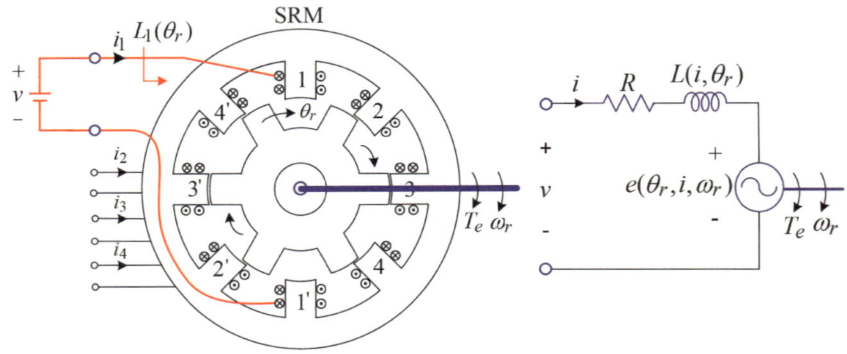

Figure 2.
Configuration of a four-phase SRM and its phase equivalent circuit.

2.2.2 Torque equation

The developed torque T_{ei} of an SRM can be derived using energy or coenergy as

$$T_{ei} = \left.\frac{\partial W_c(i, \theta_r)}{\partial \theta_r}\right|_{i=\text{constant}} = \frac{1}{2} i_i^2 \frac{\partial L(i, \theta_r)}{\partial \theta_r} \triangleq k_t i_i^2 \tag{4}$$

where $k_t \triangleq 0.5\partial L(i, \theta_r)/\partial \theta_r$. Accordingly, the composite electromagnetic developed torque and the mechanical equation of a N-phase SRM drive can be written as

$$T_e = \sum_{i=1}^{N} T_{ei} = \frac{1}{2}\sum_{i=1}^{N} i_i^2 \frac{\partial L(i, \theta_r)}{\partial \theta_r} = T_L + B\omega_r + J\frac{d\omega_r}{dt} \tag{5}$$

where T_e is the total electromagnetic torque, N is the phase number, T_L is the load torque, J is the total moment of inertia, and B is the total damping ratio.

2.2.2.1 Switched-reluctance generator

In contrast to SRM drive, SRG converts mechanical input power P_i into electrical output power P_g as

$$T_e = T_i - B\omega_r - J\frac{d\omega_r}{dt}, \quad P_e = T_e\omega_r, \quad P_g = P_e - P_c - P_{cu} \tag{6}$$

where T_i is the input shaft torque, P_e is the electromagnetic developed power, P_{cu} is the SRM copper loss, and P_c is the SRM core loss.

Comments: From the above governing voltage and torque equations, one can be aware of the following facts: (i) the SRM back-EMF is dependent on winding inductance slope, current, and rotor speed; (ii) the current response and the developed torque performance are affected by the back-EMF, especially under higher speed and/or heavier load; (iii) tuning the commutation shift angle properly can improve the winding current tracking performance. As the speed is further increased, the DC-link voltage boosting approach must be applied instead; (iv) T_{ei} is proportional to the square of current. Therefore, it possesses highly developed torque like a series DC motor; (v) for operating as SRG, the generating characteristics are more sensitive to winding current waveform and commutation instant setting for the negative back-EMF; and (vi) the observed torque can be obtained from Eq. (5) for achieving direct torque control.

2.3 Motor and generator operation modes

The phase winding inductance profile versus rotor position and current waveforms under motor mode and generator mode are sketched in **Figure 3**. According to Eq. (4), arranging the excitation under the region with positive or negative slope of winding inductance, an SRM can be operated as a motor or a generator.

2.4 Dynamic models

The dynamic modeling and dynamic control affairs can be referred to [2, 10, 25]. The standard SRM drive control belongs to cascade structure consisting inner current-loop and outer speed-loop. **Figure 4(a)** shows the hypothesized control block of an SRM drive under PWM control, wherein the related variables are defined as phase current command $i_i^* \triangleq I_c S_i$, $I_c \triangleq$ current command magnitude, and $S_i \equiv$ i-th phase switching function and $i_i \triangleq H_i(\bullet) i_i^*$ is assumed with $H_i(\bullet)$ being the i-th phase closed-loop current tracking transfer function.

Figure 4(b) shows the detailed phase winding control block, wherein denotes the current sensing factor. Obviously, the winding current tracking control is significantly affected by the back-EMF and nonlinear winding inductance. **Figure 4(c)** illustrates the speed-loop control block, wherein $\bar{k}_t$ denotes the average torque generating constant. By assuming ideal current control with $i_i \approx I_c$, the dynamic torque generating constant is found from Eq. (5) by linearization process at an operation point ($i_i \approx I_{c0}$) as

$$T_{e0} = \frac{1}{2}\frac{\partial L(I_{c0}, \theta_r)}{\partial \theta_r} I_{c0}^2 \tag{7}$$

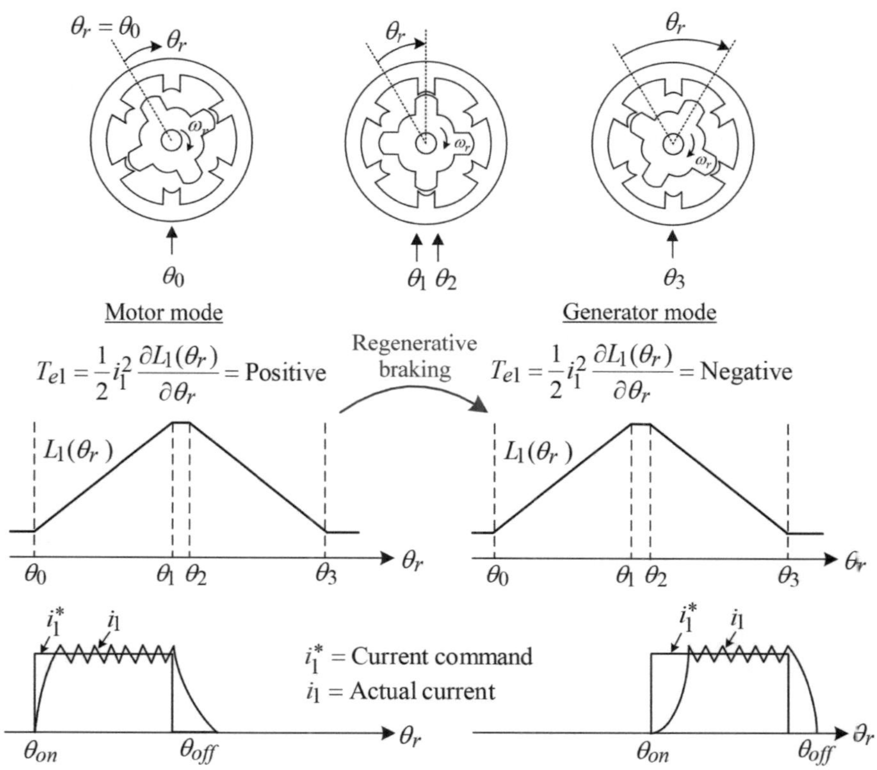

Figure 3.
Idealized phase winding inductance $L_1(\theta_r)$ and currents of an SRM in motor mode and generator mode.

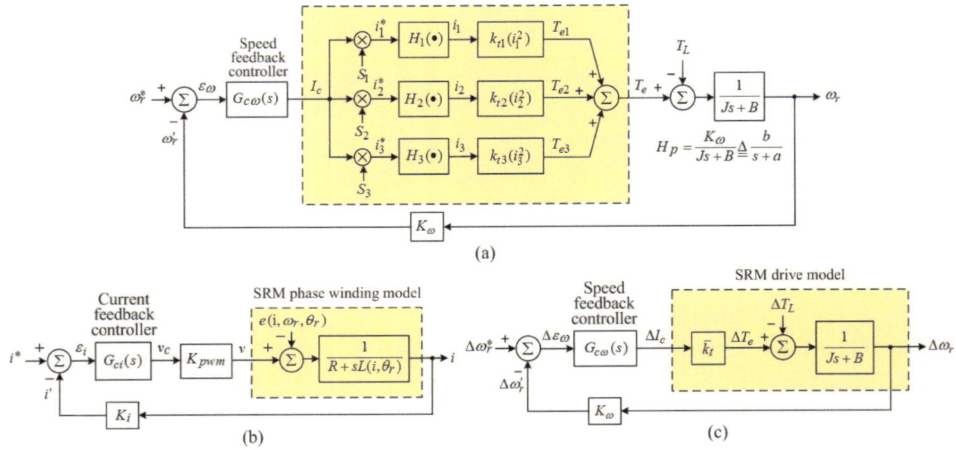

Figure 4.
Control blocks of an SRM drive with current-controlled PWM scheme: (a) hypothesized control block diagram; (b) per-phase inner current control loop; (c) outer speed control loop.

$$T_e = \frac{1}{2}\frac{\partial L(I_c, \theta_r)}{\partial \theta_r} I_c^2 = \frac{1}{2}\frac{\partial L((I_{c0} + \Delta I_c), \theta_r)}{\partial \theta_r}(I_{c0} + \Delta I_c)^2 \tag{8}$$

$$\Delta T_e = T_e - T_{e0} \approx \frac{\partial L(I_{c0}, \theta_r)}{\partial \theta_r} I_{c0} \Delta I_c \triangleq \overline{k}_t \Delta I_c \tag{9}$$

with the dynamic torque generating constant being

$$\overline{k}_t \triangleq \frac{\partial L(I_{c0}, \theta_r)}{\partial \theta_r} I_{c0} \tag{10}$$

Obviously, the torque generating constant is not a constant, but rather it is varied with the changing operating conditions. The robust controls are needed to yield better performance.

And the dynamic model depicted in **Figure 4(c)** can be represented from Eq. (5) as

$$H_p(s) = \frac{K_\omega}{Js + B} \triangleq \frac{b}{s + a}, \quad a \triangleq \frac{B}{J}, \quad b \triangleq \frac{K_\omega}{J} \tag{11}$$

where K_ω denotes the speed sensing factor.

2.5 SRM converters

There already have many existing converters for SRM drive [22–24]. Among these ones, the asymmetric bridge converter shown in **Figure 5(a)** has the most flexible PWM switching capability, and it is also the most generally adopted one for SRM drive and SRG. Two diodes and two switches are required in one phase. For each phase, the lower switch conducts commutation switching, while the upper switch is in charge of PWM switching.

Figure 5(b) sketches the typical winding current waveforms of SRM and SRG. **Figure 5(c)–(e)**, respectively, shows the schematics and current paths of three operation modes: (i) mode-1, excitation mode; (ii) mode-2, soft freewheeling mode;

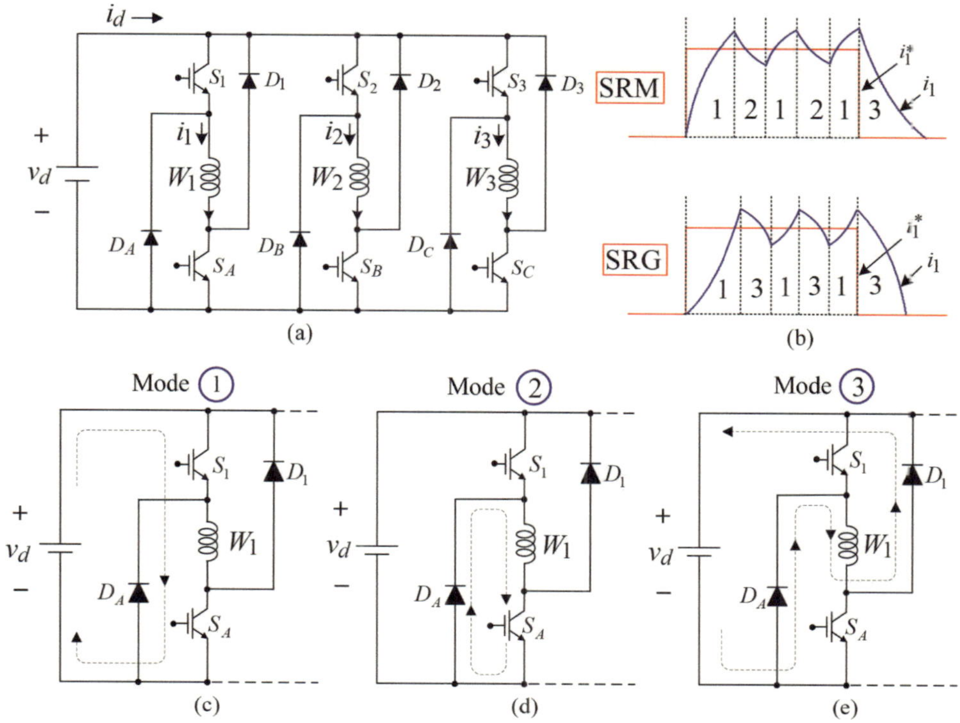

Figure 5.
Asymmetric bridge converter: (a) schematic; (b) SRM and SRG typical winding current waveforms; (c) excitation mode; (d) soft freewheeling mode; (e) demagnetization mode (or hard freewheeling mode).

and (iii) mode-3, demagnetization mode or hard freewheeling mode. For a current-controlled PWM (CCPWM) scheme, hard freewheeling operation will cause the faster demagnetization owing to the negative demagnetizing voltage rather than zero voltage being applied. For a SRG, the hysteresis CCPWM scheme with hard freewheeling (mode-3) is normally applied to counteract the effects of back-EMF on the winding current response.

2.6 Equivalent circuit parameter estimation and performance test of SRM drive

Figure 6 shows the suggested test facilities for establishing an SRM drive: (a) stationery test equipment for measuring the key motor parameters and variables, (b) the SRM drive running characteristics test environment using eddy current brake, and (c) the alternative SRM drive running characteristics test using load SPMSG as dynamic load. Since the accurate eddy current brake and torque meter are not available, this alternative of loading test is worth of adopting. However, the surface-mounted permanent-magnet synchronous generator (SPMSG) must be properly set, and it should be noted that the motor efficiency is both speed and load dependent.

2.6.1 DC test

Powering the stator winding with different values of the DC voltage leads to various corresponding values of the current. By calculating the sets of readings and averaging their results, a value of stator winding resistance R is reasonably obtained.

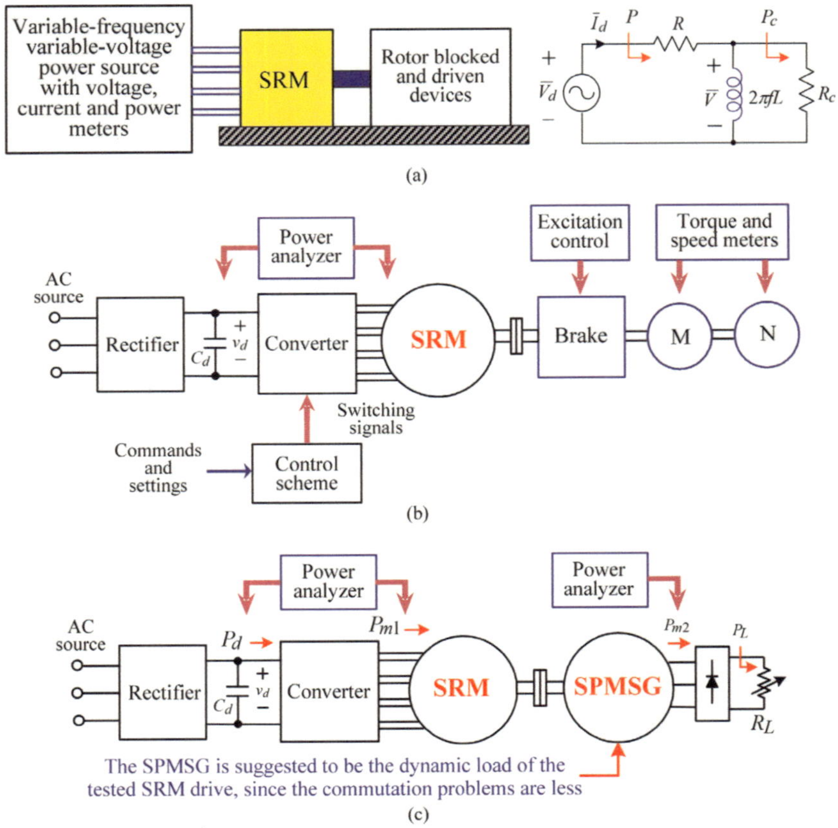

Figure 6.
Test facilities for SRM drive: (a) stationery test equipment and the equivalent circuit of SRM phase winding under blocked rotor; (b) SRM drive running characteristics test using eddy current brake; (c) SRM drive running characteristics test using load SPMSG as dynamic load.

2.6.2 Blocked rotor test

2.6.2.1 AC test

With the rotor being blocked at a specific position, the back-EMF is zero from Eq. (3). Hence the equivalent circuit of SRM phase winding under blocked rotor can be expressed in **Figure 6(a)**. The phase winding is excited with a variable-voltage variable-frequency AC source. At each frequency with fixed current level, the input power P, frequency f, and RMS values of the input voltage V_d and current I_d are recorded. Similarly, readings are taken at different constant currents. The same procedure is repeated with an increment of rotor position for the next rotor position till one rotor pole pitch is covered, namely, from the aligned to unaligned position of the rotor with respect to the excited stator. From the above sets of readings, the phase winding inductance at a specified excitation current, frequency, and rotor position is obtained following the well-known AC estimation approach.

With the measured P, f, V_d, and I_d, the phase winding inductance can be obtained from **Figure 6(a)** as

$$L = \left[2\pi f\sqrt{(I_d/V)^2 - (1/R_c)^2}\right]^{-1} \tag{12}$$

where

$$\overline{V} = \overline{V}_d - \overline{I}_d R \tag{13}$$

$$P_c = P - I_d^2 R, R_c = V^2 / P_c \tag{14}$$

with $\overline{V}_d \equiv V_d \angle 0°$, $\overline{I}_d \equiv I_d \angle \phi$, $\overline{V} \equiv V \angle \delta$, P_c, and R_c being the core loss and its equivalent resistance.

2.6.2.2 LCR meter test

For simplicity, the above measurement procedure is conducted using LCR meter at various rotor positions under different frequencies. However, the current-dependent characteristics of winding inductance cannot be obtained, since the LCR meter is small-signal excitation.

Measured example: The measured three-phase winding inductance profiles of a three-phase SRM (12/8, 380 V, 2000 rpm, 2.2 kW) using the LCR meter (HIOKI 3532-50 LCR HiTESTER) at 42 Hz and 267 Hz corresponding the speed of 315 rpm and 2000 rpm are shown in **Figure 7**. From **Figure 7**, one can observe that the three winding inductances are slightly unsymmetrical and have frequency-dependent characteristic. The winding DC resistance is measured using DC excitation method as $R = 1.43\Omega$.

2.6.3 No-load test

2.6.3.1 AC test

The SRM is run forcibly at a constant speed. By exciting the SRM phase winding with AC voltage (or current), one can measure the Hall signal and rotor position modulated voltage (or current) to observe the adequacy of Hall sensor installation.

2.6.3.2 Constant current injection test

From Eq. (3) one can find that the back-EMF of a SRM is zero at no load. The constant current injection method is proposed to measure its back-EMF. The constant current source is produced by voltage-to-current converter using the power operational amplifier OPA 548 shown in **Figure 8**. The load PMSG coupled to the test SRM (4-phase, 8/6, 48 V/2.3 kW, 6000 rpm, DENSEI Company, Japan) is powered by the commercial inverter and turned at a constant speed. The measured

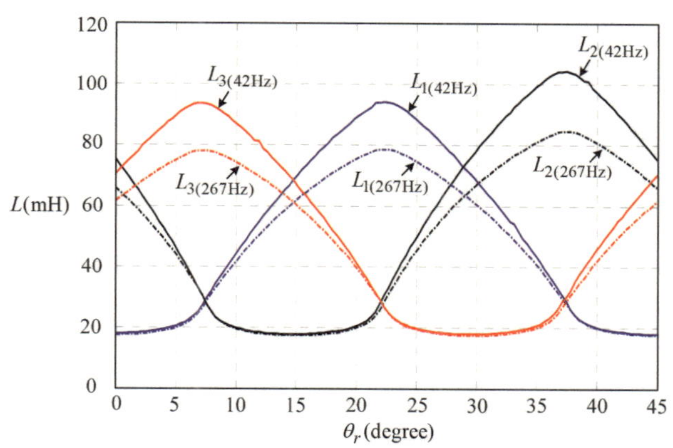

Figure 7.
Measured winding inductance profiles of an example three-phase SRM.

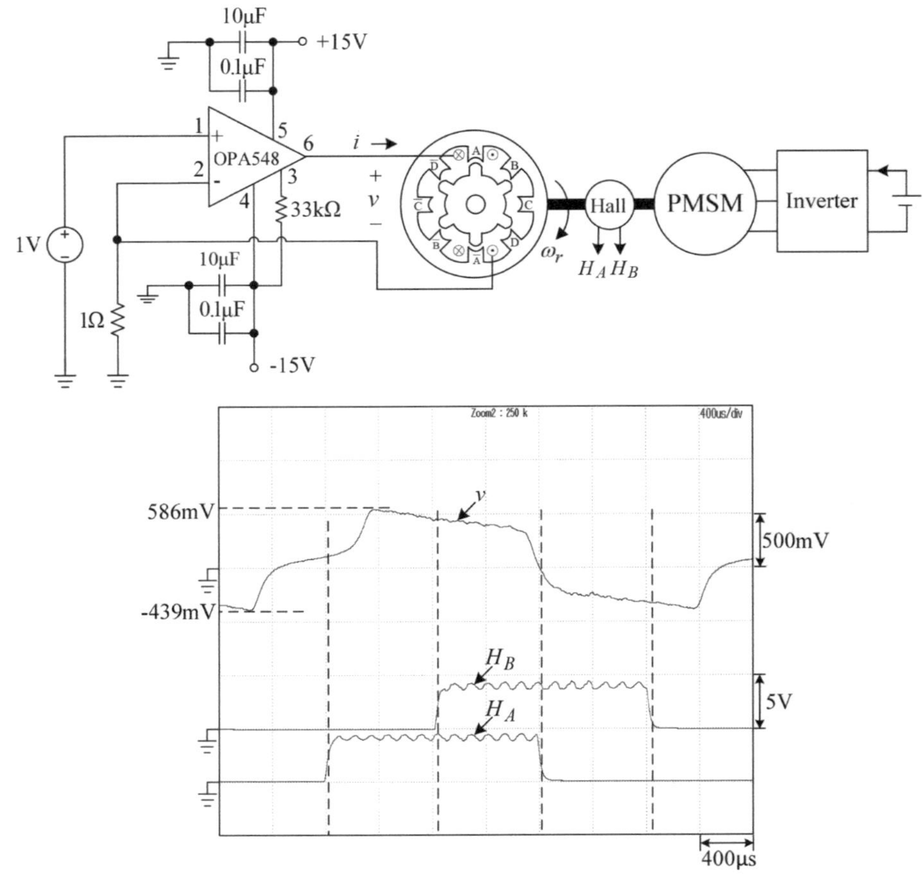

Figure 8.
Measured winding terminal voltage v and Hall signals of an example SRM by constant current injection at 0.979A and 3000 rpm.

winding terminal voltage v and quadrature Hall signals by the constant current injection at $i = 0.979$A and $\omega_r = 3000$rpm are also plotted in **Figure 8**. The measured voltage v is approximately equal to the back-EMF for the negligibly small winding resistance drop. From the measured results, one can be aware of the correctness of Hall sensor installation of this example SRM.

Comments: If the driven device for the test SRM is not available, one can turn a rope wound on the motor shaft.

3. Possible front-end converters

3.1 DC/DC front-end converters

The typical system configuration of a battery-powered SRM drive with DC/DC front-end converter is shown in **Figure 9**. The equipped DC/DC converter may possess some merits: (i) the selection of battery voltage is more flexible; (ii) the boostable and well-regulated DC-link voltage can enhance the motor driving performance, especially in high speed and/or heavy load. If the regenerative braking with energy recovery battery charging is desired, the bidirectional DC/DC converter must be employed. **Figure 10(a)** and **(b)** shows two bidirectional front-end DC/DC converters. For the latter, the DC-link voltage can be made smaller or

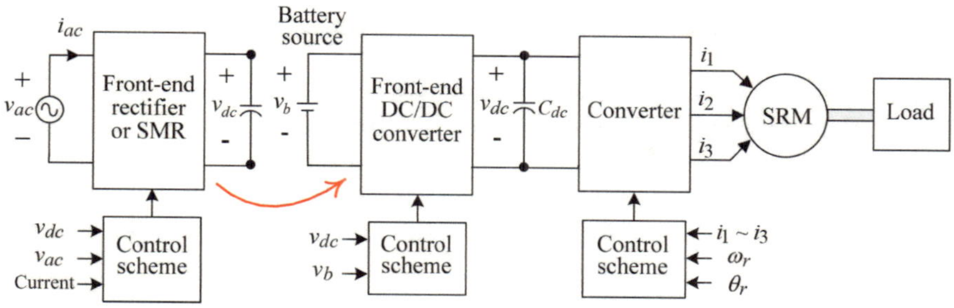

Figure 9.
A SRM drive with front-end DC/DC converter or SMR.

higher than the battery voltage. The motor drive may possess better performance over wider speed range.

3.2 AC/DC front-end converters

For AC power (utility power)-fed SRM drive as indicated in **Figure 9**, one can replace the DC/DC converter with a suited type of boost switch-mode rectifier (SMR). The SMRs can provide boosted and well-regulated DC-link voltage to enhance the motor drive driving characteristics with good line drawn power quality. **Figure 11(a)** and **(b)** shows a boost unidirectional SMR and a boost/buck bidirectional SMR. The latter possesses higher flexibility in voltage transfer ratios.

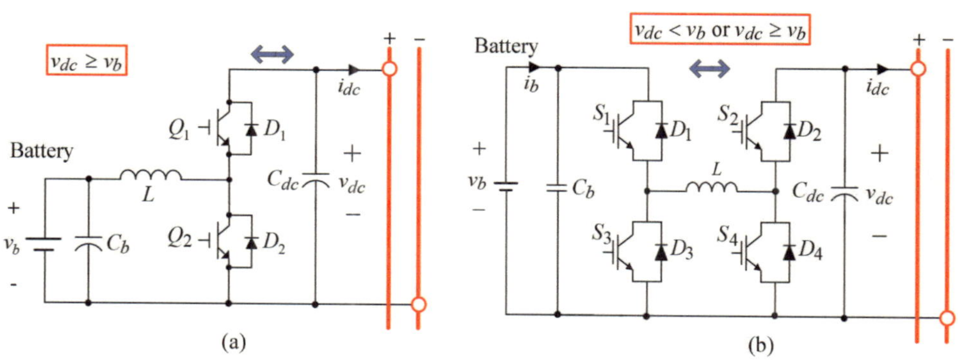

Figure 10.
Two typical bidirectional interface DC/DC converters: (a) one-leg boost/buck converter; (b) H-bridge four-quadrant converter.

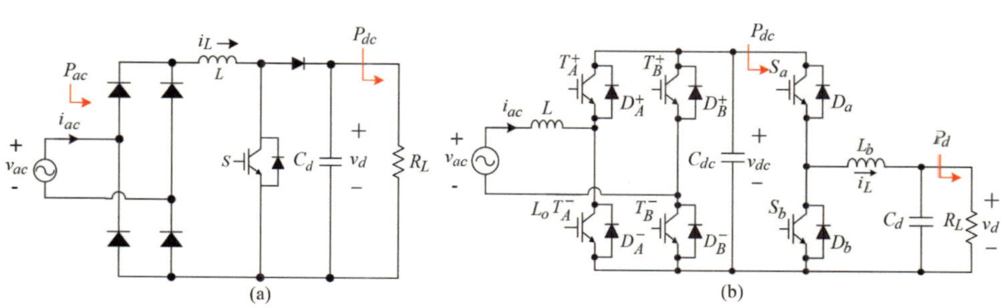

Figure 11.
Two boost type SMRs: (a) unidirectional SMR; (b) bidirectional SMR.

4. Some key issues of SRM and SRG

Some key issues affecting the operating characteristics of an SRM drive and an SRG system are depicted in **Figures 12** and **13**. From Eqs. (3) and (5), one can be aware that commutation angle setting and shifting are the critical factors affecting the SRM and SRG winding current tracking characteristics. Under higher speed and/or heavier load, the DC-link voltage boosting must be adopted. This task can be

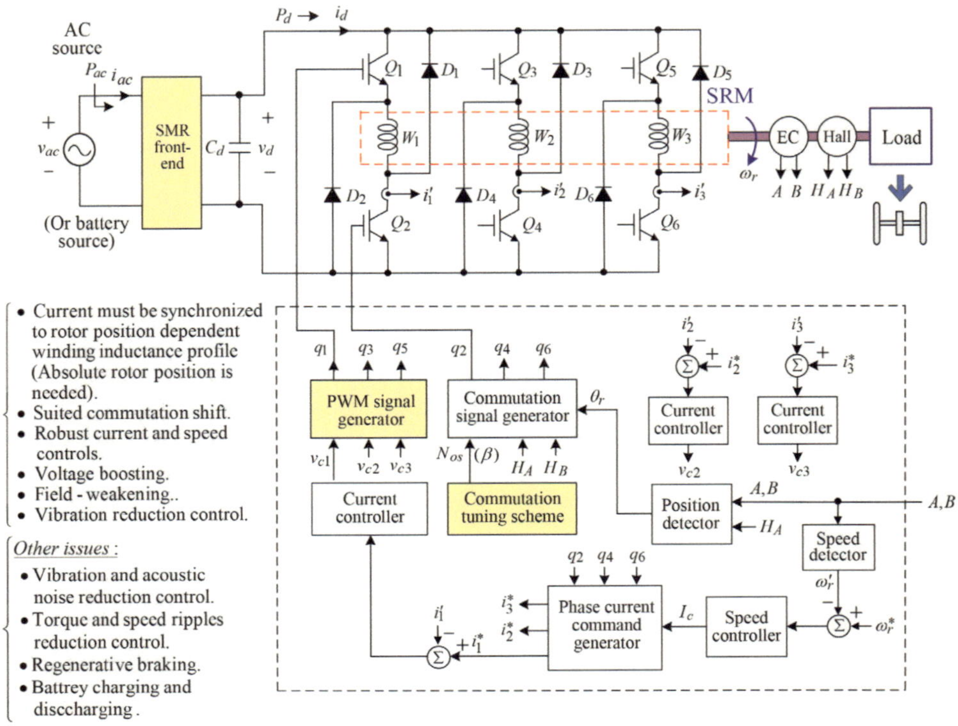

Figure 12.
Key issues affecting the performance of an SRM drive.

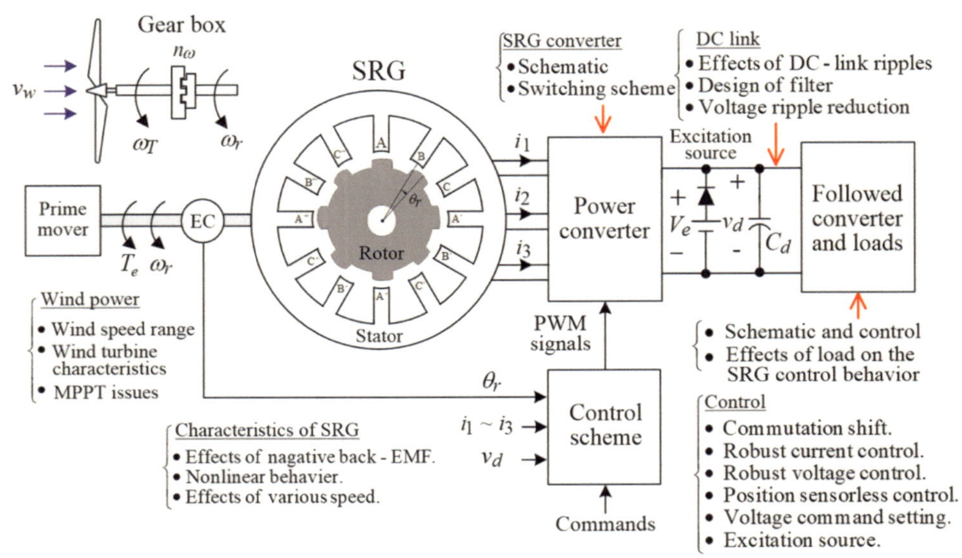

Figure 13.
Key issues affecting the performance of an SRG system.

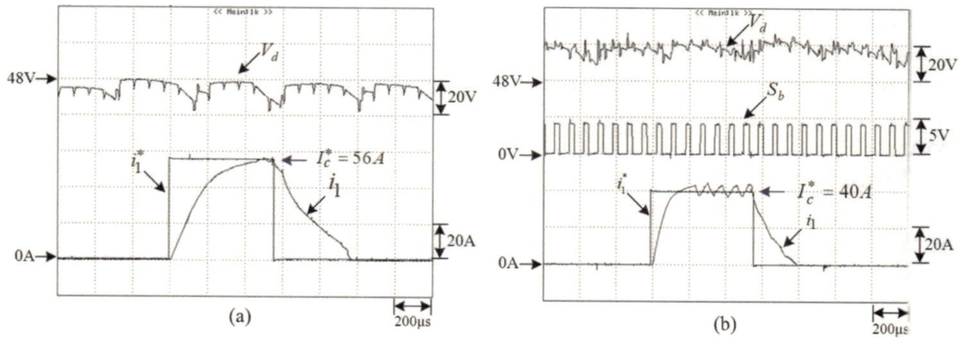

Figure 14.
Measured DC-link voltage V_d, boost switching signal S_b, phase-1 winding current i_1, and its command i_1^* at ($\omega_r = 6000$ rpm, $R_L = 6.3\ \Omega$): (a) $V_b = 45$ V, $V_d = 45$ V (fixed); (b) $V_b = 40$ V, $V_d = 64$ V.

fulfilled by adding a proper front-end DC/DC converter for battery-powered drive or SMR for utility-powered cases.

To comprehend the effectiveness of voltage boosting approach, a battery (48 V)-powered SRM drive [26] is equipped with a one-leg boost/buck interface converter as shown in **Figure 10(a)**. **Figure 14(a)** and **(b)**, respectively, shows the measured DC-link voltage V_d, boost switching signal, phase-1 winding current i_1, and its command i_1^* at ($\omega_r = 6000$ rpm, $R_L = 6.3\Omega$, R_L denotes the load resistance of the lad PMSG) under ($V_b = 45$ V, $V_d = 45$ V) and ($V_b = 40$ V, $V_d = 64$ V). Significant improved winding current response by boosting the DC-link voltage can be observed from the results.

5. Example SRG system: a grid-connected SRG-based microgrid

5.1 System configuration

Figure 15 shows the circuit of the grid-connected SRG-based microgrid with bidirectional 1P3W isolated inverter [27]. The wind SRG is followed by an interleaved DC/DC boost converter to establish the 400 V microgrid common DC bus.

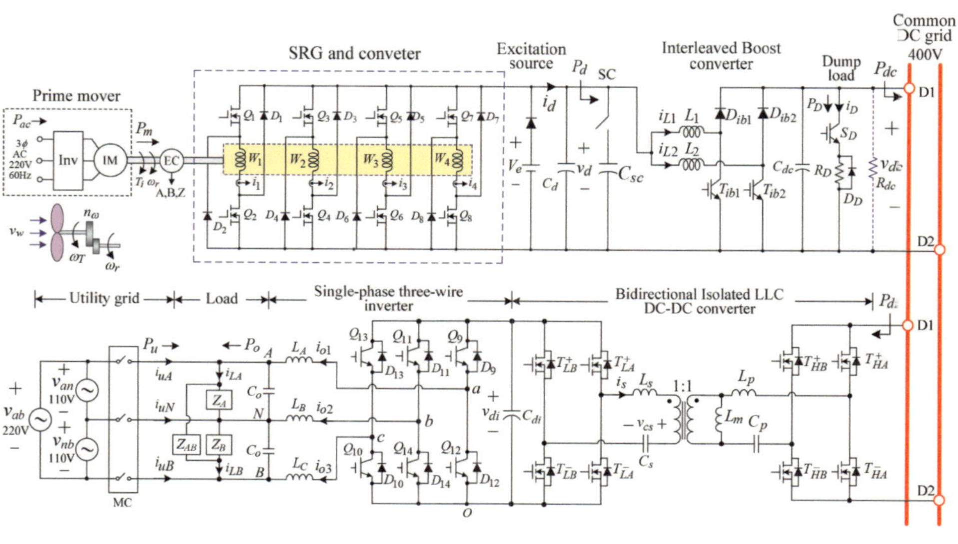

Figure 15.
Schematics of the grid-connected SRG-based microgrid with bidirectional 1P3W isolated inverter.

The bidirectional isolated 1P3W load inverter with 60 Hz 110 V/220 V AC output voltages is served as a test load. A supercapacitor (SC) placed at the SRG output provides improved voltage regulation caused by fluctuated wind speed. The other energy storage devices and other constituted parts are neglected due to the limit of scope.

The major features of the established SRG system are summarized as (1) SRG: four-phase, 8/6, 48 V, 6000 rpm, 2.32 kW (DENSEI company, Japan); (2) asymmetric bridge converter: constructed using MOSFET IRLB4030PBF (100 V/180A (continuous), 730A (peak)); (3) SC: 48 V/66F; and (4) excitation source: $V_e = 12$ V.

The control scheme of wind SRG is shown in **Figure 16**. It consists of an outer-loop voltage controller and an inner-loop HCCPWM scheme. In addition, the dynamic shift controller (DSC) is designed to automatically make the commutation shift according to the average voltage tracking error $\overline{\varepsilon}_v$. On the other hand, since the back-EMF of SRG is proportional to the rotor speed and winding current, the voltage command v_d^* is determined according to rotor speed ω_r and winding current command.

5.2 Some experimental results

5.2.1 SRG-based microgrid

The established wind SRG-powered DC microgrid is evaluated first. **Figure 17 (a)** shows the measured microgrid DC-link voltage v_d, current command i_1^*, and phase-1 winding current i_1 of the SRG under ($\omega_r = 6000$ rpm, $R_{dc} = 200\ \Omega$). And **Figure 17(b)** plots the measured output voltage v_{dc}, DC-link current i_d, and inductor currents (i_{L1}, i_{L2}) of the interleaved converter. As the results, the interleaved DC/DC boost interface converter can establish the microgrid common DC bus voltage (400 V) from the SRG output (48 V) successfully.

To evaluate the performance of the microgrid under changed SRG driving speed, at $v_d^* = 48$ V and $v_{dc}^* = 400$ V, the measured ω_r', v_d, and v_{dc} under varying driven speed $\omega_r = 6000 \rightarrow 5000 \rightarrow 6000$ rpm at $R_{dc} = 320\ \Omega$ are shown in **Figure 18**. From the results, one can deduce that the developed DC microgrid owns well-regulated common DC bus voltage under varied SRG-driven speed.

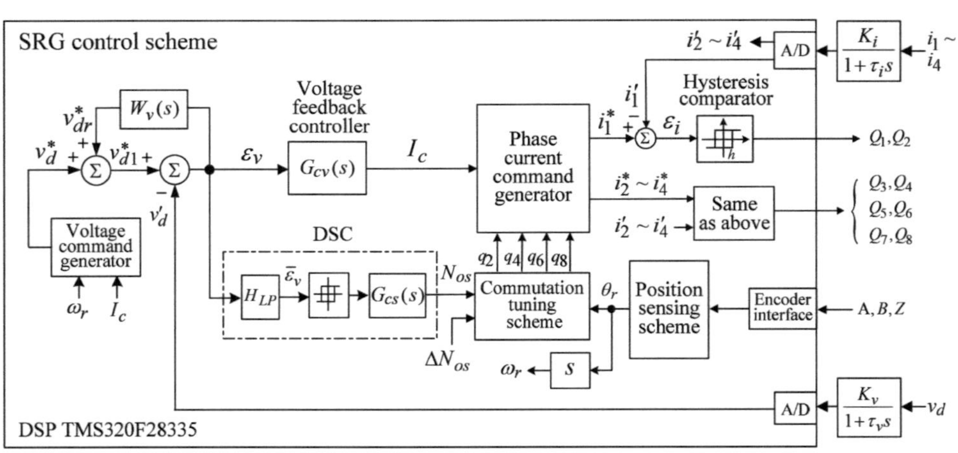

Figure 16.
Control scheme of the wind SRG.

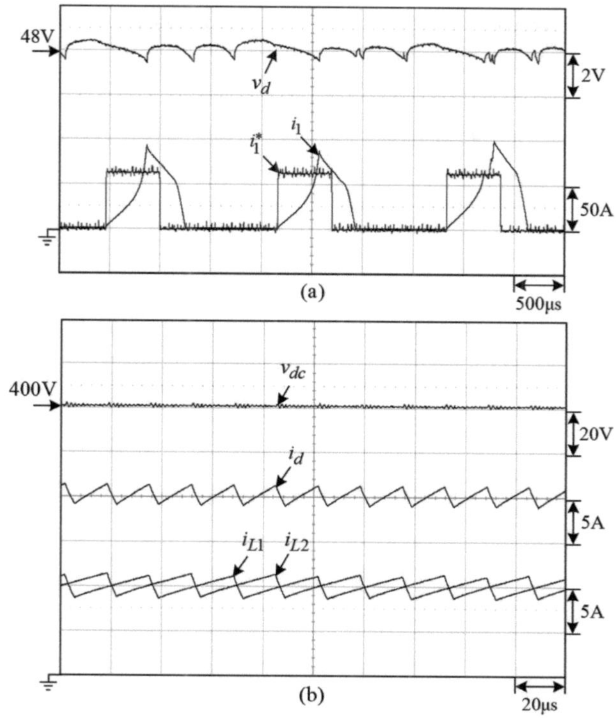

Figure 17.
Measured results of the developed SRG-based DC microgrid at ($\omega_r = 6000$ rpm, $R_{dc} = 200\ \Omega$): (a) v_d, i_1^* and i_1 of the SRG stage; (b) v_{dc}, i_d, i_{L1}, and i_{L2} of the interleaved boost interface converter.

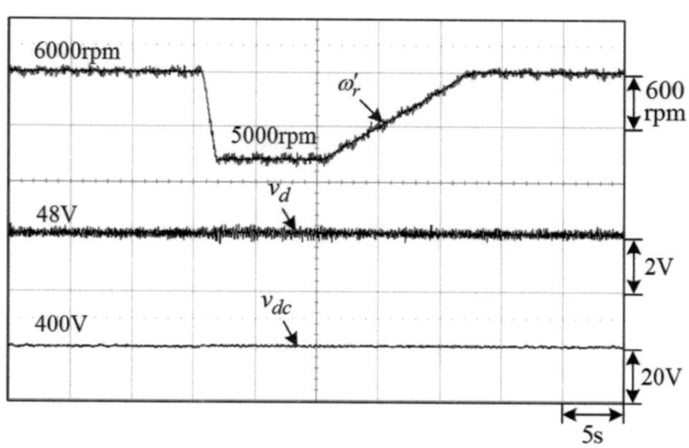

Figure 18.
Measured ω_r', v_d, and v_{dc} of the established SRG-based DC microgrid under varying driven speed $\omega_r = 6000 \rightarrow 5000 \rightarrow 6000$ rpm at $R_{dc} = 320\ \Omega$.

5.2.2 Wind SRG-powered DC microgrid with bidirectional 1P3W inverter

5.2.2.1 Microgrid-to-home (M2H) mode

The 1P3W inverter is operated under autonomous M2H mode, and SRG is driven at 6000 rpm. The measured DC-link voltage v_d and the phase-1 winding current i_1 using PR controller are shown in **Figure 19(a)**. And **Figure 19(b)** shows the DC-link current i_d, the inductor currents (i_{L1}, i_{L2}), and the common DC bus voltage v_{dc} of the interleaved boost converter. Well-regulated common DC bus voltage v_{dc} (400 V) is established by the designed interleaved boost converter.

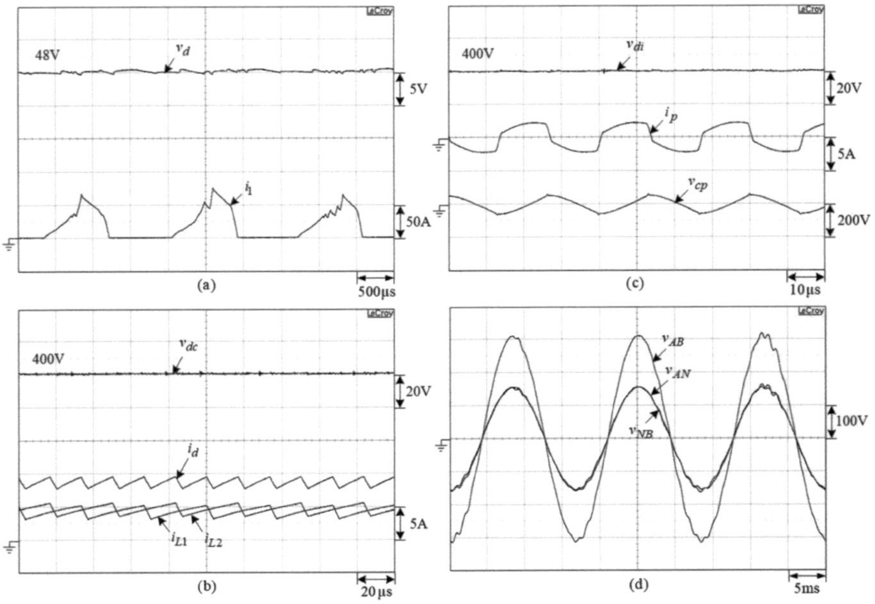

Figure 19.
Measured results of the established DC microgrid with 1P3W load inverter at $\omega_r = 6000$ *rpm and the loads* ($Z_A = 132.25\ \Omega$, $Z_B = 132.25\ \Omega$, $Z_{AB} = 484\ \Omega$)*: (a)* (v_d, i_1)*; (b)* (v_{dc}, i_d, i_{L1}, i_{L2})*; (c)* (v_{di}, i_p, v_{cp})*; (d)* (v_{AB}, v_{AN}, v_{NB}).

Figure 19(c) shows the measured secondary-side voltage v_{di}, the primary-side resonant capacitor voltage v_{cp}, and the primary-side resonant current i_p of the bidirectional LLC resonant converter. And the measured steady-state waveforms v_{AB}, v_{AN}, and v_{NB} at $Z_A = 132.25\ \Omega$, $Z_B = 132.25\ \Omega$, and $Z_{AB} = 484\ \Omega$ are shown in **Figure 19(d)**. The results show that normal M2H operation is achieved.

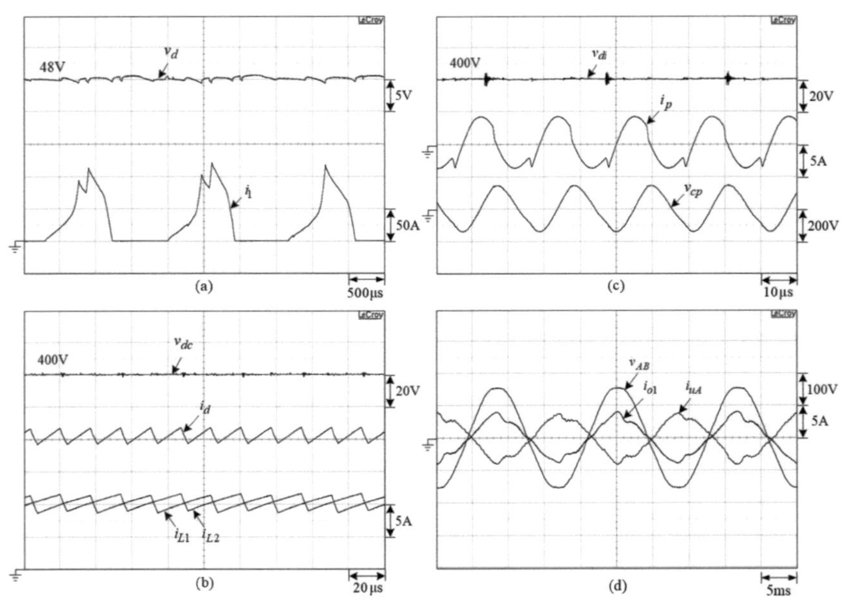

Figure 20.
Measured results of the established DC microgrid with 1P3W load inverter in M2G mode at the power command ($P_u^* = -800$ W) *under the loads* ($Z_A = Z_B = Z_{AB} = \infty$)*: (a)* ($v_d$, i_1)*; (b)* (v_{dc}, i_d, i_{L1}, i_{L2})*;* *(c)* (v_{di}, i_p, v_{cp})*; (d)* (v_{AB}, i_{o1}, i_{uA}).

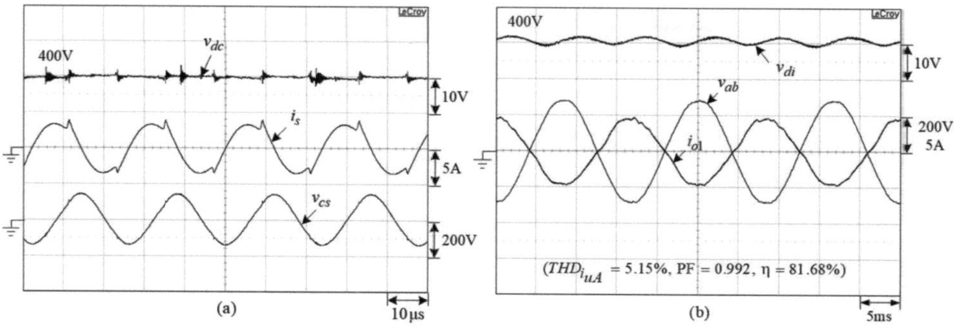

Figure 21.
Measured results of the HF-isolated single-phase three-wire load inverter in G2M charging mode with $R_{dc} = 200\Omega$*: (a)* $(v_{dc},\ i_s,\ v_{cs})$*; (b)* (v_{di}, v_{ab}, i_{01})*.*

5.2.2.2 Microgrid-to-grid (M2G) mode

In M2G mode, the SRG is driven at 6000 rpm and the isolated 1P3W inverter is employed. The discharging power command is set as $P_u^* = -800\ \text{W}$; **Figure 20 (a)–(d)** shows the measured steady-state waveforms of $(v_d,\ i_1)$, $(v_{dc},\ i_d,\ i_{L1},\ i_{L2})$, $(v_{di},\ i_p, v_{cp})$, $(v_{AB},\ i_{o1},\ i_{uA})$ of the SRG stage, the interleaved boost converter, the bilateral LLC resonant converter, and the 1P3W inverter under the loads $(Z_A = Z_B = Z_{AB} = \infty)$. According to the experimental results, the M2G operation is achieved with low distortion.

5.2.2.3 Grid-to-microgrid (G2M) mode

The DC bus voltage $v_{dc} = 400\ \text{V}$ is established by the bidirectional 1P3W isolated inverter from the mains. And a test load resistor R_{dc} is placed across the DC bus. The measured $(v_{dc},\ i_s,\ v_{cs})$ and (v_{di}, v_{ab}, i_{o1}) at $R_{dc} = 200\ \Omega$ are shown in **Figure 21(a)** and **(b)**. The normal G2M operation can also be observed from the results.

6. Example SRM drive: a battery-/SC-powered EV SRM drive

6.1 System configuration

The power circuit of the established EV SRM drive is shown in **Figure 22(a)** [28]. It consists of a battery bank and a SC bank with their bidirectional DC/DC converters, a SRM drive, a test load, and a dynamic brake leg. **Figure 22(b)** shows the control scheme of the SRM drive, whereas the control schemes of battery, SC, and dynamic brake are neglected here.

The major features of the developed EV SRM drive are (1) SRM: 3-phase, 12/8, 550 V, 1500 rpm, 2.2 kW; and (2) asymmetric bridge converter—it is constructed using two three-phase IGBT modules CM100RL-12NF (Mitsubishi Company). The switches (Q_1, Q_5, Q_9) are in charge of PWM switching control, and (Q_4, Q_8, Q_{12}) are commutation switches.

The proposed control scheme of the developed SRM drive shown in **Figure 22 (b)** consists of the outer speed-loop, the inner current-loop, and a dynamic commutation tuning (DCT) scheme. In the proposed control scheme, the basic feedback controller is augmented with an observed back-EMF current feed-forward controller (CFFC) and a robust current tracking error cancelation controller (RCECC) with

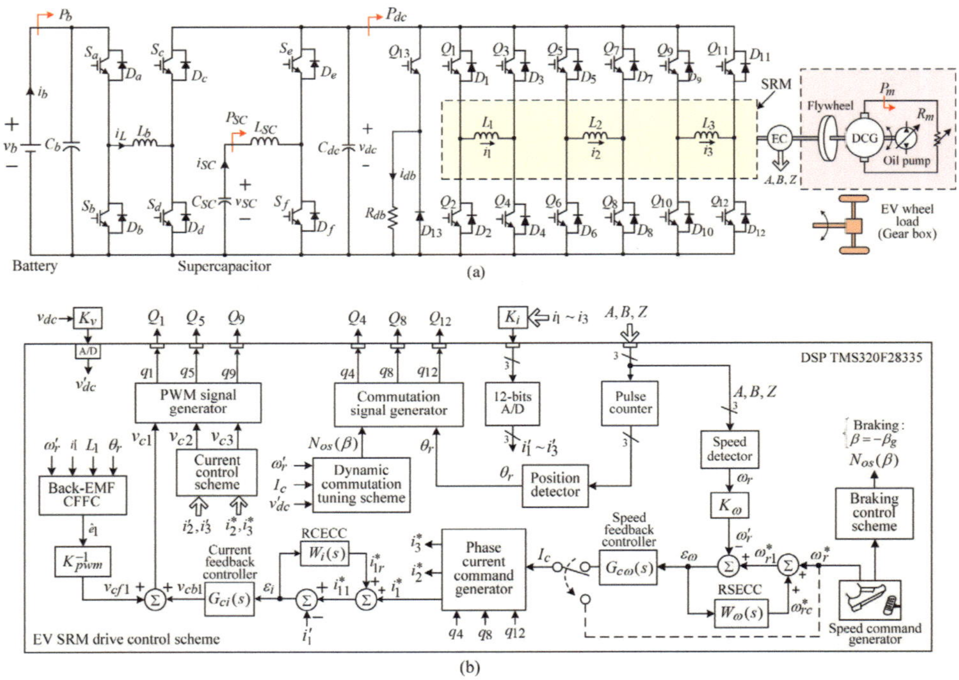

Figure 22.
The EV SRM drive: (a) schematic; (b) SRM drive control scheme.

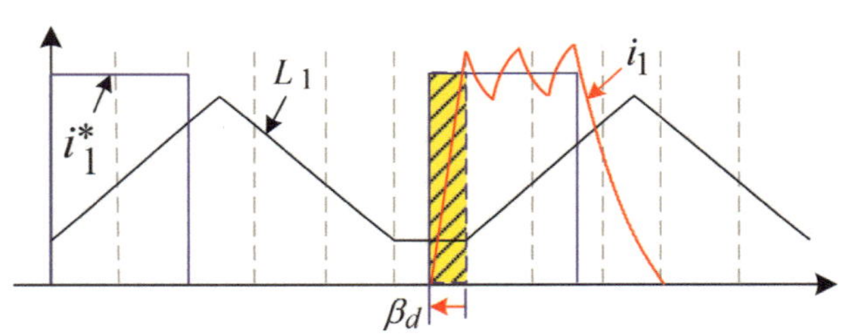

Figure 23.
The proposed dynamic commutation tuning mechanism.

the robust weighting factor $W_i(s) = W_i/(1 + \tau_i s)$, $W_i = 0.4$, $\tau_i = 2.6525 \times 10^{-4} s$. The DCT scheme is proposed here to conduct the commutation shift automatically. As shown in **Figure 23**, in order to let the winding current be the commanded value within the advanced shift angle in the minimum and constant inductance region, from the voltage equation listed in Eq. (3), one can obtain

$$v_{dc} = Ri + e(i, \omega_r, \theta_r) + L(i, \theta_r)\frac{di}{dt} \cong L_{\min}\frac{I_c}{\Delta T_\beta} \tag{15}$$

where ΔT_β denotes the advanced shifting time interval within which the winding current being linearly risen to I_c, where $R \cong 0$, $\partial L(i, \theta_r)/\theta_r = 0$, $e(i, \omega_r, \theta_r) = 0$, and $L(i, \theta_r) = L_{\min}$ are assumed. Thus, the required dynamic commutation advanced shift angle β_d can be derived from Eq. (15):

$$\beta_d = \Delta T_\beta \omega_r = \frac{L_{\min}}{v_{dc}} I_c \omega_r \tag{16}$$

6.2 Some experimental results

6.2.1 Winding current responses

Figure 24(a)–(c) shows the measured (i_1^*, i_1) at ($V_{dc} = 550$ V, $R_m = 320$ Ω, $\omega_r = 1000$ rpm) by the PI control augmented with the back-EMF CFFC and the RCECC, respectively. Obviously, the effectiveness of the proposed current control methods can be observed from the results. **Figure 25(a)** and **(b)** shows the measured (i_1^*, i_1) by all controls at ($V_{dc} = 550$ V, $R_m = 37.33$ Ω, $\omega_r = 2000$ rpm) without and with DCT. One can observe that applying DCT can let the winding current tracking control be improved.

6.2.2 Regenerative braking

By exciting the winding under the negative winding inductance slope region, the SRM will be operated as an SRG. The SRM is initially driven to 2000 rpm under ($V_{dc} = 550$ V, $\omega_r = 2000$ rpm, $R_m = 61.3$ Ω). Then the speed command is set from $\omega_r^* = 2000$ rpm to 0 rpm with 600 rpm/s decelerating rate. The measured speed command ω_r^*, speed ω_r', winding current command I_c, and DC-link voltage v_{dc} are shown in **Figure 26(a)**. **Figure 26(b)** confirms the successful SRG operation during regenerative braking from the winding current waveform.

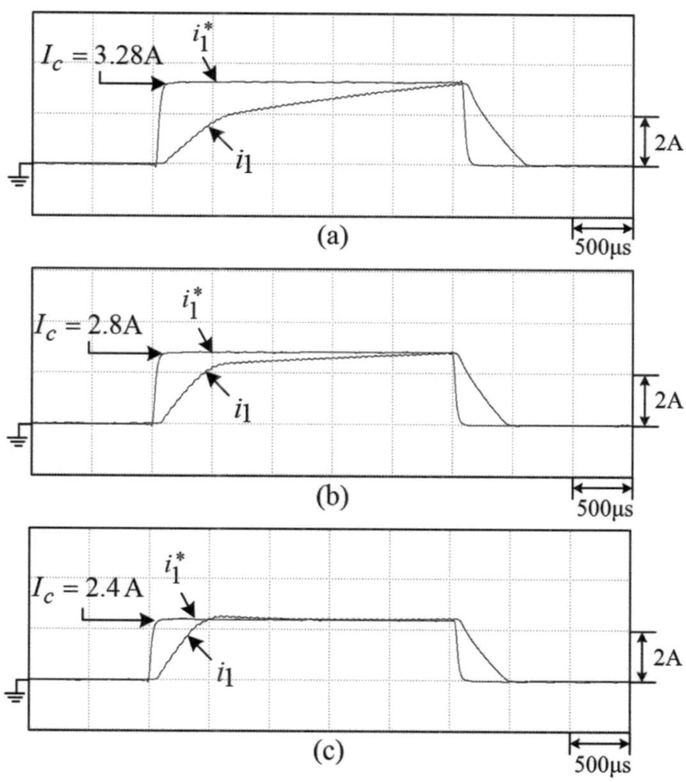

Figure 24.
Measured (i_1^*, i_1) of phase-1 winding at ($V_{dc} = 550$ V, $R_m = 320$ Ω, $\omega_r = 1000$ rpm): (a) PI only; (b) PI and back-EMF CFFC; (c) PI, back-EMF CFFC, and RCECC.

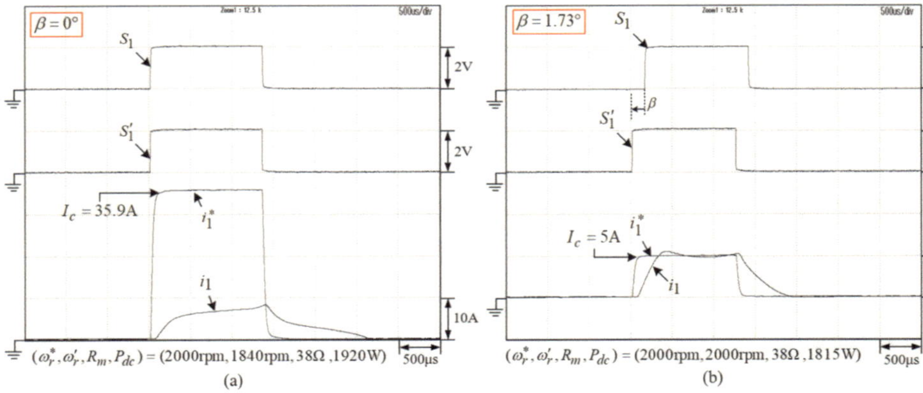

Figure 25.
Measured (i_1^*, i_1) of phase-1 winding at $(V_{dc} = 550\ \ V, R_m = 37.33\ \ \Omega, \omega_r = 2000\ \ rpm)$ by all controls without and with dynamic commutation tuning: (a) without DCT, $\beta = 0°$; (b) with DCT, $\beta = 1.73°$.

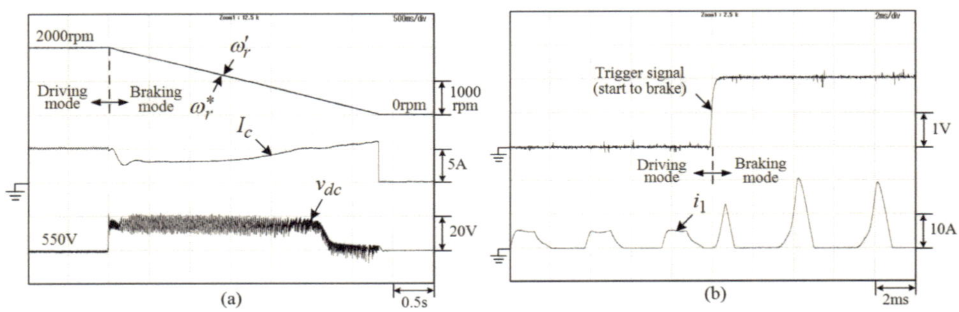

Figure 26.
Measured results during braking by letting the speed command be changed from $\omega_r^* = 2000$ rpm to 0 rpm with 600 rpm/s falling rate: (a) speed command ω_r^*, speed ω'_r, winding current command I_c, and DC-link voltage v_{dc}; (b) trigger signal and phase-1 winding current i_1.

6.2.3 Performance evaluation of the established EV SRM drive

6.2.3.1 Battery only

First, let the SRM drive be powered by battery only; the measured $(\omega'_r, \omega_r^*, v_b, i_b, v_{dc})$ at $(V_b = 156\ \ V, V_{dc} = 550\ \ V, R_m = 455\ \ \Omega)$ due to the speed command setting of $\omega_r^* = 0$ rpm $\rightarrow$ 2000 rpm $\rightarrow$ 1500 rpm $\rightarrow$ 2000 rpm $\rightarrow$ 1000 rpm $\rightarrow$ 2000 rpm $\rightarrow$ 0 rpm with rising rate and falling rate both 300 rpm/s are shown in **Figure 27**. Normal speed tracking and battery discharging/charging characteristics are seen from the results.

6.2.3.2 Battery and supercapacitor

The measured $(\omega'_r, \omega_r^*, v_b, i_b, v_{sc}, i_{sc}, v_{dc}, i_{db})$ of the battery/SC hybrid energy-powered EV SRM drive at $(V_b = 156\ \ V, V_{SC} = 100\ \ V, V_{dc} = 550\ \ V, R_m = 455\ \ \Omega)$ due to the speed command setting of $\omega_r^* = 0$ rpm $\rightarrow$ 2000 rpm $\rightarrow$ 1500 rpm $\rightarrow$ 2000 rpm $\rightarrow$ 1000 rpm $\rightarrow$ 2000 rpm $\rightarrow$ 0 rpm with rising rate and falling rate both 300 rpm/s are shown in **Figure 28**. Compared to **Figure 27**, much smaller battery discharging currents are yielded thanks to the assistance of SC.

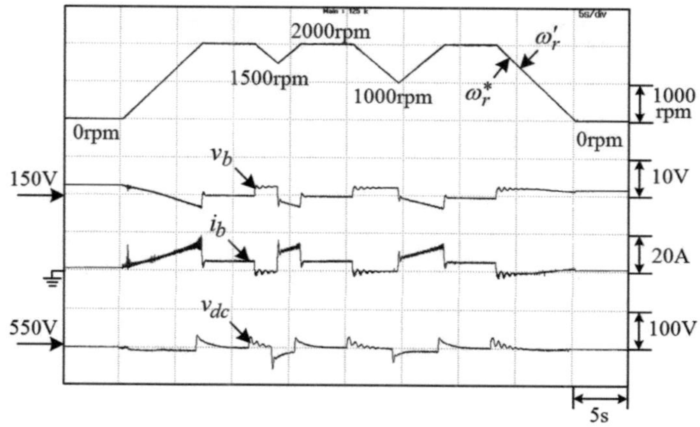

Figure 27.
The EV SRM drive powered by battery only at $V_b = 156$ V, $V_{dc} = 550$ V, $R_m = 455$ Ω.

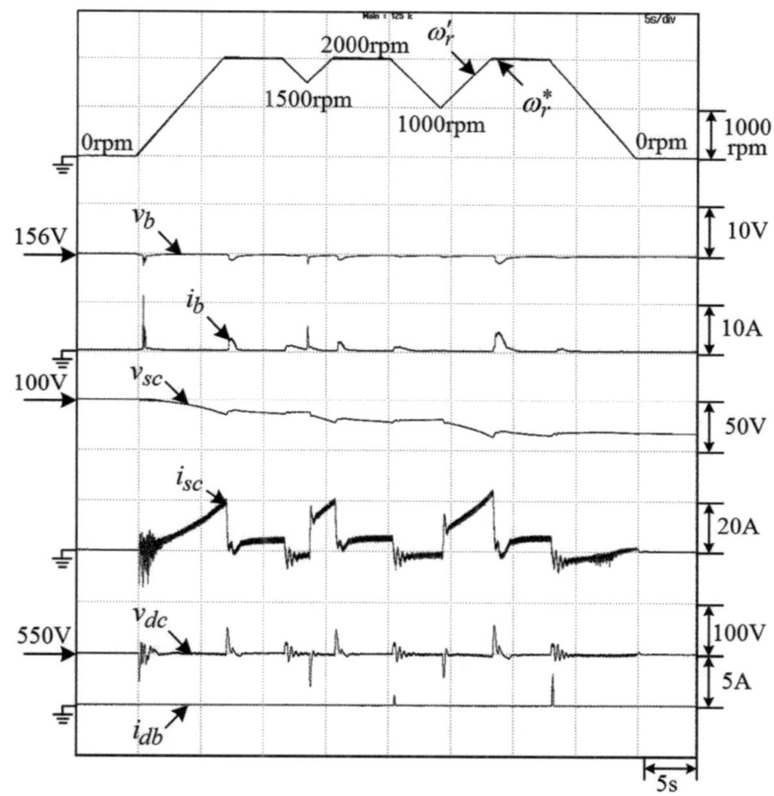

Figure 28.
The EV SRM drive powered by battery and SC at $V_b = 156$ V, $V_{SC} = 100$ V, $V_{dc} = 550$ V, $R_m = 455$ Ω.

6.2.3.3 Varied DC-link voltage

The DC-link voltage setting approaches used for comparison are shown in **Figure 29**, which are listed below:

1. Fixed DC-link voltage: $V_{dc} = 550$ V $(0 \le \omega_r \le 2000$ rpm)

2. Varied DC-link voltage:

i. Boosted DC-link voltage: $V_b = 156$ V, $V_{dc} = 156$ V $(0 \le \omega_r \le 500$ rpm); $156 < V_{dc} \le 550$ V $(500 < \omega_r \le 2000$ rpm)

ii. Lowered and boosted DC-link voltage: $V_b = 156$ V, $V_{dc} = 100$ V $(0 \le \omega_r \le 500$ rpm); $100 < V_{dc} \le 550$V $(500 < \omega_r \le 2000$rpm)

Figure 30(a)–(c) shows the measured results of the developed EV SRM drive under three DC-link voltage setting approaches. And the energies of battery are

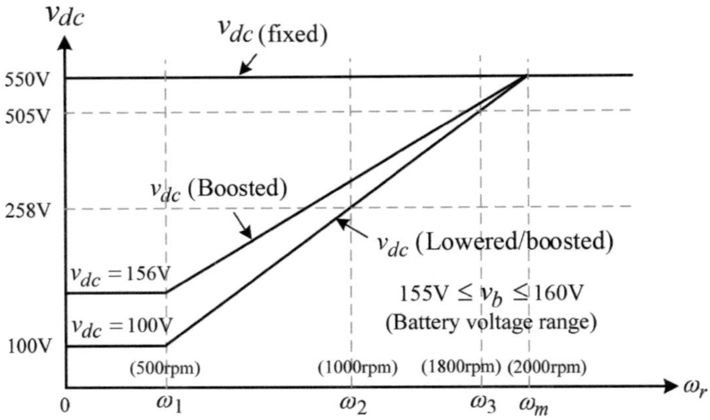

Figure 29.
Voltage profiles of DC-link versus motor speed.

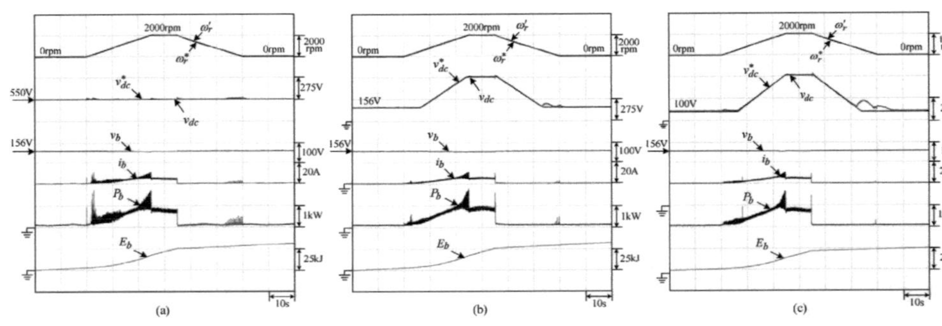

Figure 30.
Measured results of the SRM drive powered by the battery due to ramp speed command change with changing rate of 80 rpm/s at (R_m = 322 Ω) and (ω_r = 0 rpm → 2000 rpm → 0 rpm): (a) fixed voltage (V_{dc} = 550 V); (b) boosted voltage ($V_b \le V_{dc}$); (c) with lowered/boosted voltage ($V_b \ge V_{dc}$ or $V_b \le V_{dc}$).

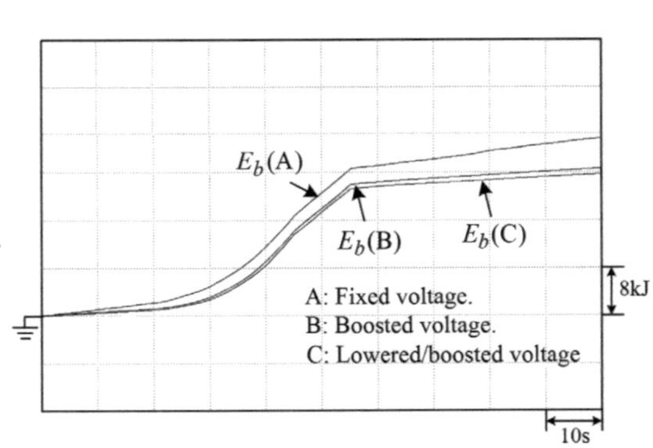

Figure 31.
Measured energies of battery of the developed standard SRM drive powered by battery with three DC-link voltage setting approaches due to ramp speed command change with changing rate of 80 rpm/s at (R_m = 322 Ω) and (ω_r = 0 rpm → 2000 rpm → 0 rpm).

compared in **Figure 31**. From **Figure 31**, the effectiveness in energy saving by applying the lowered/boosted DC-link voltage over the boosted and the fixed ones can be observed.

7. Conclusions

Compared to other machines, SRM is more difficult to control for yielding satisfactory operating characteristics. This chapter has presented some basic and key issues for SRM operated as motor and generator. These include structural features, governing equations and dynamic model, SRM converter, some front-end DC/DC converters and SMRs, some key motor parameters estimation and experimental performance evaluation of an SRM drive, commutation scheme, dynamic controls, reversible and regenerative braking operation controls, operation control and tuning issues of SRM and SRG, etc. Finally, two example SRM plants are presented to demonstrate the affairs being described, including a wind SRM-based microgrid and an EV SRM drive.

Author details

Chang-Ming Liaw*, Min-Ze Lu, Ping-Hong Jhou and Kuan-Yu Chou
Department of Electrical Engineering, National Tsing Hua University, Hsinchu, Taiwan, ROC

*Address all correspondence to: cmliaw@ee.nthu.edu.tw

References

[1] Sen PC. Principles of Electric Machines and Power Electronics. 3rd ed. New Jersey: John Wiley & Sons, Inc; 2014

[2] Miller TJE. Switched Reluctance Motors and their Control. Oxford: Clarendon Press; 1993

[3] Y Z, Shang F, Brown IP, Krishnamurthy M. Comparative study of interior permanent magnet, induction, and switched reluctance motor drives for EV and HEV applications. IEEE Transactions on Transportation Electrification. 2015; **1**(3):245-254

[4] Bartolo JB, Degano M, Espina J, Gerada C. Design and initial testing of a high-speed 45-kW switched reluctance drive for aerospace application. IEEE Transactions on Industrial Electronics. 2017;**64**(2):988-997

[5] Namazi MM, Nejad SMS, Tabesh A, Rashidi A, Liserre M. Passivity-based control of switched reluctance-based wind system supplying constant power load. IEEE Transactions on Industrial Electronics. 2018;**65**(12):9550-9560

[6] Santos Neto PJD, Santos Barros TAD, Paula MVD, Souza RRD, Filho ER. Design of computational experiment for performance optimization of a switched reluctance generator in wind system. IEEE Transactions on Energy Conversion. 2018;**33**(1):406-419

[7] Miller TJE. Optimal design of switched reluctance motors. IEEE Transactions on Industrial Electronics. 2002;**49**(1):15-27

[8] Schulz SE, Rahman KM. High-performance digital PI current regulator for EV switched reluctance motor drives. IEEE Transactions on Industry Applications. 2003;**39**(4):1118-1126

[9] Lin Z, Reay D, Williams B, He X. High-performance current control for switched reluctance motors based on on-line estimated parameters. IEEE Transactions on Electric Power Applications. 2010;**4**(1):67-74

[10] Hwu KI, Liaw CM. Quantitative speed control for SRM drive using fuzzy adapted inverse model. IEEE Transactions on Aerospace and Electronic Systems. 2002;**38**(3):955-968

[11] Hwu KI, Liaw CM. Intelligent tuning of commutation for maximum torque capability of a switched reluctance motor. IEEE Transactions on Energy Conversion. 2003;**18**(1):113-120

[12] Hu KW, Chen YY, Liaw CM. A reversible position sensorless controlled switched-reluctance motor drive with adaptive and intuitive commutation tuning. IEEE Transactions on Power Electronics. 2015;**30**(7):3781-3793

[13] Huang HN, Hu KW, Wu YW, Jong TL, Liaw CM. A current control scheme with back-EMF cancellation and tracking error adapted commutation shift for switched-reluctance motor drive. IEEE Transactions on Industrial Electronics. 2016;**63**(12):7381-7392

[14] Chan S, Bolton HR. Performance enhancement of single-phase switched reluctance motor by DC link voltage boosting. IEE Proceedings—Electric Power Applications. 1993;**140**(5): 316-322

[15] Hwu KI, Liaw CM. DC-link voltage boosting and switching control for switched reluctance motor drives. IEE Proceedings—Electric Power Applications. 2000;**147**(5):337-344

[16] Chang HC, Liaw CM. Development of a compact switched-reluctance motor drive for EV propulsion with voltage

boosting and PFC charging capabilities. IEEE Transactions on Vehicular Technology. 2009;**58**(7):3198-3215

[17] Chai JY, Chang YC, Liaw CM. On the switched-reluctance motor drive with three-phase single-switch switch-mode rectifier front-end. IEEE Transactions on Power Electronics. 2010;**25**(5):1135-1148

[18] Torrey DA. Switched reluctance generators and their control. IEEE Transactions on Industrial Electronics. 2002;**49**(1):3-14

[19] Chang YC, Liaw CM. On the design of power circuit and control scheme for switched reluctance generator. IEEE Transactions on Power Electronics. 2008;**23**(1):445-454

[20] Santos Barros TAD, Santos Neto PJD, Filho PSN, Moreira AB, Filho ER. An approach for switched reluctance generator in a wind generation system with a wide range of operation speed. IEEE Transactions on Power Electronics. 2017;**32**(11): 8277-8292

[21] Rahmanian E, Akbari H, Sheisi GH. Maximum power point tracking in grid connected wind plant by using intelligent controller and switched reluctance generator. IEEE Transactions on Sustainable Energy. 2017;**8**(3): 1313-1320

[22] Vukosavic S, Stefanovic VR. SRM inverter topologies: A comparative evaluation. IEEE Transactions on Industry Applications. 1991;**27**(6): 1034-1049

[23] Barnes M, Pollock C. Power electronic converters for switched reluctance drives. IEEE Transactions on Power Electronics. 1998;**13**(6): 1100-1111

[24] Cabezuelo D, Andreu J, Kortabarria I, Ibarra E, Garate I. SRM converter topologies for EV application: State of the technology: proceedings of IEEE International Symposium on Industrial Electronics. Edinburgh, UK; 19-21 June 2017;**2017**:861-866

[25] Pereira1 M, Araújo RE. Analysis and design of a speed controller for switched reluctance motor drive. U.Porto Journal of Engineering. 2019;**5**(1):46-58

[26] Chang HC. Development of battery powered switched-reluctance motor drives [Ph.D. dissertation]. ROC: Dept. Electric Eng., National Tsing Hua Univ.; 2010

[27] Jhou PH. A wind switched-reluctance generator based grid-connected micro-grid [Master thesis]. Hsinchu, ROC: Department of Electrical Engineering, National Tsing Hua University; 2017

[28] Chou KY. A batter/supercapacitor powered EV SRM drive with bidirectional isolated charger [Master thesis]. Hsinchu, ROC: Department of Electrical Engineering, National Tsing Hua University; 2017

Chapter 11

A Review of Classic Torque Control Techniques for Switched Reluctance Motors

Marcelo Vinícius de Paula, Tárcio André dos Santos Barros and Pedro José Dos Santos Neto

Abstract

In this chapter, the most relevant electronic torque control methods established in the current literature are addressed. The electronic torque control methods are divided into two technical topologies. The first one aims to control the average torque produced by the switched reluctance machine and is ideal for applications that require wide speed range of operation. The second topology enrolls the instantaneous torque production and is required in low torque ripple applications. Different torque control methods are addressed, for instance, direct average torque control, current profiling through torque sharing functions, direct torque control, and direct instantaneous torque control. Detailed information regarding the working principle and implementation of the methods is presented. Mathematical simulations are conducted in Matlab/Simulink® environment to elucidate the methods. Finally, a comparison of all addressed methods regarding the torque ripple minimization capabilities is presented.

Keywords: instantaneous torque control, average torque control, torque ripple minimization, direct average torque control, current profiling, torque sharing function, direct torque control, direct instantaneous torque control

1. Introduction

Switched reluctance machines (SRMs) main advantages are their simple structure, without coils, and rare materials in the rotor, low cost, ruggedness, brushless operation, operation in high temperature environments, high reliability, fault tolerance, wide speed range of operation, high efficiency and high torque, and power density. These advantages enable SRMs as competitive to more traditional electric machines, such as induction motors, permanent magnet synchronous machines (PMSM), and DC machines in many applications [1, 2].

Besides the advantages, SRMs are strongly dependent in electronic converters for its drive. Also, automatic control of these machines is complex due to nonlinear characteristics. Furthermore, the efficiency and power factor of SRM are generally lower than those of PMSM as a result of higher losses in SRM. The switched reluctance machines have doubly salient structure and nonlinear characteristics, thus, these machines present elevated torque ripple and acoustic noise. These are the predominant disadvantages encountered in SRMs [3, 4].

The scientific community focuses on mitigating these disadvantages through electronic control or through structural changes in the machine. For instance, in [5], the authors analyze the effect of different frames and ribs structure topologies over the acoustic noise production. The methods that target the structural changes consist in manipulating the magnetic flux linkage in such way that the reluctance variation between cycles is smaller, thus resulting in shallower torque ripple. On the other hand, the speed range of operation is diminished because the torque per ampere ratio is also diminished.

In contrast, the electronic control methods aim to minimize the torque ripple by manipulating a set of control parameters, such as electric phase current, turn on angle, turn off angle, and excitation voltage, which regulate the machine operation. These parameters may be controlled directly or indirectly. The electronic control methods will be focused in this chapter.

In order to understand the working principle of the control methods, some other concepts must be addressed initially. In SRMs, the phase inductance (L) varies according to the rotor position (θ) because of the reluctance variation caused mostly by the air-gap changes as the rotor revolves. The inductance profile for a generic four-phase ideal SRM phase is highlighted in **Figure 1(a)**. The pointed angles in the abscissa axle refer to the relative position between a point in the rotor and a point in the stator, where θ_{as} is the start of alignment position, θ_a is the aligned position, θ_{ms} is the start of misalignment position, and θ_m is the misaligned position. These positions and their respective inductance are also pointed in the SRM physical structure in **Figure 1(b)**. The aligned position occurs when the entirety of the rotor pole is bounded by the stator pole, and in this position, the inductance is at its maximum and the reluctance at its minimum. The misaligned position occurs when the entirety of the rotor pole is off the stator pole bounds [6].

Because of its working principle, SRMs require an electronic converter as a driver. The classic and most used topology is the asymmetric half bridge (AHB) [6, 7], which demand two power switches and two power diodes for its construction. There are several other proposed topologies in the literature for specific applications. In [8], the SRM phases are used as components of a charging station. In [9], the authors embed the SRM and AHB components as part of a converter with voltage boosting and power factor correction capabilities. The research presented in [10, 11] apply the SRM windings and AHB components for flexible energy conversion and multi-operation mode of the SRM. The authors in [12] propose an integrated SRM and drive system designed for hybrid energy storage systems composed of batteries and supercapacitors.

The schematic circuit for a three phase AHB is presented in **Figure 2**. This topology may be applied for any number of phases and enables operation in all four

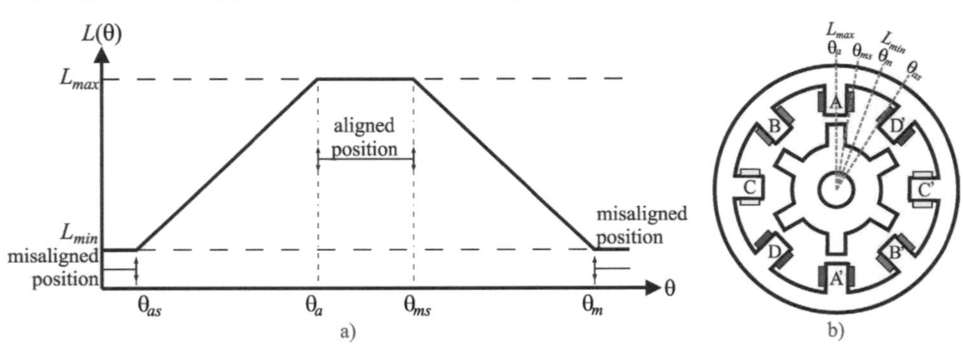

Figure 1.
Inductance variance according to rotor position: (a) inductance profile for one electric cycle of one SRM phase; (b) physical structure of a 8/6 SRM with main rotor position highlighted.

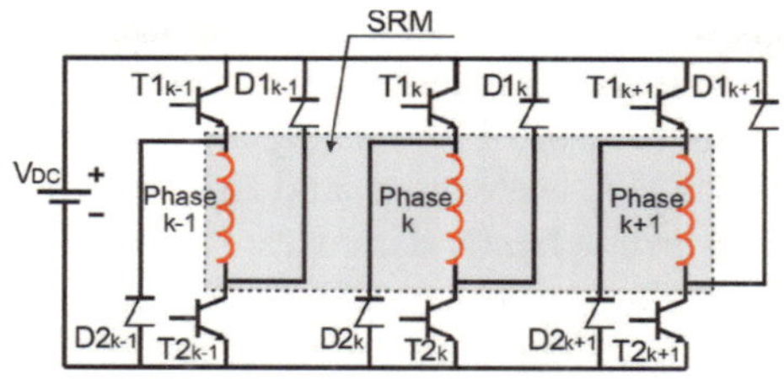

Figure 2.
Schematic circuit of the asymmetric half bridge converter for SRM drive.

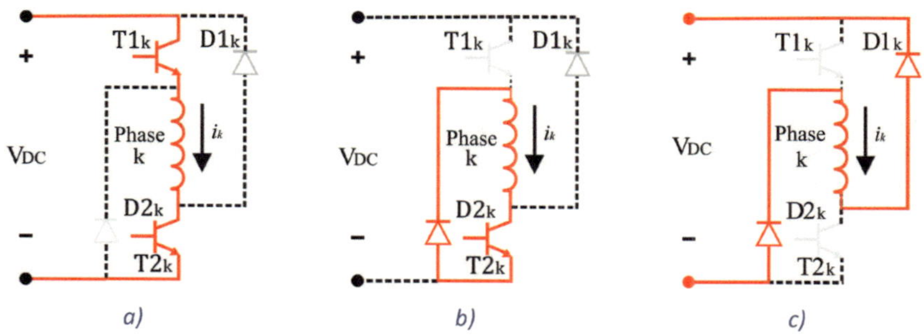

Figure 3.
Voltage states for the AHB converter: (a) magnetization or positive state (1), (b) free-wheeling or null state (0), and (c) demagnetization or negative state (−1).

quadrant of the torque-speed plane. The AHB support independent voltage control for all the SRM phases. Three voltage levels are possible according to the power switches states.

Figure 3 shows all three possible voltage states that may be applied to the SRM phases using an AHB converter. In case both switches are conducting, positive voltage is applied to the phase terminals, and this state is known as positive state or magnetization (1) as depicted in **Figure 3(a)**. If only one switch is conducting, null voltage will be applied to phase terminals and the electric current will flow through the circuit formed by the conducting switch, power diode, and SRM phase winding. This state is named free-wheel or null state (0) (**Figure 3(b)**). In case both switches are not conducting, negative voltage is applied to the SRM phase, characterizing the demagnetization or negative state (−1) as depicted in **Figure 3(c)**.

The choice of the driving angles (θ_{on} and θ_{off}) has great influence over the electronic control method applied. Some researches focus on the influence of these angles. The research presented in [13, 14] is focused on optimal angles for applications in current hysteresis control, while in [15], the authors focus the analysis on the voltage single pulse operation. The angle selection may have different objectives, for example, maximizing the efficiency, minimizing torque ripple or minimizing acoustic noise. The majority of the studies present optimizations aiming to minimize torque ripple [16, 17] and/or minimize the losses [18]. For this chapter, the focus will be on minimizing the torque ripple.

When the SRM is working at low speeds and current hysteresis control may be applied, at the beginning of the conduction period, the inductance value is almost constant, and consequently, a small amount of torque is produced. When the stator and rotor poles start to overlap, the torque production rises, thus when the rotor position overcomes the turn-off angle, a negative voltage must be applied to the phase terminals to guarantee that the electric current flowing through the phase reaches zero before the inductance starts to decrease. The magnetic flux linkage rise is constant because of the quazi-linear inductance variation. At the end of the commutation, the magnetic flux returns to zero because of the negative voltage applied. After that, zero voltage is applied to the phase.

On the other hand, when the SRM is working at high speeds, the back electromotive force (back EMF) is elevated, and the excitation voltage is not enough to guarantee appropriate hysteresis current control. Thus, voltage single pulse control is applied. At the beginning of the conduction interval, the electric current rises because there is voltage being applied to the phase terminals after the rotor position overcome the turn on angle (θ_{on}). At certain point, the back EMF rises and the electric current can no longer increase, thus current hysteresis is inapplicable. The back EMF is directly proportional to the electric current and phase inductance. At certain speed, known as the base speed, the back EMF becomes the predominant term in the voltage equation (Eq. (1)), and then, the produced torque must be controlled by adjusting the θ_{on} and θ_{off} angles. In Eq. (1), V_{DC}, R, i, ω, and $L(\theta, i)$ are the source voltage, the phase resistance, the phase electric current, the rotor angular speed and the angular position, and electric current dependent phase inductance, respectively.

$$V_{DC} = Ri + L(\theta, i)\frac{di}{dt} + i\omega\frac{dL(\theta, i)}{d\theta} \tag{1}$$

2. Torque production in switched reluctance machines

The SRM motion is directly linked to the minimum reluctance principle, that is, when the stator pole winding is energized, the closest rotor pole tends to move toward the energized stator pole, and during this process, the reluctance is minimized and the inductance is maximized. The coordination of this process leads to rotational acceleration. The electromagnetic torque produced by an SRM phase may be expressed as in Eq. (2), where T_e is the electromagnetic torque.

$$T_e = \frac{1}{2}i^2\frac{\partial L(i, \theta)}{\partial \theta} \tag{2}$$

From Eq. (2), one can infer that: (1) the torque production does not depend on the direction of the electric current because of their quadratic relation; (2) the direction of the torque produced depends exclusively on the inductance derivative according to the rotor position. The inductance depends on the electric current and on the rotor position; and (3) the SRM is able to operate in all four torque-speed quadrants [19].

The total instantaneous torque produced by the SRM can be expressed as the sum of the individual torque produced by the phases as in Eq. (3), where N is the number of phases and $T_{e,k}$ is the electromagnetic torque produced by the phase k. Each phase torque can be calculated as Eq. (2).

$$T_{tot} = \sum_{k=1}^{N} T_{e,k} \tag{3}$$

Thus, the average developed torque $\left(\overline{T_{tot}}\right)$ in a period is given by the time integral of the total torque produced in the same period as in Eq. (4), where τ is one revolution period.

$$\overline{T_{tot}} = \frac{1}{\tau}\int_0^{\tau} T_{tot}dt \tag{4}$$

As aforementioned, due to the SRM working principle and driving methods, strong torque undulations occur. These undulations are known as torque ripples and one way of measuring this phenomenon is through the ratio between peak to peak torque and the average torque produced in one cycle as in Eq. (5), where $\max(T_e)$ and $\min(T_e)$ are the maximum and minimum torque produced in one electric period, respectively.

$$T_{ripple} = \frac{\max(T_e) - \min(T_e)}{\overline{T_{tot}}} \quad (5)$$

3. Average torque control

The main aspect of average torque control (ATC) techniques is to keep a constant value of reference current throughout an electric cycle. Therefore, the major task of the controller is to adjust the control variables, for instance driving angles (θ_{on} and θ_{off}) and electric current reference (i^*), for each operation point (T, ω) in the torque-speed plane. The basic structure for ATC is presented in **Figure 4**.

There are two main methods for implementing the ATC [20]. The first consists of maintaining the driving angles while varying the electric current reference in each cycle, and the realization of this method is straightforward, but the operation region is limited and no efficiency optimization may be achieved. In the second method, both the electric current and the driving angles are controlled. Thus, although the control system becomes more complex and expensive, a good performance can be achieved, and the machine can operate in the entire speed range.

Studies focus on methods that enable real time variation of all control variables. This can be achieved online, through formulation and controllers or through genetic algorithms, or it can be achieved offline, using lookup tables [20, 21].

There are different combinations of control variables that attain the same control requirements for each operation point, such as minimizing the torque. It is possible, through cumbersome simulations, to create lookup tables containing the optimum values for each of these parameters according to the control objective.

Different optimization objectives may be conflicting, for example, minimizing torque ripple and maximizing efficiency. Thus, for a multi-objective optimization, a weighting process must be added to the optimization [22]. In this chapter, only torque ripple minimization is considered as objective.

Therefore, for the tables, creation diverse simulations were realized with the SRM modeling process developed in [23]. An open loop simulation of the system is implemented, while the control parameters and the load torque are varied covering all the operation region of the SRM (One Factor at a Time – OFAT optimization).

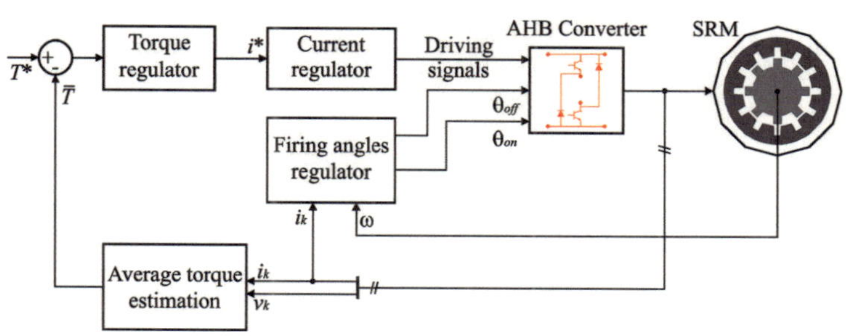

Figure 4.
Closed-loop average torque control.

After each simulation, the data are stored. The stored data incorporate the steady state rotational speed, the control variables used, and the torque ripple. By comparing the data, a combination of control parameters is selected, which enable the SRM to operate in the entire torque × speed plane (T, ω) with the minimum torque ripple. The selected parameters are stored in a lookup table.

Graphic representation of the lookup tables with the optimum values for the turn-on angle, turn-off angle, and reference current from the simulations are presented in **Figure 5**. As expected, both the turn-on and the turn-off angles must be advanced according to the rotational speed and load torque rise. On the other hand, the optimal electric current rises according to the increase in speed and load torque.

In some applications, the excitation voltage must also be considered as control parameter. The addition of one control parameter increases exponentially the number of simulations. To avoid this problem, Inderka [20] proposed a closed-loop average torque control method known as direct average torque control (DATC). The method is based on comparing the reference torque to the average torque produced. However, measuring the average torque with a torque transducer is expensive. To overcome this downside, the average torque may be estimated through the machine terminal quantities.

3.1 Average torque estimation

Considering SRMs with two or more phases, two phases may contribute to torque generation during commutation. Thus, the average torque estimation becomes dependent on the flux linkage derivative estimation, which may be calculated with terminal quantities, that is, phase current and phase voltage.

The torque production principle in SRM is similar to the torque production principle of solenoids. Assuming that the mechanical work incremented in a period is proportional to the rate of change in coenergy [6], the mechanical energy (W_{mec}) can be expressed as in Eq. (6), where $\phi(\theta, i)$ is the angular position and phase current dependent flux linkage. With this equation, the average torque $(\overline{T})$ can be calculated as a function of the number of phases (N_k) and number of rotor poles (N_r), as in Eq. (7). Also, Eq. (1) can be rearranged to express the flux derivative (Eq. (8)).

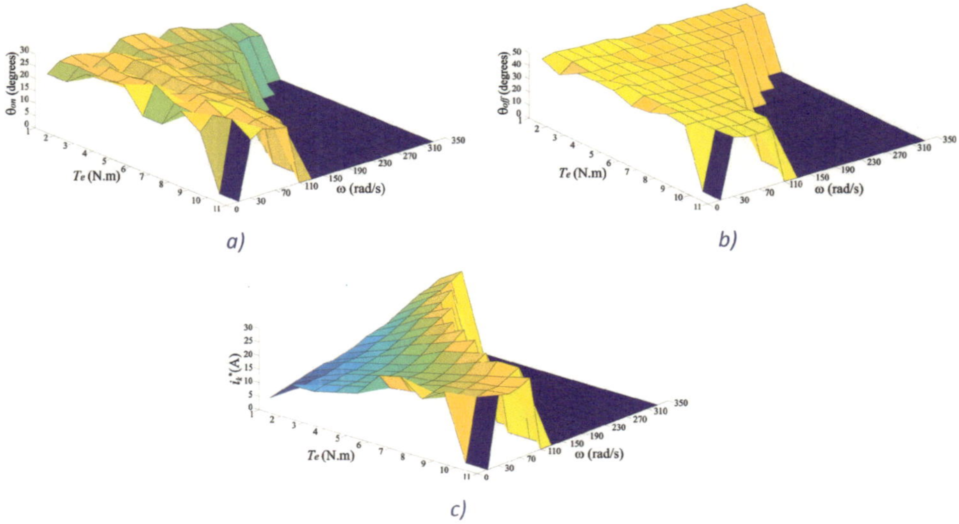

Figure 5.
Graphic representation of the lookup tables with the optimum control parameters values for torque ripple minimization: (a) $\theta_{on}(\omega, T)$, (b) $\theta_{off}(\omega, T)$, and (c) $i_k^*(\omega, T)$.

$$W_{mec} = \oint i \frac{d\phi(\theta, i)}{dt} dt \tag{6}$$

$$\overline{T} = \frac{N_k N_r}{2\pi} W_{mec} \tag{7}$$

$$\frac{d\Phi(\theta, i)}{dt} = V_{DC} - Ri \tag{8}$$

The average value of torque produced by the phases does not differ much from one another in each cycle due to phase symmetry, thus enabling to use only one average torque estimator. Note that in practice, the use of only one torque estimator reduces the cost of the project, but also can decrease the speed and quality of the controller. **Figure 6(a)** depicts the structure used to estimate the average torque of one SRM phase according to Eqs. (6)–(8). Only the current and voltage quantities are required. For implementation, it is not necessary to calculate the derivative of flux, since it is given by the difference between the excitation voltage and the resistance voltage drop. Note that the integrator must be reset every time the current returns to zero in order to avoid the drifting problem during practical implementation. Also, the sample and hold logic must be used to hold the mechanical energy value at the instant the current reaches zero. **Figure 6(b)** presents the waveform of the quantities used to estimate the average torque produced in one cycle.

3.2 Direct average torque control

In total, the DATC method relies on three lookup tables, one for reference electric current of the phase $i_k^*(T, \omega)$, one for the turn-on angle $\theta_{on}(T, \omega)$, and one for the turn-off angle $\theta_{off}(T, \omega)$. The block diagram structure for the DATC is presented in **Figure 7**. An external speed loop is responsible to generate the reference torque through a PI controller. A feedforward is added to the control system and generates a new reference torque. The lookup tables output the control parameter values. A hysteresis current regulator drives the AHB converter according to the control parameters.

Since the average torque is estimated only once per electric cycle, the DATC method requires low sampling ratio. This imply in lower cost for the controller mass production. On the other hand, the controller response to reference torque variations tend to be slow. To cope with that, one can use a torque estimator per SRM phase and recalculate the average torque at the beginning of each phase electric cycle. This solution increases the computational effort and cost of the system.

In order to verify the method, a computational model was developed in Matlab/Simulink® environment. The mathematical model of the SRM was developed as

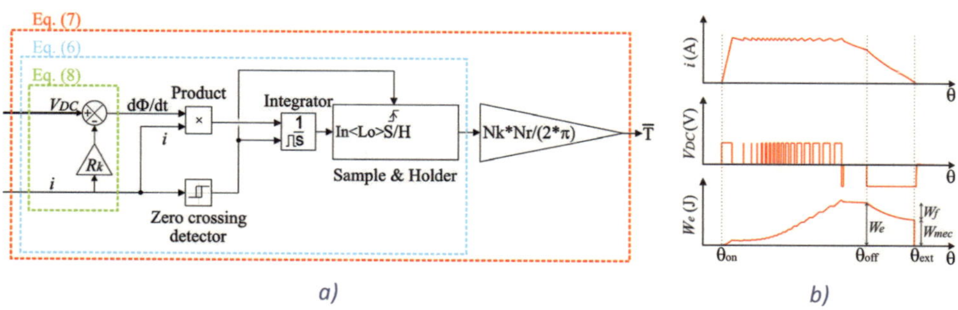

Figure 6.
Average torque estimation: (a) structure based on aforementioned equations; (b) time domain representation used in the mechanical energy estimation.

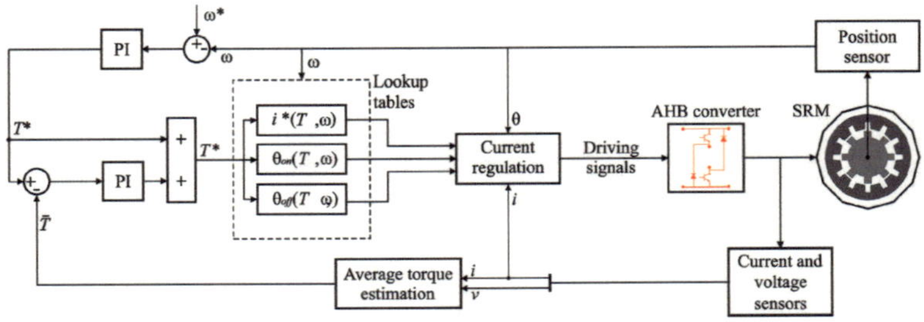

Figure 7.
Block diagram for the direct average torque control.

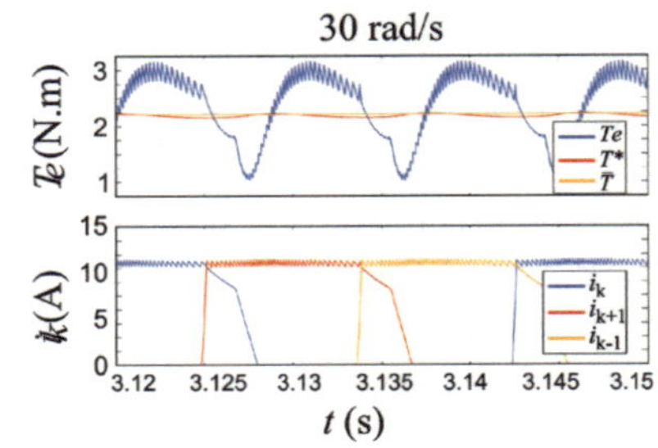

Figure 8.
DATC simulation results for setup S_1.

explained in [23]. The parameters for the SRM used for the simulation tests are presented in Appendix A. The base speed of the SRM is 954.93 *rpm* or 96 *rad/s*. The excitation voltage for the tests was maintained in 80 *V*. A constant load torque of 2 *N.m* was used for the tests.

For this test and for the following tests, two types of simulation are performed. The first simulation (*S*1) is performed for a constant load of 2 *N.m* and a reference speed of 30 *rad/s* (below base speed). For the second simulation (*S*2), the load is maintained in 2 *N.m* and the reference speed is increased to 130 *rad/s* (above base speed). The same simulations setup will be used in future subsections for the other torque control methods presented, where they will be referred to as *S*1 and *S*2, respectively.

The simulation results for the *S*1 and *S*2 setups using the DATC method are presented in **Figures 8** and **9**, respectively. The torque ripple in *S*1 conditions is 69.4%, while in *S*2 conditions, the torque ripple is 58.64%. Note that in *S*1, when the rotational speed is 30 *rad/s*, the electric current is limited by the hysteresis controller, while in *S*2, when the rotational speed is 130 *rad/s*, there is almost no chopping

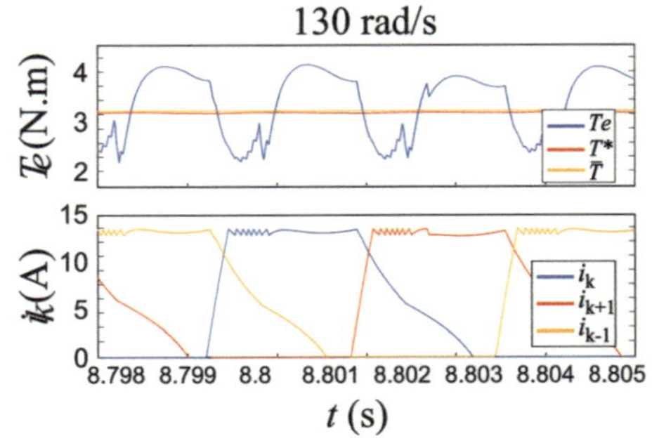

Figure 9.
DATC simulation results for setup S_2.

in the current waveform because the back EMF is high and preclude the current rise. The results show a high torque ripple in both simulation setups. This occurs because the DATC main objective is to control the average torque produced, and this way, the instantaneous torque is not controlled, leading to a high torque ripple response.

4. Instantaneous torque control

The biggest difference between ATC and instantaneous torque control (ITC) is that for the ITC methods, the electric current is regulated at every sampling period, while for ATC methods, the electric current is maintained constant through the entire excitation period. Therefore, the torque must be measured/estimated at every sampling period for ITC [24].

Another relevant factor is the operation region for each type of controller. The ATC methods may be applied through the entire speed range of operation. In contrast, the ITC methods have a shortened operation region and can only be applied while the controller is capable of controlling the current with accuracy. This becomes impractical when the speed grows above base speed because the back EMF forces the electric current in the opposite direction of the desired current.

Regarding the ITC methods, there are three main techniques present in the literature, which are the direct torque control (DTC), current profiling through torque sharing functions (TSF), and direct instantaneous torque control (DITC). Each of these techniques has its own derivations. Nonetheless, all of them depend on the instantaneous torque value. However, as in the case of the ATC, measuring the instantaneous torque may be expensive. Thus, to lower the price of the drive system, an instantaneous torque estimation technique may be employed.

4.1 Instantaneous torque estimation

Measuring the instantaneous torque is often expensive. Thus, sensorless torque estimation is preferable over torque transducer provided that the estimated torque accuracy is satisfactory.

The SRM model used in the simulations requires two lookup tables, one that returns the electric current according to flux linkage and electric current ($i(\Phi,\theta)$) and the other that returns the electromagnetic torque according to the electric current and rotor position ($T(i,\theta)$).

If the model is accurate, the torque lookup table may be used to estimate the electromagnetic torque produced by the SRM given the rotor position and the electric current in the phase winding. The electromagnetic torque estimation may also be performed using other lookup tables. For example, the electromagnetic torque may be estimated through the magnetic flux linkage and the rotor position with a $T(\Phi,\theta)$ table. This method also depends on a precise position sensor. If no position sensor may be used, the electromagnetic torque can also be estimated through the electric current and the voltage in the phase winding. With these two quantities, the flux linkage may be calculated online and used as input to a $T(i,\Phi)$ lookup table.

4.2 Current profiling technique through torque sharing functions

Given the relation between electromagnetic torque and electric current, it is possible to indirectly control the instantaneous torque produced through current control. On the other hand, the nonlinearities of the inductance profile must be accounted. Thus, to maintain the produced torque constant, the current profile presents a nonlinear waveform that varies according to the SRM characteristics, which is hard to interpolate.

The current profiling technique consists of controlling the electric current around a desired reference. The reference current is generated aiming to minimize the torque ripple or maximize the system efficiency. Some current profiling techniques are based on finite element analysis (FEM) [3, 4], and other methods are based on generating a reference electric current from desired torque profiles named torque sharing functions (TSF). The TSF may be generated and/or compensated online or offline. The TSF are responsible for dividing the produced torque equally through the phases and may present different formats. The most common formats are linear, sinusoidal, cubic, and exponential. In some cases, modifications are applied to these TSF by performing optimization techniques with different objectives.

Considering that the controller is fully capable of regulating the electric current, the torque production may be distributed among the phases in such a way that the sum of the torque produced by the phases is equal to the reference torque (T^*).

4.2.1 Torque sharing functions

The reference torque and rotor position are used as inputs of the TSF. The TSF outputs the reference torque of each phase (T_k^*). The torque division follows the profile defined in Eq. (9), where θ_{on}, θ_{off}, θ_{al}, and θ_{ov} are the turn-on, turn-off, alignment, and overlapping angles, respectively. The functions that rule the profile while the phase is entering and leaving the commutation period are defined as $f_{up}(\theta)$ and $f_{down}(\theta)$, respectively. The TSF is divided into five sectors according to angular position.

$$T_k^*(\theta) = \begin{cases} 0, \&0 \le \theta \le \theta_{on} & sector\ 1 \\ T^* f_{up}(\theta), \&\theta_{on} \le \theta \le \theta_{on} + \theta_{ov} & sector\ 2 \\ T^*, \&\theta_{on} + \theta_{ov} \le \theta \le \theta_{off} - \theta_{ov} & sector\ 3 \\ T^* f_{down}(\theta), \&\theta_{off} - \theta_{ov} \le \theta \le \theta_{off} & sector\ 4 \\ 0, \&\theta_{off} \le \theta \le \theta_{al} & sector\ 5 \end{cases} \tag{9}$$

In the literature, there are many functions to describe the rise and descend of the reference torque during commutation. The equations referring to the most commonly found TSF are presented hereafter along with the expected waveform for these functions. In the equations, the indexes *lin*, *sin* , *cub*, and *exp* indicate the linear TSF, the sinusoidal TSF, the cubic TSF, and the exponential TSF, respectively.

The up and down functions for the linear TSF are presented in Eqs. (10) and (11). The reference torque waveform for this TSF is presented in **Figure 10(a)**.

$$f_{up}^{lin}(\theta) = \frac{1}{\theta_{ov}}(\theta - \theta_{on}) \tag{10}$$

$$f_{down}^{lin}(\theta) = 1 - \frac{1}{\theta_{ov}}(\theta - \theta_{off}) \tag{11}$$

The equations for the sinusoidal TSF are presented following. The reference torque waveform is depicted in **Figure 10(b)**.

$$f_{up}^{sin}(\theta) = \frac{1}{2} - \frac{1}{2}\cos\left[\frac{\pi}{\theta_{ov}}(\theta - \theta_{on})\right] \tag{12}$$

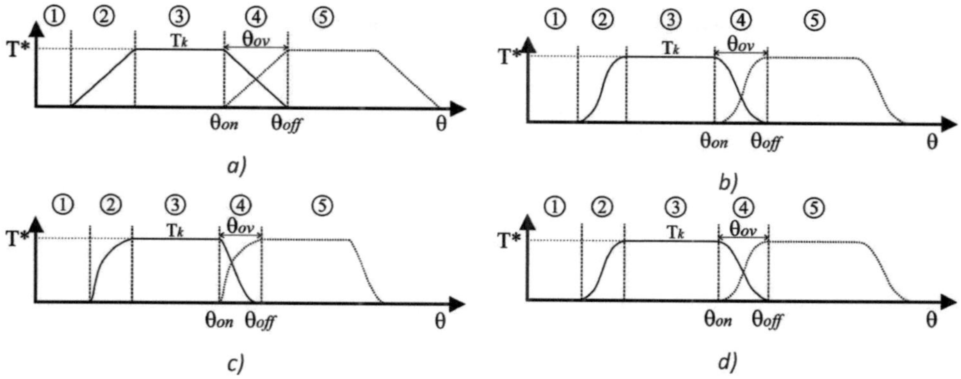

Figure 10.
TSF waveforms and sector divisions: (a) linear TSF, (b) sinusoidal TSF, (c) exponential TSF, and (d) cubic TSF.

$$f_{down}^{sin}(\theta) = \frac{1}{2} + \frac{1}{2}\cos\left[\frac{\pi}{\theta_{ov}}(\theta - \theta_{off})\right] \tag{13}$$

For the exponential TSF, the up and down functions are described by Eqs. (14) and (15), respectively, and the waveform of the expected reference torque is shown in **Figure 10(c)**.

$$f_{up}^{exp}(\theta) = 1 - \exp\left(-\frac{(\theta - \theta_{on})^2}{\theta_{ov}}\right) \tag{14}$$

$$f_{down}^{exp}(\theta) = 1 - f_{down}^{exp}(\theta - \theta_{off} + \theta_{on}) \tag{15}$$

The last TSF is the cubic. The up and down functions are represented as followed in Eqs. (16) and (17), while the waveform of the reference torque is illustrated in **Figure 10(d)**.

$$f_{up}^{lin}(\theta) = \frac{1}{\theta_{ov}}(\theta - \theta_{on}) \tag{16}$$

$$f_{down}^{lin}(\theta) = 1 - \frac{1}{\theta_{ov}}(\theta - \theta_{off}) \tag{17}$$

Observing Eqs. (10)–(17), one can perceive that the firing angles play an important role on the TSF creation. Normally, these angles are kept constant throughout the operation range. Varying the firing angles enables better performance in different operation points. Thus, for this book chapter, an OFAT optimization [22] was performed, and the optimum angles for each operation point (T, ω) were stored in lookup tables. Note that the optimum firing angles for each TSF are different. Thus, two lookup tables are necessary for each TSF, totalizing eight lookup tables.

4.2.2 Current profile creation

After obtaining the reference torque signals for each phase with the TSF, the reference current signals per phase (i_k^*) must be determined. This procedure may be accomplished by inverting the $T(i, \theta)$ lookup table into an $i(T, \theta)$ lookup table. In [25], the authors use the inverted $T(i, \theta)$ LUT to estimate electromagnetic torque and compare to a non-unity torque reference. The inverted $T(i, \theta)$ LUT is also applied in [26, 27], and in these cases, the contribution of the authors is to propose an alternative to commonly used TSFs. Alternatively, a nonlinear function of current as function of torque and rotor position can be used. The development of such function is cumbersome and requires a mathematical approach that models the inductance variation as function of electric current and rotor position. The analytical inversion function is expressed in Eq. (18), where $L_{0p}(\theta)$ is the inductance derivative for one SRM phase before saturation, that is, with low electric current, $f(\theta)$ is the function that describe the ratio change between the produced torque and the electric current according to the saturation level. More detailed explanation about the derivation of Eq. (18) may be found in [28].

The $i(T, \theta)$ lookup table obtained after inverting the $T(i, \theta)$ lookup table from the SRM model is presented in **Figure 11**. The parameters of the SRM used for the development of Eq. (18) and of the lookup table are presented in Appendix A.

$$i_k\left(T_k^*, \theta\right) = \frac{2T_k^*}{L_{0p}(\theta)}\left[\frac{f(\theta)}{2} + \sqrt{\frac{f^2(\theta)}{4} + \left(\frac{L_{0p}(\theta)}{2T_k^*}\right)^{\frac{1}{p}}}\right] \tag{18}$$

4.2.3 Simulation results

The block diagram implemented in the Matlab/Simulink® environment is represented in **Figure 12**. The speed error is processed in a PI controller to create the reference torque (T^*), and the reference torque and the rotational speed are used as input of the optimum firing angles lookup tables. The optimum angles and the reference torque are used to create the phase reference torque (T_k^*) through the TSF. The phase reference torque is then transformed in a phase reference current. Now, a hysteresis current controller drives the AHB in order to maintain the electric current following the reference.

To verify the TSF with the best torque ripple reduction capability, simulations throughout the entire speed range of operation were done, in steps of 10 rad/s. The results are presented in **Figure 13**. In the figure, a vertical line demarks the base speed ($\omega_b = 96\ rad/s$). The presented TSF returns similar torque ripple. The higher disparity occurs at 170 rad/s, where the exponential TSF have 19.59% less torque ripple than the cubic TSF. However, since the current profiling is an ITC method, it is expected that above base speed results are not acceptable and the interest speed range lies below the base speed. Below base speed, the sinusoidal TSF and the cubic TSF convey minor torque ripple. Thus, for future comparisons, the sinusoidal TSF will be employed because of its simple implementation.

The simulation results for sinusoidal TSF are presented in the following figures. The setups used in the simulations were the Simulation 1 ($S1$) and Simulation 2 (S_2) setups. The results for S_1 and S_2 setups are presented in **Figures 14** and **15**,

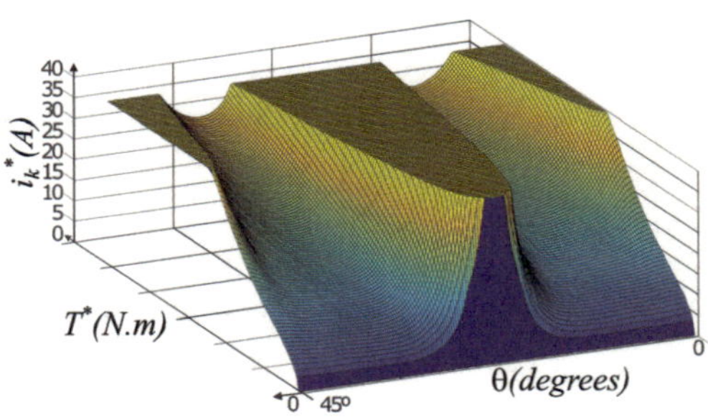

Figure 11.
$i(T, \theta)$ lookup table.

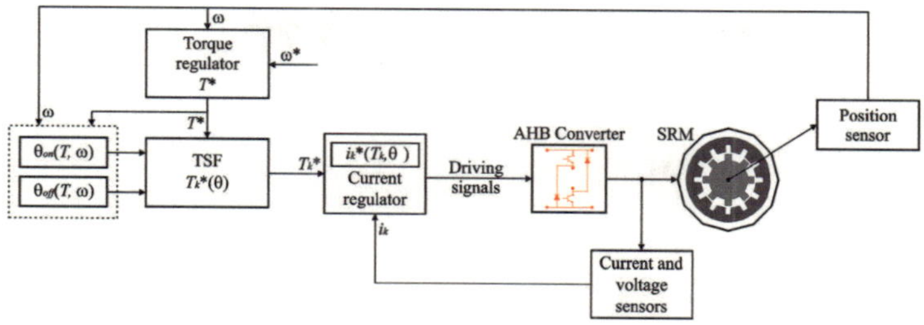

Figure 12.
Block diagram representing the current profiling through TSF with optimal driving angles.

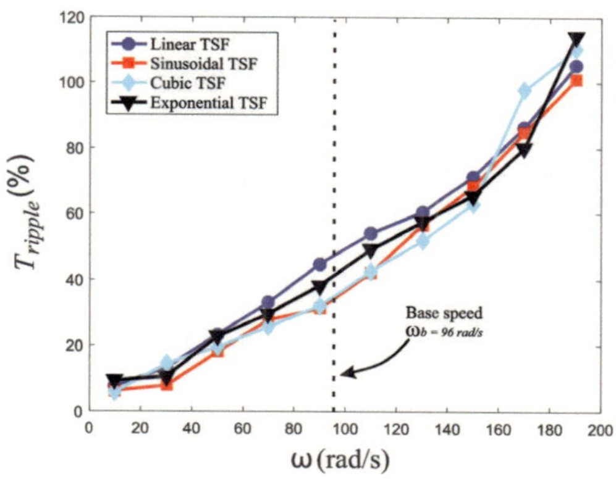

Figure 13.
Torque ripple level for each of the presented TSF in different speeds.

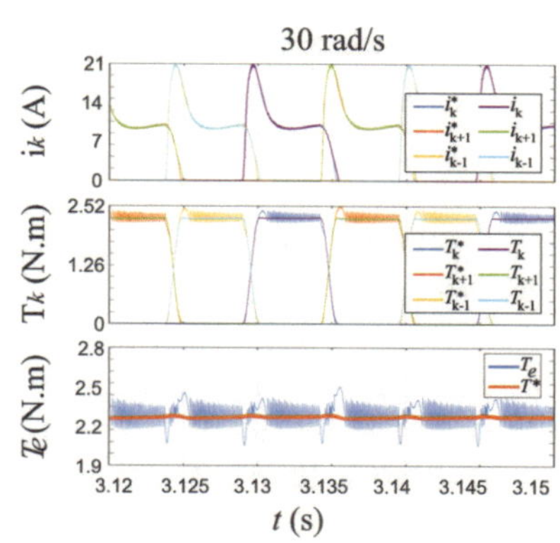

Figure 14.
Current profiling through sinusoidal TSF simulation results for setup S_1.

respectively. With S_1 setup, the current is accurately profiled, and the torque ripple is diminished, however, in S_2, the back EMF is high, and the controller is not capable of profiling the current, thus the torque ripple is considerably higher. The torque ripple, turn-on angle, and turn-off angle for each TSF using both setups are presented in **Table 1**.

When the speed rises, the error between reference torque and instantaneous torque produced by each phase rises as well. This phenomenon occurs for every TSF because there is a small error between the instantaneous current and the current reference at the phase energization and at the phase demagnetization. This current error is a consequence of the reduced maximum increase and decrease ratio of the electric current. When the speed rises, the back EMF also rises, diminishing the

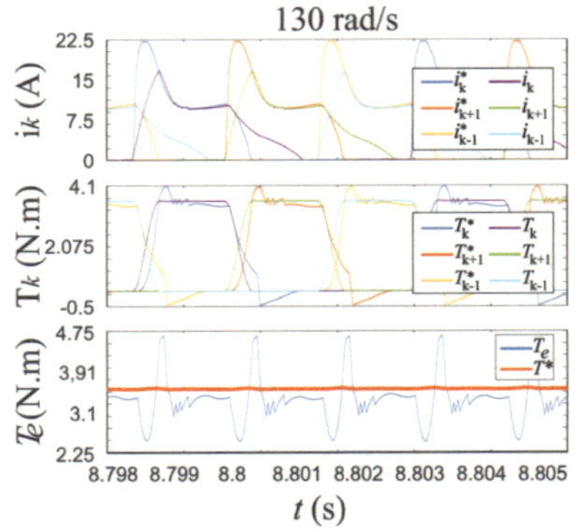

Figure 15.
Current profiling through sinusoidal TSF simulation results for setup S_2.

	Turn-on angle (°)		Turn-off angle (°)		Torque ripple (%)	
TSF	S_1	S_2	S_1	S_2	S_1	S_2
Linear	25.6	23.0	43.4	42.5	13.21	60.26
Sinusoidal	25.3	23.5	43.5	42.5	7.92	55.87
Exponential	27.3	23.6	43.5	43.0	10.7	58.12
Cubic	26.0	23.9	43.5	43.8	14.52	53.24

Table 1.
Control parameters for the different TSF using S_1 *and* S_2 *setups.*

maximum current change ratio. Then, the current error is reflected in the electromagnetic torque produced, causing the torque ripple to rise.

4.3 Direct torque control

The direct control technique (DTC) for SRM is based on the DTC technique for induction machines from the 80s and present simple dynamic and robustness. The DTC may be adapted to SRMs with any number of phases. However, three or five phases' adaptations are more recurrent in the literature than four phases' adaptations. This technique does not control the electric current, and instead, it aims to control the torque directly by applying different voltage states to the phases. The voltage states are stored in vectors and are chosen according to torque and magnetic flux linkage errors [29, 30].

The DTC uses flux linkage and electromagnetic torque values as control variables. The torque and flux conditions are used to determine which voltage state vector should be applied to the AHB in order to control the instantaneous torque produced. Therefore, to better understand the working principle of this method, the relation between excitation voltage and flux linkage and how the produced torque can be controlled through magnetic flux linkage changes must be clarified.

The vector of magnetic flux linkage in the stator ($\overrightarrow{\lambda_s}$) may be calculated as the time integral of the difference between the excitation voltage ($\overrightarrow{V_{DC}}$) and the voltage drop in the resistance ($R_s \overrightarrow{i_s}$) minus the initial flux linkage value ($\overrightarrow{\lambda_0}$), as described in Eq. (19).

$$\overrightarrow{\lambda_s} = \int_0^{\tau} \left(\overrightarrow{V_s} - R_s \overrightarrow{i_s} \right) dt - \overrightarrow{\lambda_0} \tag{19}$$

Generally, the voltage drop in the stator resistance is much smaller than the excitation voltage and then may be disregarded. In this case, the variation of flux linkage in the stator depends only on the voltage vector applied and can be described as in Eq. (20), indicating that a voltage vector applied to a phase causes a variation on the flux linkage of the same direction.

$$\Delta\overrightarrow{\lambda_s} = \overrightarrow{V_s}\,\Delta t \tag{20}$$

To calculate the flux linkage vector in the stator, one must know the flux linkage from each SRM phase and parse them to their stationary component. For a three-phase SRM, the Clarke transform may be used to calculate the stationary components of the magnetic flux linkage vector (λ_α and λ_β) through Eqs. (21) and (22), respectively.

$$\lambda_\alpha = \lambda_a - \lambda_b\cos 60^o + \lambda_c\cos 60^o \tag{21}$$

$$\lambda_\beta = \lambda_b\sin 60^o - \lambda_c\sin 60^o \tag{22}$$

If the stationary components of flux linkage in the stator are know, it is possible to calculate the total flux linkage as:

$$\lambda_s = \sqrt{\lambda_\alpha^2 + \lambda_\beta^2} \tag{23}$$

Furthermore, the stationary components may be used to estimate the rotor position through Eq. (24), enabling a position sensorless application.

$$\theta = arctan\left(\frac{\lambda_\beta}{\lambda_\alpha}\right) \tag{24}$$

Given the relation presented above, the DTC aims to maintain the flux linkage and torque values constrained in hysteresis bands. Variations in the produced torque are similar in direction to the variations in the flux linkage vector in relation to the rotor position, hence, for a torque increase, the flux linkage vector must be advanced from the rotor position vector, while for a torque decrease, the flux linkage vector must be delayed from the rotor position vector. These flux linkage variations are named flux acceleration and flux deceleration, respectively.

4.3.1 Voltage state vectors definition

Consider a three-phase SRM 12/8 poles with an AHB as the driver converter. As mentioned in the introduction, three different voltage states may be applied to each SRM phase. Thus, the voltage state vector (U_x) has three columns and one row as: $U_x = (u_A u_B u_C)$, where u_A, u_B, and u_C are the voltage states applied to phase A, phase B, and phase C, respectively. Now, to understand the vectors creation, first, it is important to understand the voltage sector division.

In the considered SRM, the phases are spaced in 120 electric degrees as in **Figure 16**. Tracing concentric axels over the phases, three axels are obtained. Each of them may be divided into a positive part, located at the axel portion that overlap the phase, and a negative part, on the other half of the axel, for example, for phase A, we obtain the A^+ portion and the A^- portion. The axels divide the vector space in *six* sectors of 60° each, named S_0, S_1, S_2, S_3, S_4, and S_5 as observed in **Figure 16**.

It is possible to define the state vectors for the DTC from the sectors. In total, since it is possible to apply 3 different voltage states per phase and there are 3

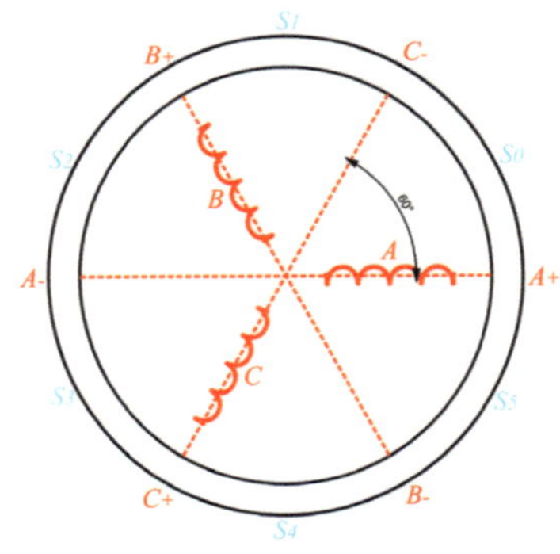

Figure 16.
Sectors division for a three-phase SRM.

phases, there are 27 possible state combinations. However, some of the combinations must be avoided, such as keep the three phases turned on at the same time, remaining only six combinations of interest for this case. Each of the remaining vectors are located at the center of each sector and are named U_0, U_1, U_2, U_3, U_4, and U_5 as in **Figure 17**. Each vector has three coordinates that represent the relation between the vector position and the axel portion of each phase, for example, the vector U_0 is located at the center of the sector S_0. The sector S_0 is bounded by the positive portion of axel A (A^+) and by the negative portion of axel C (C^-) and is not bounded by any portion of axel B, hence, U_0 is defined as $(1, 0, -1)$.

Table 2 summarizes the voltage states applied to the SRM phase windings according to the vector. Note that in **Figure 17**, there are two circles, one more internal and one more external. These circumferences represent the inferior and superior hysteresis bands for the flux linkage, respectively. The flux linkage vector in the stator must be maintained within this bandwidth through the application of the state vectors.

The choice of which vector must be applied depends on two conditions, one regarding the error between the reference torque and the instantaneous torque and other regarding the error between reference flux linkage and instantaneous flux linkage in the stator. Both conditions are implemented employing hysteresis

Vector	U_0	U_1	U_2	U_3	U_4	U_5
States	$(1, 0, -1)$	$(0, 1, -1)$	$(-1, 1, 0)$	$(-1, 0, 1)$	$(0, -1, 1)$	$(1, -1, 0)$

Table 2.
Voltage vectors to be applied in the AHB converter.

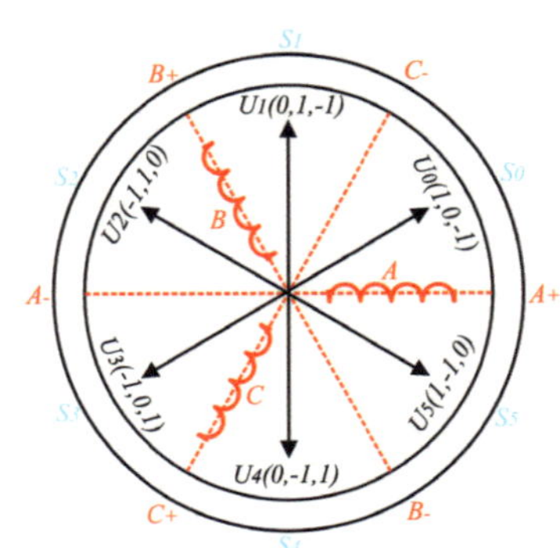

Figure 17.
Voltage state vectors definition for the DTC method.

controllers. The flux hysteresis controller receives the flux error and indicates if it necessary to increase or decrease the flux magnitude, while the torque hysteresis controller receives the torque error and indicates whether the flux vector must be accelerated or decelerated in order to control the torque production. The voltage vectors applied to make the stator flux linkage vector complete one clockwise cycle are detailed in **Figure 18** (considering that it is necessary to raise the torque). In case it is necessary to reduce the electromagnetic torque, the stator flux linkage must be rotated counterclockwise.

Table 3 compiles the logic behind the voltage state vector choice according to the sector (stator flux linkage vector position) and to the torque and flux conditions. For better understanding, considering the case in which the flux linkage vector is located at the S_4 sector, it is demanded to escalate the torque (instantaneous torque reached the inferior hysteresis band) and to reduce flux linkage (stator flux linkage reached the superior hysteresis band, that is, the external circle of **Figure 18**). In such case, the vector U_0 must be applied in order to keep the acceleration of the flux vector (increase torque) and lessen the amplitude (decrease flux), as depicted in **Figure 18**.

4.3.2 Variable magnetic flux linkage reference

The DTC requires a flux linkage reference for the hysteresis control realization. In the literature, diverse researches apply a constant reference throughout the entire operation speed range [29, 30]. However, observing the characteristics of the SRM, it is perceptible that for high values of speed (above base speed), where a high

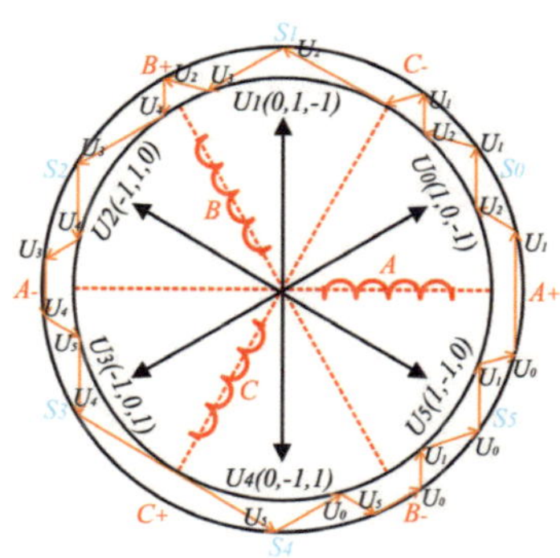

Figure 18.
State vectors defined and representation of a complete cycle of the flux linkage in the stator for a three-phase SRM.

		T&λ condition			
Sector	**Angular position**	**λ↑T↑**	**λ↑T↑**	**λ ↑ T ↑**	**λ↑T↑**
S_0	$0-60^o$	$U_1(0,1,-1)$	$U_5(1,-1,0)$	$U_2(-1,1,0)$	$U_4(0,-1,1)$
S_1	$60-120^o$	$U_2(-1,1,0)$	$U_0(1,0,-1)$	$U_3(-1,0,1)$	$U_5(1,-1,0)$
S_2	$120-180^o$	$U_3(-1,0,1)$	$U_1(0,1,-1)$	$U_4(0,-1,1)$	$U_0(1,0,-1)$
S_3	$180-240^o$	$U_4(0,-1,1)$	$U_2(-1,1,0)$	$U_5(1,0,-1)$	$U_1(0,1,-1)$
$\mathbf{S_4}$	$\mathbf{240-300^o}$	$U_5(0,1,-1)$	$U_3(-1,0,1)$	$\mathbf{U_0(1,0,-1)}$	$U_2(-1,1,0)$
S_5	$300-360^o$	$U_0(1,0,-1)$	$U_4(0,-1,1)$	$U_1(0,1,-1)$	$U_3(-1,0,1)$

Table 3.
DTC switching table for an three-phase SRM.

electric current is required, the flux linkage behavior becomes nonlinear according to the saturation level causing the torque ripple to increase according to the rotational speed increase. To overcome this drawback, a variable flux linkage reference may be used.

In certain operation speed, the electromagnetic torque developed starts to present drops, causing the torque ripple growth according to speed. This speed is defined as the critical speed (ω_{crit}). To estimate the critical speed, an open loop simulation with constant operation speed must be performed while observing the electromagnetic torque waveform. According to the speed rise, it is possible to note dips in the instantaneous torque waveform [29]. For the SRM used in this book chapter studies, the critical speed (ω_{crit}) is approximately 37.5 rad/s.

The flux linkage reference used here is divided into three cases, one for speeds below the critical speed, the second for speeds between the critical and the base speed, and the third for speeds above the base speed. For the first case, the flux reference is obtained with a PI controller that processes the speed error. For the second case, a function that describes the optimum flux decaying according to the speed increase is implemented ($\lambda_{dec}(\omega)$). In the third case, the flux linkage reference is maintained constant at the nominal value.

The decaying flux linkage function is established using an OFAT optimization to find the optimum flux linkage values for each operation speed. Then, the optimum values are interpolated to create the $\lambda_{dec}(\theta)$ function. The function is a sum of sinusoids presented in Eq. (25). The coefficients used in the equation are displayed in **Table 4**.

$$\lambda_{\text{dec}}(\omega) = \sum_{\text{i}=1}^{4} A_i \sin\left(B_i\omega + C_i\right) \tag{25}$$

Thus, the flux linkage reference ($\lambda^*(\omega)$) for the entire speed range of the DTC operation is described as following:

$$\lambda^*(\omega) = \begin{cases} PIcontrolleroutput, 0 \leq \omega < \omega_{crit} \\ \lambda_{dec}(\omega), \omega_{crit} \leq \omega < \omega_b \\ 0.0875, \omega_b \leq \omega \end{cases} \tag{26}$$

4.3.3 Simulation results

The block diagram for the DTC implementation is shown in **Figure 19**. The error between reference and instantaneous speed is processed through a PI controller to generate the reference torque. The flux linkage reference is according to Eq. (26). The torque reference is compared to the instantaneous torque, and the error is processed by a hysteresis controller to determine the torque condition. The same happens for the flux linkage to return the flux condition. Clarke transform is employed to estimate position and to calculate the total flux linkage in the stator. The sector selector block informs the sector according to the estimated position. The

Coefficients	A_1	B_1	C_1	A_2	B_2	C_2
Values	0.128	0.02826	−0.4405	0.04907	0.06595	−0.02546
Coefficients	A_3	B_3	C_3	A_4	B_4	C_4
Values	0.01876	0.1314	−1.89	0.01024	0.01658	4.76

Table 4.
A_i, B_i, C_i, and D_i coefficients for the $\lambda_{dec}(\theta)$ function.

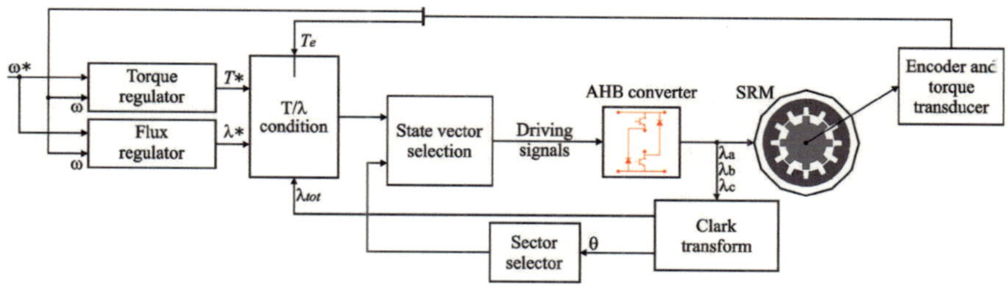

Figure 19.
Block diagram of the DTC.

sector and the conditions are then used to choose the voltage state vector according to **Table 3**.

The DTC controls the electromagnetic torque produced and the flux linkage in the stator. Considering the stationary frame ($\alpha\beta$-reference frame), the phase of the flux linkage vector must be accelerated or decelerated to produce torque, while the magnitude of the vector must be maintained constant. Plotting the α flux component (λ_α) against the β flux component (λ_β) should result in a circular profile as depicted in **Figure 20**.

The results obtained using the DTC with the S_1 and S_2 conditions are presented in **Figures 21** and **22**, respectively. The figures contain the waveforms of the flux linkage of each phase, instantaneous and reference magnitude of the flux linkage in

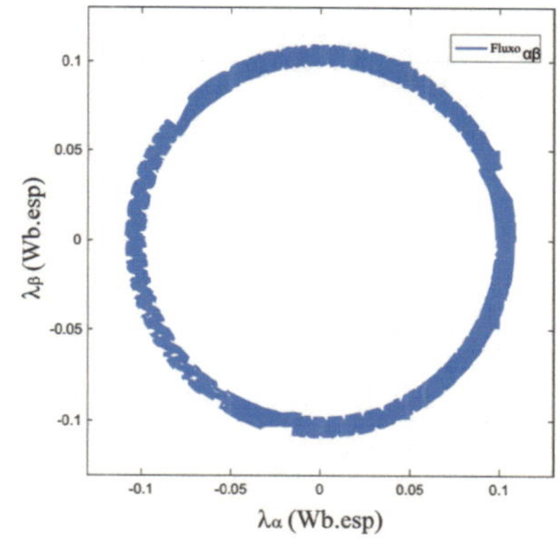

Figure 20.
Flux linkage in stationary coordinates for the DTC.

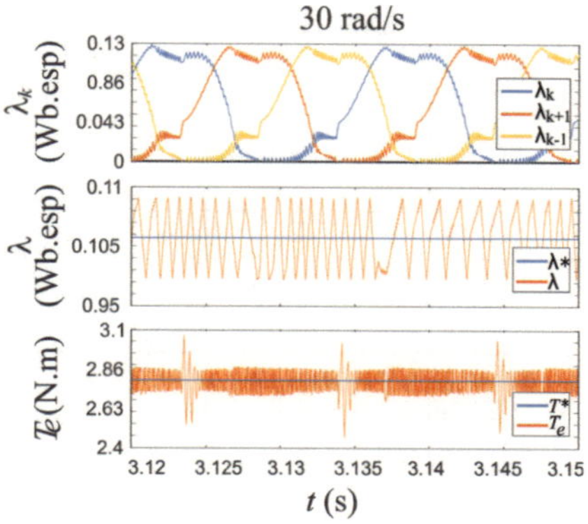

Figure 21.
DTC simulation results for setup S_1.

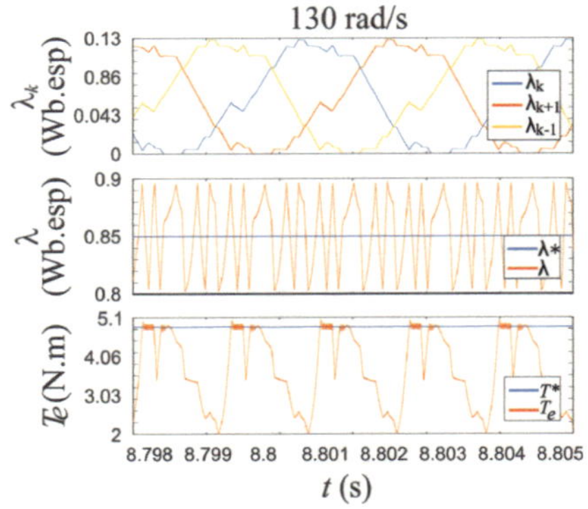

Figure 22.
DTC simulation results for setup S_2.

the stator (in stationary frame), and the instantaneous and reference electromagnetic torque. The torque ripple returned in both conditions is 24.41%, for the S_1 simulation, and 72.84%, for the S_2 simulation.

4.4 Direct instantaneous torque control

The direct instantaneous torque control (DITC) was first envisioned by Inderka [31]. The closed-loop control system uses a combination of double band current hysteresis controllers to generate the driving signals for each SRM phase. Differently from the current profiling through TSF, the DITC controls the torque directly.

The method dynamically divides the torque between two phases during commutation. The control system coordinates the AHB switch states according to an enabled conduction condition signal and to the instantaneous torque produced by each phase.

For better understanding of the DITC, first, some concepts must be clarified. In **Figure 23**, the electric currents for the three SRM phases are depicted. Two different intervals may be defined. One is the commutation interval, defined as the period in which two phases are simultaneously in conduction and both phases are responsible for the torque production. The other one is the single conduction interval, defined as the period in which only one phase is in conduction and is responsible for the entire torque production.

Furthermore, during the commutation interval, the two phases in conduction may be defined as incoming phase and outgoing phase. The outgoing phase is the

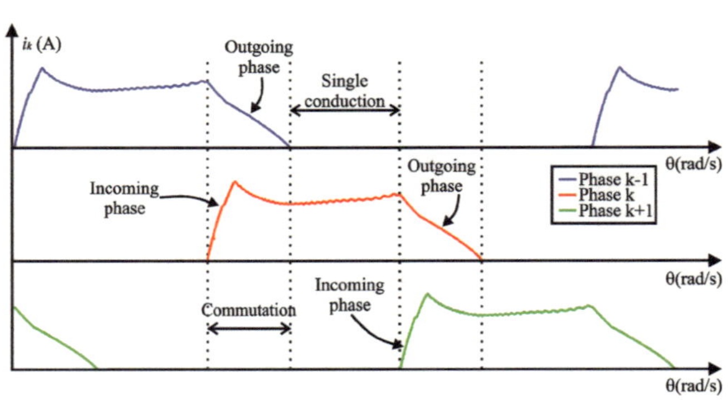

Figure 23.
Electric current in phases when DITC is applied and concepts definition.

phase that is in the end of the conduction period and is leaving commutation, while the incoming phase is the phase that is entering the conduction period.

The DITC classifies the phases, indicating whether the phase is enabled to enter conduction. The phase position must be in the interval between the turn-on angle and the turn-off angle ($\theta_{on} \leq \theta \leq \theta_{off}$) to be considered enabled to enter conduction. If this condition is satisfied by only one phase, single conduction is characterized, and the control system changes the voltage state of this phase to control the torque produced. If two phases satisfy the condition, phase commutation is characterized, and the voltage states of both phases must be individually controlled in such way that the sum of the produced torque by the phases equals to the total reference torque. The control strategies used in both cases are minutely explained hereafter.

4.4.1 State machines for direct instantaneous torque control

During commutation, there are two independent control logics simultaneously implemented. One for the incoming phase and one for the outgoing phase. The incoming phase is preferred for the torque production because the outgoing phase has already started the demagnetization process. Thus, the outgoing phase must be energized only if the incoming phase is not capable of producing the required reference torque by itself.

The following three state machines elucidate the working principles of DITC. The state machine in **Figure 24(a)** regulates the operation of the incoming phase, and the state machine in **Figure 24(b)** regulates the operation for the outgoing phase during commutation. Also, any phase that is in single conduction mode follows the logic within the state machine in **Figure 24(c)**. The conditions for voltage state change are indicated in the state machines, where T_{err} is the error between instantaneous phase torque and reference phase torque, *thss* is the superior torque limit, *thsm* is the superior medium torque limit, *thim* is the inferior medium torque limit, and *thii* is the inferior torque limit.

The waveforms of the total torque produced are presented in **Figure 25** along with the current of two commutating phases. In the figure, two regions are indicated during phase commutation. In region (1), the electromagnetic torque produced reaches the external inferior limit (*thii*), and then, the outgoing phase, which was in free-wheeling, is magnetized to augment the torque production, when the torque reaches back the *thim* limit, the phase state goes back to free-wheeling. In region (2), the torque reaches the superior torque band *thss*, thus the outgoing phase is demagnetized to reduce the torque production, and when the torque reaches back the *thsm* limit, the phase enter the free-wheeling state again.

Table 5 summarizes the working principle of the implemented CDTI. In this table, the voltage states (1, 0, and −1) are used to indicate the action. The "-" signal in the cells indicate that the phase is not accounted for torque production. The switching rules in **Table 5** guarantee three conditions aiming to achieve low torque ripple:

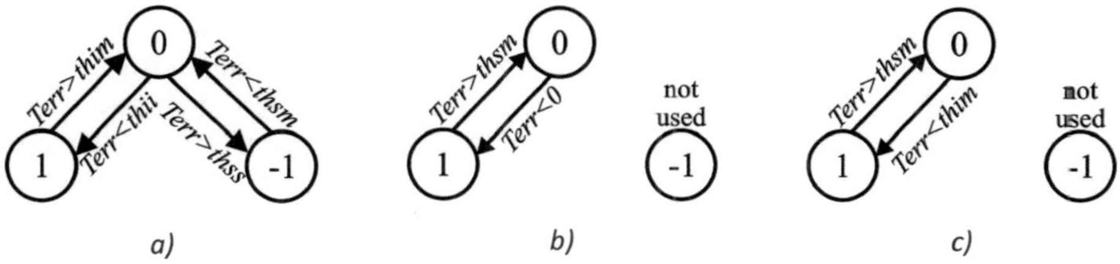

Figure 24.
State machines used in the DITC: (a) outgoing phase, (b) incoming phase, and (c) single conduction phase.

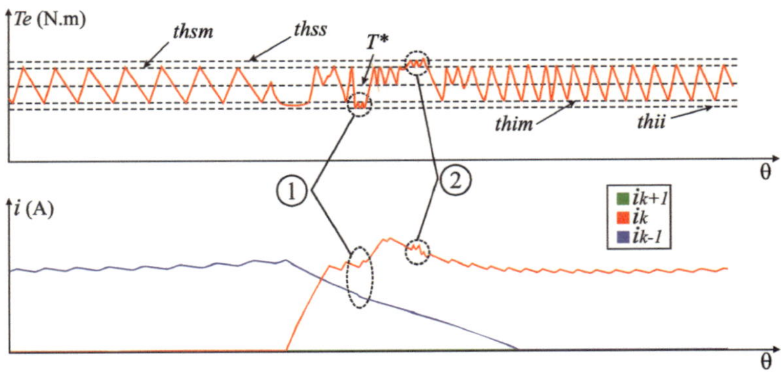

Figure 25.
Instantaneous electromagnetic torque and phase current waveforms highlighting the DITC logic.

Phase condition			State applied to phase		
$k-1$	k	$k+1$	$k-1$	k	$k+1$
0	0	0	—	—	—
1	0	0	1|0	—	—
1	1	0	1|0| − 1	1|0	—
0	1	0	−1|−	1|0	—
0	1	1	—	1|0| − 1	1|0
0	0	1	—	−1|−	1|0

Table 5.
Switch voltage state logic for DITC in a three-phase SRM.

- Avoid hard-chopping to minimize switching losses and reduce the torque ripple;
- Prioritize the incoming phase for the torque production so that the electric current flowing in the outgoing phase reaches zero before the inductance profile inclination becomes negative, thus avoiding negative torque production;
- Keep the incoming phase and the single conduction phase in the free-wheeling state for as long as possible to diminish the copper losses and the torque ripple.

4.4.2 Simulation results

The simulated system with DITC is depicted in **Figure 26**. The speed error is processed through a PI controller to return the reference torque. The reference torque is compared to the total instantaneous torque produced. The output error is used to decide the voltage state applied to the phase according to the switching state machines. The logic used depends on a condition signal generated for each phase separately. To synthesize the phase condition, the rotor position and the firing angles are required. The firing angles used are provided by a lookup table and depend on the SRM operation point (T, ω) as for the previous methods introduced.

The results obtained for the DITC in S_1 and S_2 conditions are presented in **Figures 27** and **28**, respectively. The waveforms presented are the electric current

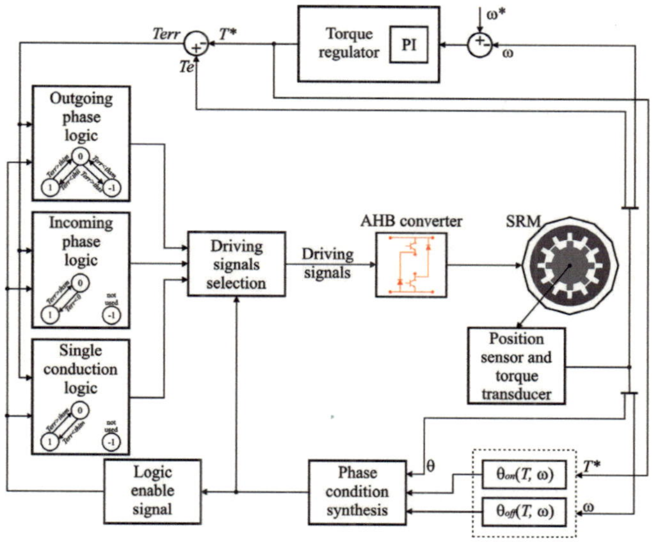

Figure 26.
Block diagram for the DITC.

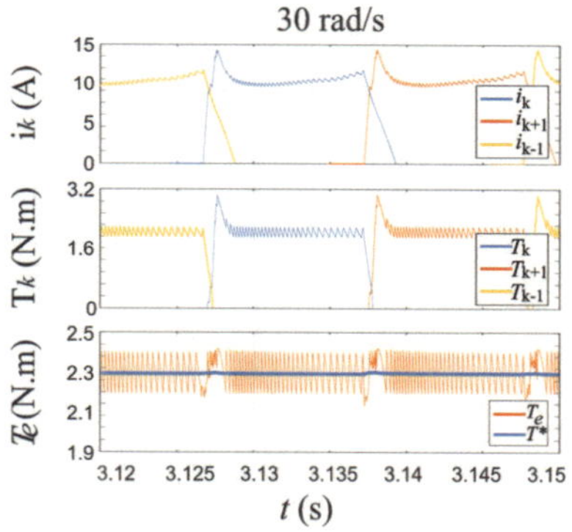

Figure 27.
DITC simulation result for setup S_1.

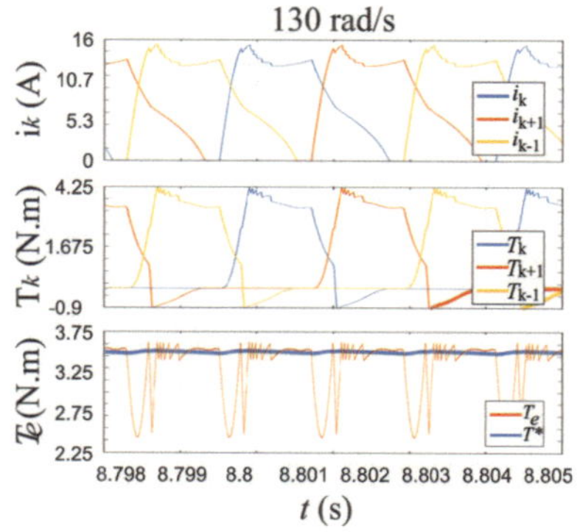

Figure 28.
DITC simulation result for setup S_2.

in the phases, individual electromagnetic torque produced by each phase, and the instantaneous and reference total electromagnetic torque produced. The turn-on angle, turn-off angle, and torque ripple for both simulations are disclosed in

Simulation	Turn-on angle (°)	Turn-off angle (°)	Torque ripple (%)
S_1	25.7	43.9	10.43
S_2	23.2	40.5	59.43

Table 6.
DITC simulation results for S_1 and S_2 conditions.

Table 6. When the speed is higher than the base speed, the current enters in the negative torque generation zone and causes torque dips as observed in **Figure 28**.

5. Results and discussion

For a better comparison of the methods behavior and performance throughout the SRM speed range of operation, simulations are performed from 10 rad/s up to the maximum possible speed with each method (even above nominal speed) at every 20 rad/s. The ripple factor at each speed for all methods presented in this chapter is calculated through Eq. (5), and the results are stored. The obtained values are grouped in **Figure 29** along with the results for the case where only a simple speed control is employed. The base speed (ω_b) is indicated as a vertical dashed line. The base speed for the studied machine is approximately 96 rad/s.

Remark that the ITC methods present better performance than the ATC method only when the electric current may be accurately controlled. Above 150 rad/s, the DATC presents better torque regulation capability, while below this speed, the DITC returned smaller torque ripple. Note that, below 30 rad/s, the current profiling with the sinusoidal TSF presented the lowest torque ripple; however, in all other speeds, the DITC surpassed the current profiling.

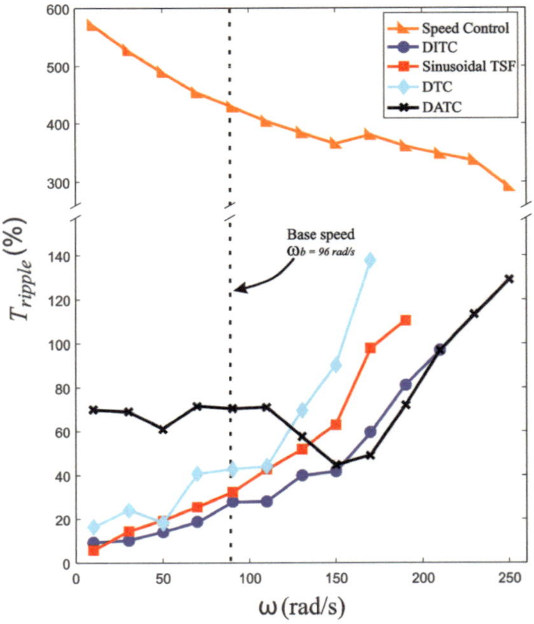

Figure 29.
Electromagnetic torque ripple for the different control methods presented throughout the entire SRM speed rage of operation.

When no torque ripple suppression method is applied, the ripple level reaches up to 570.1% at the low speed of 10 rad/s. According to the speed growth, the ripple is naturally diminished as an effect of the inertia moment. Note that, when any torque control method is employed, the torque ripple is diminished.

The preferred TSF among the studied functions are the sinusoidal one because of its easy implementation and good performance in the operation speed range of interest. When the sinusoidal TSF is compared to the other methods, it has better performance in a narrow speed range (below 30 rad/s) and then it is surpassed by the DITC. The DITC is the fittest ITC method since it naturally divides the torque production among phases and directly controls the total torque produced by the phases. However, as an ITC method, the DITC malfunctions in high speeds (above base speed), that is, torque dips appear as the speed increases until the instantaneous torque is no longer controlled. At above base speeds, the DATC method is preferred, because it considers only the average torque as target variable of the control.

6. Conclusions

In this chapter, different torque control methods were presented aiming to clarify which method is preferred for each operation point of the machine. One ATC method entitled DATC was presented, along with different ITC methods. The ITC methods addressed were the direct torque control (DTC), the current profiling through torque sharing function (TSF), and the direct instantaneous torque control (DITC). The most common TSF were used for the comparison, which are the linear, sinusoidal, exponential, and cubic TSF.

Simulations with each method were performed throughout the entire SRM speed range of operation. The DITC is the method that returned better torque ripple minimization capabilities bellow base speed, while above base speed, the DATC was the fittest method. With the presented simulations, the reader is able to understand how the SRM behavior changes according to the control method employed and operation point.

Some modifications and/or adaptations may be applied to the methods in order to improve the performance of the control system toward the chosen objective. For instance, the DATC method uses three LUT, which may be optimized according to the application and control objective. This feature makes this method very flexible.

The current profiling through torque sharing functions may be improved by creating torque profiles that account the difference between magnetization and demagnetization dynamics and how much each phase contributes to total torque produced. The DTC method can be improved by augmenting the number of sectors considered in the control strategy. The DITC technique may be modified by inserting variable turn-on and turn-off angles that can be optimized for different control objectives, such as minimize torque ripple, maximize efficiency or minimize acoustic noise, which also makes this method deeply adaptable.

Acknowledgements

This book chapter was a result of researches supported by the São Paulo Research Foundation, FAPESP under Grant 17/21640-9 and by the Brazilian National Council for Scientific and Technological Development - CNPq under Grant 142597/2018-0.

Appendix: switched reluctance motor parameters

This appendix presents the parameters of the SRM used in this chapter.

- Nominal power: 1.5 kW;
- Nominal speed: 1500 rpm;
- Number of poles: 12/8 (stator/rotor);
- Operation voltage: $30 - 80\ V$;
- Winding resistance: 0.3 Ω;
- Maximum flux linkage: 0.22 $Wb.esp$;
- Inertia moment (obtained through deceleration test): 0.002 kgm^2;
- Minimum phase inductance: 1.675 mH;
- Maximum phase inductance: 13.88 mH.

Author details

Marcelo Vinícius de Paula[1*], Tárcio André dos Santos Barros[1]
and Pedro José Dos Santos Neto[2]

1 School of Mechanical Engineering, University of Campinas, Campinas, Brazil

2 School of Electrical and Computer Engineering, University of Campinas, Campinas, Brazil

*Address all correspondence to: mvpaula@outlook.com

References

[1] Hu KW, Yi PH, Liaw CM. An ev srm drive powered by battery/supercapacitor with g2v and v2h/v2g capabilities. IEEE Transactions on Industrial Electronics. 2015;**62**(8): 4714-4727

[2] Bilgin B, Emadi A, Krishnamurthy M. Comprehensive evaluation of the dynamic performance of a 6/10 srm for traction application in phevs. IEEE Transactions on Industrial Electronics. 2013;**60**(7):2564-2575

[3] Mikail R, Sozer Y, Husain I, Islam M, Sebastian T. Torque ripple minimization of switched reluctance machines through current profiling. In: 2011 IEEE Energy Conversion Congress and Exposition. 2011. pp. 3568-3574

[4] Mikail R, Sozer Y, Husain I, Islam M, Sebastian T. Four quadrant torque ripple minimization of switched reluctance machine through current profiling with mitigation of rotor eccentricity problem and sensor errors. IEEE Transactions on Industry Applications. 2015;**5**:2097-2104. DOI: 10.1109/TIA.2014.2365715

[5] Castano SM et al. Acoustic noise analysis of a high-speed high-power switched reluctance machine: Frame effects. IEEE Transactions on Energy Conversion. 2016;**31**:69-77. DOI: 10.1109/TEC.2015.2470079

[6] Krishnan R. Switched Reluctance Motor Drives: Modeling, Simulation, Analysis, Design, and Applications. New York: CRC Press LLC; 2001. 398 p. ISBN 9780849308383

[7] Miller TJE. Switched Reluctance Motors and their Control. Oxford: Oxford Press; 1993. ISBN 0-19-859387-2

[8] Hu Y et al. Split converter-fed srm drive for flexible charging in ev/hev applications. IEEE Transactions on Industrial Electronics. 2015;**62**: 6085-6095. DOI: 10.1109/TIE.2015.2426142

[9] Chang HC, Liaw CM. Development of a compact switched-reluctance motor drive for ev propulsion with voltage-boosting and pfc charging capabilities. IEEE Transactions on Vehicular Technology. 2009;**58**:3198-3215. DOI: 10.1109/TVT.2009.2017546

[10] Hu Y et al. Modular tri-port high-power converter for SRM based plug-in hybrid electrical trucks. IEEE Transactions on Power Electronics. 2017;**33**:3247-3257. DOI: 10.1109/TPEL.2017.2701784

[11] Gan C et al. New integrated multilevel converter for switched reluctance motor drives in plug-in hybrid electric vehicles with flexible energy conversion. IEEE Transactions on Power Electronics. 2017;**32**: 3754-3766. DOI: 10.1109/TPEL.2016.2583467

[12] Lu S, Corzine KA, Ferdowsi M. A new battery/ultracapacitor energy storage system design and its motor drive integration for hybrid electric vehicles. IEEE Transactions on Vehicular Technology. 2007;**56**: 1516-1523. DOI: 10.1109/TVT.2007.896971

[13] Mademlis C, Kioskeridis I. Performance optimization in switched reluctance motor drives with online commutation angle control. IEEE Transactions on Energy Conversion. 2003;**18**:448-457. DOI: 10.1109/TEC.2003.815854

[14] Xue XD et al. Optimal control method of motoring operation for srm drives in electric vehicles. IEEE Transactions on Vehicular Technology. 2010;**59**:1191-1204. DOI: 10.1109/TVT.2010.2041260

[15] Kioskeridis I, Mademlis C. Maximum efficiency in single-pulse controlled switched reluctance motor drives. IEEE Transactions on Energy Conversion. 2005;**20**:809-817. DOI: 10.1109/TEC.2005.853738

[16] Vujicic VP. Minimization of torque ripple and copper losses in switched reluctance drive. IEEE Transactions on Power Electronics. 2012;**27**:388-399. DOI: 10.1109/TPEL.2011.2158447

[17] De Paula M, Barros TAS, Santos No PJ, Ruppert FE. Estudo comparativo de técnicas de minimização de ondulações de torque para motores de relutância variável trifásicos. In: Proceedings of the Congresso Brasileiro de Automática (CBA2018). Brazil; 2018. ISSN 2525–8311

[18] Dowlatshahi M et al. Copper loss and torque ripple minimization in switched reluctance motors considering nonlinear and magnetic saturation effects. Journal of Power Electronics. 2014;**14**:351-361. DOI: 10.6113/JPE.2014.14.2.351

[19] Brauer HJ, Hennen MD, Doncker RWD. Control for polyphase switched reluctance machines to minimize torque ripple and decrease ohmic machine losses. IEEE Transactions on Power Electronics. 2012;**27**:370-378. DOI: 10.1109/TPEL.2011.2161332

[20] Inderka RB, Doncker RWAAD. High-dynamic direct average torque control for switched reluctance drives. IEEE Transactions on Industry Applications. 2003;**39**:1040-1045. DOI: 10.1109/TIA.2003.814579

[21] Hannoun H, Hilairet M, Marchand C. Design of an srm speed control strategy for a wide range of operating speeds. IEEE Transactions on Industrial Electronics. 2010;**57**: 2911-2921. DOI: 10.1109/TIE.2009.2038396

[22] Neto PJ d S, Barros TA d S, de Paula MV, de Souza RR, Filho ER. Design of computational experiment for performance optimization of a switched reluctance generator in wind systems. IEEE Transactions on Energy Conversion. 2018;**33**:406-419. DOI: 10.1109/TEC.2017.2755590

[23] Dos Santos Barros TA, dos Santos Neto PJ, de Paula MV, Moreira AB, Nascimento Filho PS, Ruppert Filho E. Automatic characterization system of switched reluctance machines and nonlinear modeling by interpolation using smoothing splines. IEEE Access. 2018;**6**:26011-26021. DOI: 10.1109/ACCESS.2018.2825607

[24] Petrus V. Switched reluctance motors for electric vehicle propulsion: comparative, numerical and experimental study of control schemes [thesis]. Technical University of Cluj-Napoca & Universitánd Libre de Bruxellesa; 2013

[25] Hyeung S, Kiwoo P, Kyobeum L. A non-unity torque sharing function for torque ripple minimization of switched reluctance generators in wind power systems. Energies. 2015;**8**:11685-11701. DOI: 10.3390/en81011685

[26] Lee DH et al. A simple nonlinear logical torque sharing function for low-torque ripple sr drive. IEEE Transactions on Industrial Electronics. 2009;**56**: 3021-3028. DOI: 10.1109/TIE.2009.2024661

[27] Ye J, Bilgin B, Emadi A. An extended-speed low-ripple torque control of switched reluctance motor drives. IEEE Transactions on Power Electronics. 2015;**30**:1457-1470. DOI: 10.1109/TPEL.2014.2316272

[28] Vujicic VP. Modeling of a switched reluctance machine based on the invertible torque function. IEEE Transactions on Magnetics. 2008;**44**: 2186-2194. DOI: 10.1109/TMAG.2008.2000663

[29] Zhao X, Xu A, Zhang W. Research on dtc system with variable flux for switched reluctance motor. CES Transactions on Electrical Machines and Systems. 2017;**1**:199-206. DOI: 10.23919/TEMS.2017.7961342

[30] Xu AD, He K, Cao Y. Torque ripple reduction of srm using optimized voltage vector in dtc. In: 2016 IEEE Vehicle Power and Propulsion Conference (VPPC). [S.l.: s.n.]. 2016. pp. 1-5

[31] Inderka RB, Doncker RWAAD. Ditc-direct instantaneous torque control of switched reluctance drives. IEEE Transactions on Industry Applications. 2003;**39**:1046-1051. DOI: 10.1109/TIA.2003.814578

Chapter 12

Average Rated Torque Calculations for Switched Reluctance Machines Based on Vector Analysis

Aleksas Stuikys and Jan Sykulski

Abstract

An average rated torque estimation for generally saturable switched reluctance (SR) machines based on vector analysis is described. The proposed analytical method enables the switched reluctance machine designers to compute quickly and relatively accurately the rated torque of the machine. This approach offers simplicity, accuracy, and intuitive insight characteristic to analytical solutions of magnetically nonlinear problems otherwise achievable only with time-consuming computer-based numerical simulation tools. The suggested analytical methodology, therefore, offers immediate answers regarding the rated torque performance at the early stages of the machine sizing and design process. In this chapter, the switched reluctance machine rated torque calculation is derived based on the analytically estimated flux-linkage characteristic map and the knowledge of the DC bus voltage of the machine. It is further demonstrated that the proposed analytical rated torque calculations, based on vector analysis, enable construction of highly accurate instantaneous phase current profiles using a graphical method and thus aiding intuition and providing valuable insight into the nonlinear switched reluctance machine operation and control requirements. The proposed method will be found particularly suitable for those studying the nonlinear design and control of switched reluctance machine technologies for electric vehicle traction and industrial applications.

Keywords: analytic average torque calculation, vector analysis, nonlinear switched reluctance machines

1. Introduction

Since the dawn of the popularity of the switched reluctance (SR) motors [1], many diverse developments of this machine's technology have resulted, including theoretical analysis, and the volume of the published literature on these topics is enormous. For this reason it would be impossible to encompass the entire range of published methods of analysis and design of SR motors, let alone to effectively utilize them in the machine design process. Therefore the aim of this work is to provide the readers with the essential analytical formulae and methodologies in order to allow them to analyze quickly and relatively accurately a desired SR machine for a particular application.

Since the principle of operation of SR machines does not require any permanent magnets for the production of torque, such machines offer an excellent alternative electric motor technology at this moment in time in the context of the cost and environmental impact uncertainties of permanent magnet materials [2]. Usually, the SR motor will be designed to operate from a fixed DC voltage supply, such as a battery, and the stator windings of the machine will be of the concentrated type. A typical structure of an SR electric motor or generator is shown in **Figure 1**.

In **Figure 1** the depicted machine contains 18 stator poles and 12 rotor poles; therefore this particular SR machine configuration is abbreviated as an 18/12 SR motor. Therefore, with the knowledge of these variables, we are able to compute the number of "steps" the rotor will make in a single revolution as it gets aligned to each of the energized stator poles, as in Eq. (1):

$$\textit{number of rotor steps} = m \cdot N_r \tag{1}$$

where m is the number of phases and N_r is the number of rotor poles. For the 18/12 configuration, Eq. (1) gives 36 steps. Therefore, the rotor will make 36 steps from one stator pole to the next in one complete revolution.

For the SR machine to operate effectively, it requires a capable high power and switching frequency power electronic converter in order to shape the phase current waveforms to be fed into the SR machine phase windings. The most general power electronic converter structure used for SR machines, known as the asymmetric half-bridge converter, is shown diagrammatically in **Figure 2**.

The power electronic converter for the three-phase 18/12 SR machine is assumed to be connected to a fixed magnitude DC bus voltage V. Each phase of the machine is connected to two power transistor switches, *T1* and *T2*, and two power diodes, *D1* and *D2*, in an asymmetric h-bridge configuration. The transistor switches are active devices, and therefore in order to control their conduction states and switching speed, the electronic signals are used: *sig1* and *sig2*. The power diodes are passive components, and therefore they do not require electronic signals in order to work. The capacitor C is connected across the DC bus supply voltage lines to be used as an energy dump capacitor.

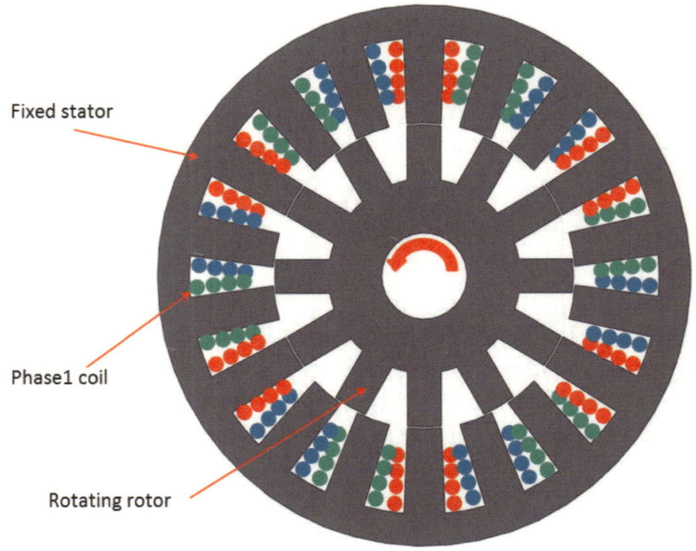

Figure 1.
Cross-sectional view of the electromechanical structure of a three-phase 18/12 switched reluctance machine.

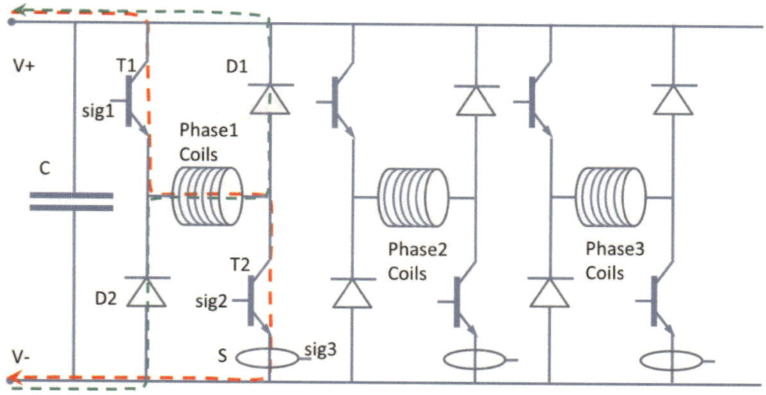

Figure 2.
H-bridge asymmetric power electronic converter for the three-phase 18/12 SR machine.

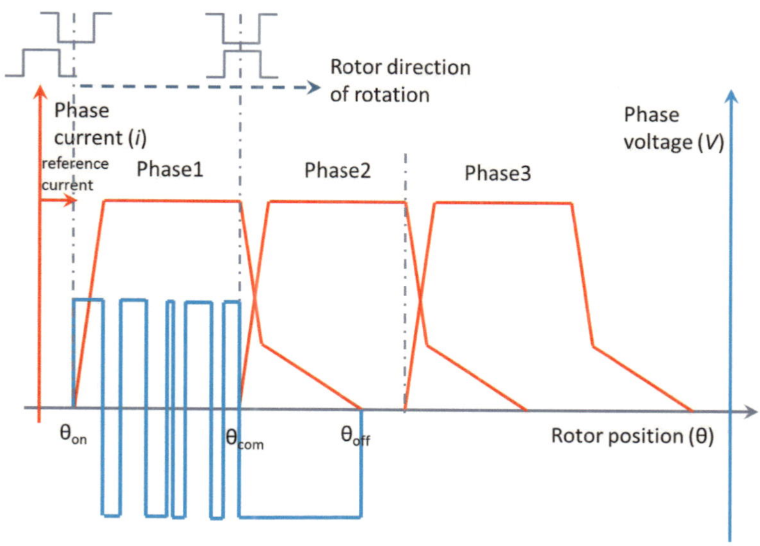

Figure 3.
Phase current profiles in the phase windings resulting from the voltage switching strategy of the asymmetric h-bridge converter.

The sequence of operation of the h-bridge converter in terms of phase current profiles as shown in **Figure 3**—for the motoring mode of the SR machine—is as follows. Assuming the hard chopping voltage switching strategy to begin with, when the rotor pole is furthest away from the stator pole, the two power transistor switches are latched so as to enable the conduction of the current. This action results in the electric current flow from the positive supply terminal, through the phase windings, to the negative terminal, as represented by the red dashed line in **Figure 2**, with the arrow indicating the direction of the flow. Once the current has reached some predetermined reference value, say peak rated current, the signal from the current sensor *S* is sent to the electronic controller (not shown) of the SR machine which, in turn, sends signals to the power transistors, *T1* and *T2*, in order to switch them off, so as to stop the flow of the current. This last action, however, does not stop the circulation of current in the phase winding immediately since the magnetic field, created by the currents, prevents the current from reducing to zero instantly. Moreover, exactly at the same time, the power diodes start to conduct current in the phase winding, as shown by the green dashed line in **Figure 2**, with the arrow indicating the direction of the current flow. From this operation we deduce that the current in the phase winding does not alternate and is therefore a direct current (DC). The resulting current profile is shown schematically in **Figure 3**.

From **Figure 3** it may be seen that once the current reaches some specified value, the electronic controller has to check the value of the measured current by the current sensor *S* in the winding, so as to keep it constant up until the commutation angle θ_{com}, at which point both transistors are switched off. The current profile from θ_{com} to θ_{off} is a result of the power diodes' action, represented by the green dashed line in **Figure 2**. At this stage it becomes clear why the dump capacitor *C* is used in h-bridge converters: because the magnetic energy established by the current circulating in the winding is not fully used up, it has to be returned in the form of a current back into the supply; however, in order to keep the supply undisturbed, the capacitor *C* in **Figure 2** is used to absorb the returning energy. The entire switching process is repeated for the other phases of the SR machine as shown in **Figures 2** and **3**, where generally the overlap of each of the three phase current profiles will depend on the timing of the transistor switching action as well as on the operating speed of the machine and the torque requirement at that speed.

It should be noted here that the analytical rated torque calculations presented below are based on the assumption that the correct switching mode of the power electronic converter is assumed first. Furthermore, the actual performance of an SR machine can be greatly affected by the design choice of the power converter topology and the switching action, including the torque production and the energy conversion efficiency of the device [3]. Therefore, the following analytical solutions will be based on the "hard chopping" voltage switching strategy as in **Figure 3** for the asymmetric half-bridge converter presented in **Figure 2**.

The particular SR motor considered in this chapter is taken from [4]. The main reason behind the decision to use a published SR machine design was that it had been extensively optimized, its performance was analyzed, and it was actually built and tested in order to verify the results. Finally, the published work should be of reproducible quality, and therefore the chosen SR machine design makes an excellent choice when comparing the validity of the proposed analytical solution with the solutions obtained from extensive computer simulations and carefully controlled experimental measurements.

The chosen 50 kW SR motor parameters to be used in the present analysis are given in **Table 1**.

By application of the open license finite element method [5] to the magnetic analysis of the chosen SRM2 design, the aligned and the unaligned rotor positions are analyzed in terms of the magnetic flux distributions in the magnetic circuit of the device, as shown in **Figures 4** and 5.

Figures 4 and 5 are used to describe the fundamental theory of SR machines in terms of the phase flux-linkage.

Therefore, if we consider the amount of magnetic flux Φ, measured in units of weber, being established in the stator pole around which the number of conductors *N* is wound in the concentrated manner as in **Figure 1**, then the flux-linkage thus created is given in Eq. (2):

$$\Psi = N \cdot \Phi (V \cdot s) \tag{2}$$

where instead of the intuitive units of weber-turns, we choose to use the volt-seconds (as a derived SI unit) since this unit of measurement will be very handy when we explain the significance of the flux-linkage in the SR machine analysis section. Furthermore, the phase flux-linkage computed in Eq. (2) is multiplied by the number of the coils connected in series, N_s, to obtain the total flux-linkage for that particular phase.

SR motor design variables and test results	
Design parameters	**50 kW SRM2**
Outer diameter [mm]	269
Stack length [mm]	135
Air gap length [mm]	0.5
Iron core material	10JNEX900
Wire diameter [mm]	0.6
Wire turns [turns]	17
Wire parallel turns [turns]	22
Slot fill factor [%]	57.0
Stator pole arc angle β_s [degrees]	10.5
Rotor pole arc angle β_r [degrees]	11
Rated angular speed [rpm]	1200
Current, peak [A]	320
Max. current density [A/mm^2]	33 (33)
RMS current (@1200 rpm) [Arms]	204 (206)
Max. torque (@1200 rpm) [Nm]	415 (400.4)
Number of phases	3
DC bus voltage [V]	500
Max. power [kW]	50

() denotes test results.

Table 1.
50 kW SR machine parameters, adapted from [4].

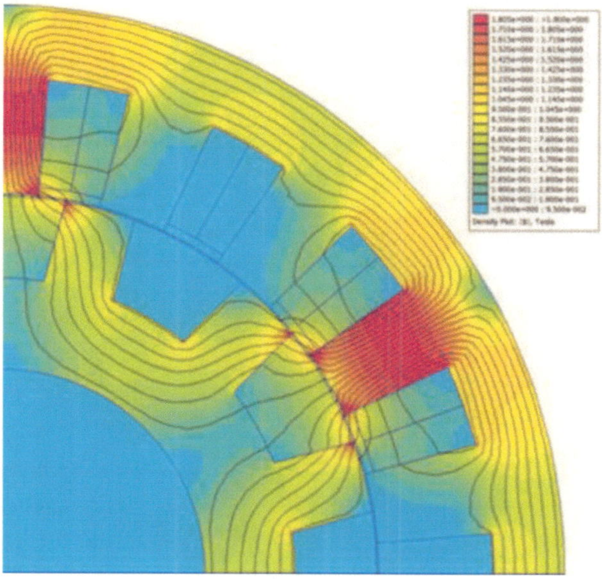

Figure 4.
Magnetic flux distribution and flux density in the switched reluctance machine for the unaligned rotor position.

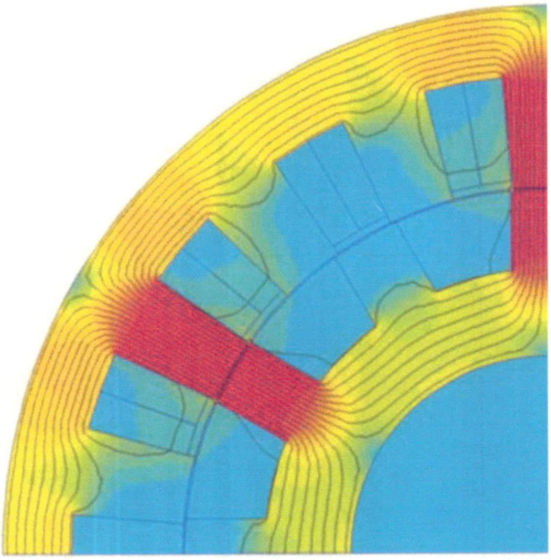

Figure 5.
Magnetic flux distribution and flux density in the switched reluctance machine for the aligned rotor position.

Using the same reasoning as above; Eq. (2) will apply for the unaligned rotor position as well as for the aligned case as shown in **Figures 4** and **5**. In fact Eq. (2) is used to compute the flux-linkage for any rotor angular position with respect to the stationary stator. One further outcome we are able to accomplish—in addition to varying the rotor angular position to compute the flux-linkage for that rotor position—is to vary the phase current that is being circulated in the concentrated winding of the stator pole. Therefore, if we increase the current from zero to its full rated peak value (see **Table 1**), for each of the rotor positions from the fully aligned to the fully unaligned, in steps of 1 mechanical degree, the complete flux-linkage map of the 18/12 SR machine will result, as plotted in **Figure 6**.

Some highly notable observations of **Figure 6a** are that for the aligned rotor position, the flux-linkage builds up very quickly for the range of currents, after which it levels off considerably; this occurs in the region of 80 (amps) called the magnetic saturation point. The unaligned rotor position flux-linkage is fairly linear for the entire range of the phase current values, especially if compared to the fully

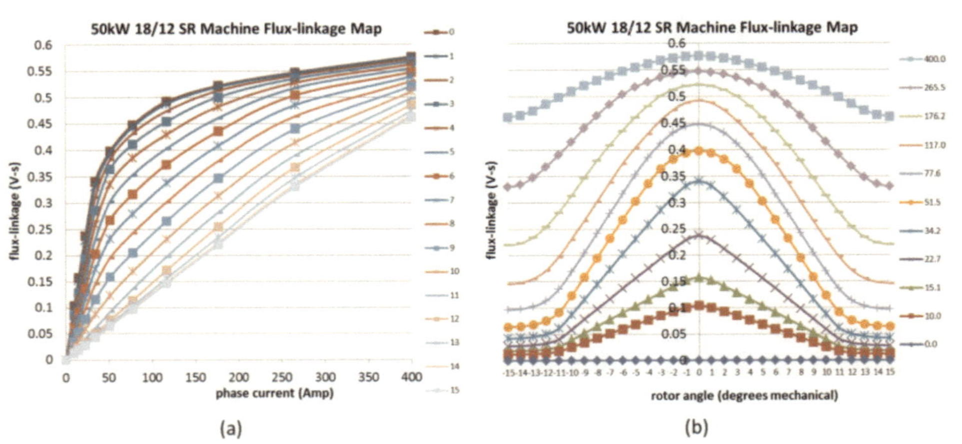

Figure 6.
Flux-linkage map for the 18/12 SR machine; (a) taking the rotor position as a parameter, (b) taking the phase current as a parameter.

aligned position. The intermediate rotor position flux-linkage curves, generally speaking, become highly nonlinear as the rotor tends toward the fully aligned position. If the flux-linkage curves with respect to the rotor position are examined taking the phase current as a parameter, as in **Figure 6b**, it may be seen that linearity is not present at all, and for a given current value, the flux-linkage value will vary with respect to all rotor positions from the fully unaligned to the fully aligned. This is true irrespective of the rotor rotation in clockwise or counterclockwise direction, as in **Figure 6b**.

The torque production of the SR machine can be described in terms of the flux-linkage map shown in **Figure 6a**. If we were to consider only the fully aligned and unaligned rotor position flux-linkage curves, as in **Figure 7**, then it would follow that as the rotor tends toward the fully aligned position, the current build-up in the phase winding would increase the stored magnetic field energy W_f. The phase current would then follow the same profile as in **Figure 3**, yet this time it is shown with respect to the flux-linkage curves in **Figure** 7 (red dashed).

As can be seen from the graphical representation, the magnetic co-energy W' is represented by the area bound by the aligned and the unaligned flux-linkage curves and therefore is given by Eq. (3) in terms of the aligned flux-linkage:

$$W' = \int_0^i \Psi di \text{ (Joules)} \tag{3}$$

Thus, generally, if the magnetic field co-energy is created as a result of the rotor moving from the unaligned to the aligned position, while the phase current is kept constant, the torque generated, T, can be computed as in Eq. (4):

$$T = \left[\frac{\partial W'}{\partial \theta}\right]_{i=constant} (N \cdot m) \tag{4}$$

Therefore, for a given SR machine, we are interested in maximizing the area W' in order to arrive at a torque dense machine design. This is possible to achieve effectively if the aligned rotor position flux-linkage curve is maximized for all phase current values, while the unaligned rotor position flux-linkage curve is minimized for a given SR machine. Furthermore, the phase current profile itself will determine the magnitude of W' created, providing the phase current locus encompasses as much of the area as possible, as shown in **Figure** 7.

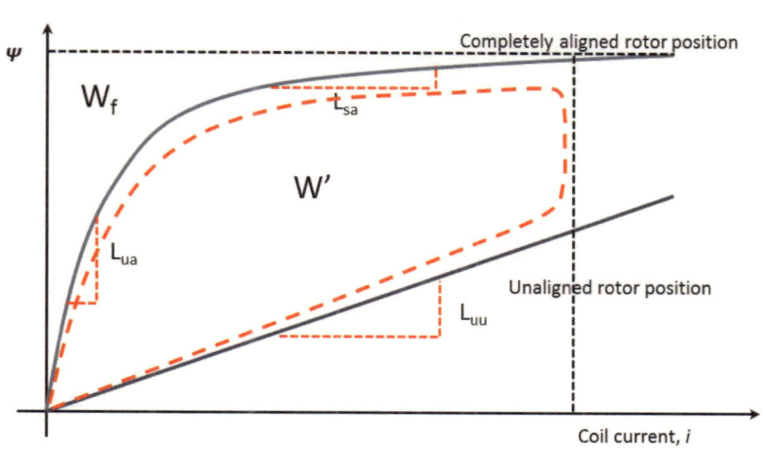

Figure 7.
Magnetic field energy W_f and co-energy W' representation.

The published and generally accepted phase torque calculations were based on the assumption of a linear machine theory, thus assuming non-saturating SR machine flux-linkage curves and their associated unsaturated aligned L_{ua} and unsaturated unaligned L_{uu} inductances, as shown in **Figure 8a**. The alternative to the linear theory is the nonlinear flux-linkage curves theory [6], with the additional saturated aligned inductance L_{sa}, in which W' is represented by the parallelogram area the phase current locus encompasses, as shown in **Figure 8b**.

From **Figure 8a** it is seen that the co-energy area W' is represented by the triangle and therefore is given as in Eq. (5):

$$W' = \frac{1}{2}\Psi \cdot i = \frac{1}{2}(L_{ua} - L_{uu})i^2 \text{ (Joules)} \tag{5}$$

where L is an inductance which defines the slope of the particular linearized flux-linkage curve from **Figure 6a** as in **Figure 8a** at a particular rotor position and the peak value i of the flat-topped current profile. The torque produced by such a non-saturating SR machine will be given as in Eq. (6):

$$T = \frac{1}{2}i^2 \frac{dL}{d\theta}(N \cdot m) \tag{6}$$

which is the same as Eq. (5) being divided by the rotor angle that is being turned.

The torque production for the saturating flux-linkage curves producing parallelogram area of magnitude W' in **Figure 8b** is given in Eq. (7) [6]:

$$W' = V\frac{c \cdot \beta s}{\omega} \cdot i\left[1 - \frac{1}{2}\frac{c}{s}\right] \text{ (Joules)} \tag{7}$$

where c is commutation factor, s is combined inductance ratio, βs is stator pole arc angle as in **Table 1**, V is the DC bus voltage, ω is rotor angular speed, and Eq. (7) is applicable only when $L_{sa} = L_{uu}$.

In general, the non-saturating SR machine is rather a special case of the generally saturating SR machine and for this reason will not deliver as high a torque value as is otherwise potentially possible with the saturating SR machine, since the area W' will be much lower for the given rated current at a specified speed of the machine. The saturating SR machine with the assumed co-energy area W' represented by the

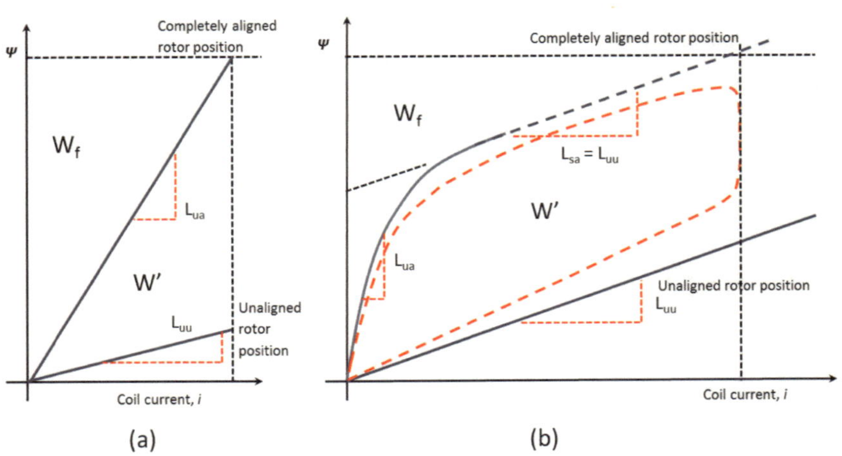

Figure 8.
Linear flux-linkage and torque production theory (a), nonlinear flux-linkage and torque production theory based on parallelogram area (b).

parallelogram area, as in **Figure 8b**, is not very accurate since in practical SR machine devices $L_{sa} \neq L_{uu}$, which is the slope in **Figure 8b**, since the performance envelope of SR machines, in terms of the inductance ratios, is limited to the following bounds [7]:

$$6 < \frac{L_{ua}}{L_{uu}} < 14 \tag{8}$$

These bounds indicate that a fully saturable SR machine having the inductance ratio of less than 6 is considered to be unsuitable as it will not be a torque dense machine. On the other hand, the inductance ratio of 10 or more in Eq. (8) will indicate exceptionally high torque density SR machine capable of high instantaneous torque production, which is most desirable. It follows that our SRM2 design, from **Figure 6**, is capable of a reasonable torque production since the inductance ratio is about 8. However, achieving the inductance ratio of more than 10 is not a simple task, calling for highly specialized computer-based multi-objective optimization routines that are computationally intensive [8], and is therefore beyond the scope of the treatment in this chapter.

Going back to the previously stated argument regarding the assumed W' based on the parallelogram area calculation with Eq. (7), from **Figure 8b** it is evident that the saturated aligned flux-linkage curve has the same slope L_{sa} as the unsaturated unaligned flux-linkage curve slope L_{uu}. Comparing these two parallel inductance flux-linkage curves in **Figure 8b** with the corresponding curves of our SRM2 design of **Figure 6a** reveals that the parallel inductance assumption is inaccurate. Therefore, in general, the parallelogram theory-based torque computations, as in Eq. (7), for the saturable SR machine can lead to underestimated values, unless some scaling factor is introduced to compensate for the discrepancy resulting from the parallelogram area assumption applied to the nonparallel inductances. It will be shown that the two methods of Eqs. (5) and (7) are both special cases of the general torque production model based on quadrilateral area computation that is proposed in this chapter.

2. General analytic theory for saturable switched reluctance machines

In this chapter we set out the proposed nonlinear theory based on vector analysis of quadrilaterals for the SR machines. By presenting the nonlinear theory, it will be demonstrated how to estimate the average rated torque at the rated angular speed of the SR machine taking into account the magnetic nonlinearity of the magnetic circuit of the machine. The nonlinear theory will also be used to construct the instantaneous current profiles at the rated angular speed of the machine as well as to estimate the average phase current and voltage. The proposed nonlinear theory is also applicable to the generator mode of the SR machine, and the presented examples will demonstrate the results. Finally, the nonlinear analytical theory enables the estimation of the power electronic converter volt-ampere requirements of the switched reluctance machine as well as the energy conversion ratio for the machine. All these estimates will be used to construct the estimated speed-torque envelopes for the analyzed saturable SR machine, thus concluding the preliminary design process of the device using the analytical theory.

2.1 Switched reluctance machine energy conversion estimation and flux-linkage map construction

Since the flux-linkage map of an SR machine is the main piece of information to be used for the torque production estimation, the already constructed complete flux-linkage map for the selected SRM2 machine design is redrawn again, this time for the aligned and the unaligned curves only, as shown in **Figure 9**.

In **Figure 9** the flux-linkage curves are first drawn from the finite element analysis-obtained data points of **Figure 6a**, assuming that $L_{sa} \neq L_{uu}$. The three linear curves are obtained by splitting the aligned flux-linkage curve in the region of the saturation point, at approximately 80 A current point in this case; however, the two parts of the split aligned flux-linkage curve are overlapping in order to facilitate more accurate linear fit. The linear fit of the unaligned flux-linkage curve also indicates that the linear fit is a very close approximation, as can be inferred from the statistical goodness of fit measure R^2, the value of which in **Figure 9** for the linear flux-linkage is very close to unity. It is noted at this point that all of the linearized fits are rather subjective; however, the aim to keep the area W' in **Figure 7** as close to the area bound by the linearized curve fits in **Figure 9** will ensure that our subsequent calculations will be accurate.

Having obtained the satisfactory linearization of the curves, as in **Figure 9**, we now define the following flux-linkage expressions (units of which can be in *volt-seconds* or *weber*) in terms of the known inductances and phase current i.

Unsaturated unaligned flux-linkage:

$$\Psi_{uu} = L_{uu} \cdot i = 0.0012072 \cdot i \tag{9}$$

Unsaturated aligned flux-linkage:

$$\Psi_{ua} = L_{ua} \cdot i_s = L_{sa} \cdot i_s + \Psi_s = 0.0071879 \cdot i_s = 0.0004948 i_s + 0.4192920 \tag{10}$$

Saturated aligned flux-linkage:

$$\Psi_{sa} = L_{sa} \cdot i_r + \Psi_s = 0.0004948 i_r + 0.4192920 \tag{11}$$

where i_r is the rated phase current, as given in **Table 1**, and is equal to 320 A for the selected SRM2 design.

The saturation current, i_s, is found from Eq. (10) since the two lines intersect:

$$i_s = \frac{\Psi_s}{L_{ua} - L_{sa}} = \frac{0.4192920}{0.0071879 - 0.0004948} \approx 62.6 \text{ (A)} \tag{12}$$

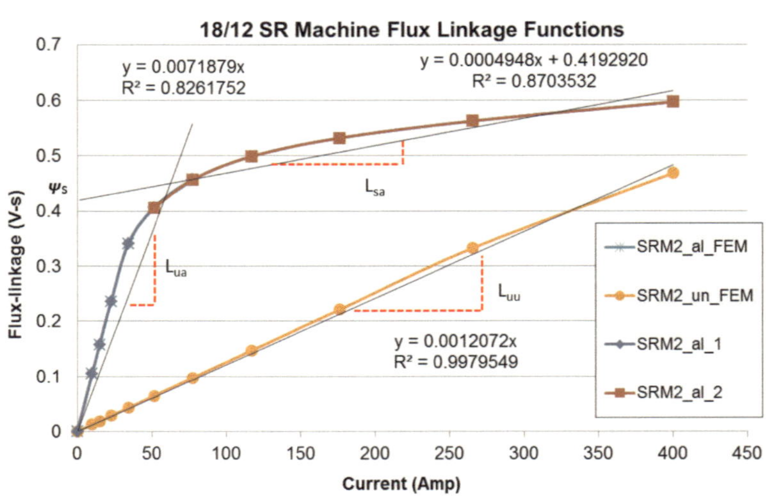

Figure 9.
Flux-linkage curves for the 18/12 SRM2, taking the rotor position as the parameter and the linearized fits.

Moreover, the aligned flux-linkage intercept, Ψ_s, is found from Eq. (11) by substituting the zero current value:

$$\Psi_s = L_{sa} \cdot 0 + \Psi_s = 0.0004948 \times 0 + 0.4192920 = 0.4192920 \quad (13)$$

Our next goal is to express mathematically the area W' bounded by the linearized curve fits in **Figure 7** in terms of the inductances and the currents. To achieve this goal, we first recognize that area W' is neither in the trapezoidal nor the parallelogram form, and therefore the simple area calculations obtained with such formulae may be inaccurate. However, we recognize also that the area W' in **Figure 7** is a general quadrilateral if we assume the flat-topped phase current profile of **Figure 3** to follow the path as shown diagrammatically in **Figure 10**.

In **Figure 10** the phase current is assumed to rise from zero to the full rated phase current value at V_1, to be maintained constant while the rotor is moving toward the fully aligned position with the stator pole and while the flux-linkage is increasing from V_1 to V_2. Just before the full alignment, at V_2, the phase current is commutated by applying full wave voltage, as in **Figure 3**, and from that point the current decays to V_3 and finally to V_O, at which point it is zero again. Such quadrilateral phase current locus (albeit very idealized and nonpractical) will encompass the total area bound by the two flux-linkage curves and will be equal to W', which in turn is used to compute the torque generated by the SR machine.

The geometric figure formed by $V_O V_1 V_2 V_3 V_O$ in **Figure 10** is the general quadrilateral, the area of which can be computed with the help of Varignon's theorem [9], which states that if we take the midpoints of each of the sides of the quadrilateral, this will result in a new quadrilateral *ABCD*—the midpoint quadrilateral called Varignon parallelogram—as shown in **Figure 10**. The Varignon parallelogram is exactly half of the area of the original quadrilateral $V_O V_1 V_2 V_3 V_O$ defining W'. In terms of the phase current trajectory vectors in **Figure 10**:

$$W' = \frac{1}{2} |((V_2 - V_O) \times (V_3 - V_1))| \text{ (Joules)} \quad (14)$$

and in terms of the coordinates of the vectors V_i in **Figure 10** and the known numeric quantities defined by Eq. (9) to (13), Eq. (14) becomes.

$$W' = \frac{1}{2} [(i_r - 0)(\Psi_{ua} - \Psi_{uu}) - (i_s - i_r)(\Psi_{sa} - 0)] \text{ (Joules)} \quad (15)$$

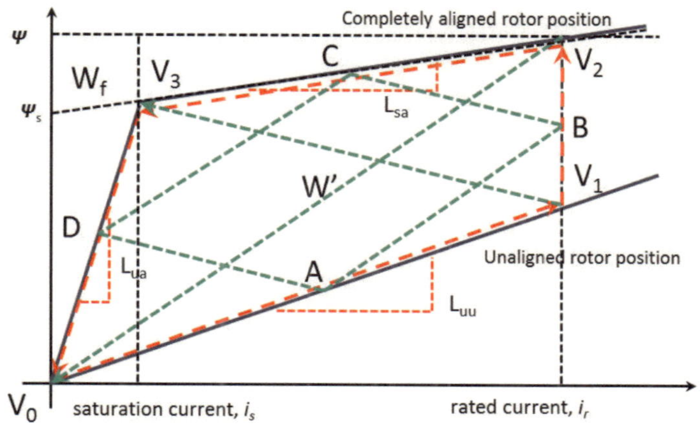

Figure 10.
Linearized flux-linkage curves for the 18/12 SRM2 and the phase current locus (dashed red).

Eq. (15) is only convenient if the flux-linkage and the current quantities are known numerically at the required points on the quadrilateral. Usually it is preferred to work with the inductances L and the rated current value, since these quantities are used for other electric machines and therefore can be more intuitive in the SR machine design process as well.

Substituting the inductance relations, Eq. (9) to (13), into Eq. (15) and simplifying yield

$$\begin{aligned} W' &= \frac{1}{2}[(L_{ua} \cdot i_s)i_r - (L_{uu} \cdot i_r)i_r - (L_{sa} \cdot i_r + \Psi_s)i_s + (L_{sa} \cdot i_r + \Psi_s)i_r] \\ &= \frac{1}{2}\left[(L_{sa} - L_{uu})i_r{}^2 + (L_{ua} - L_{sa})i_s i_r + \Psi_s i_r - \Psi_s i_s\right] \end{aligned} \tag{16}$$

Eq. (16) is not yet the preferred final formula for the area calculation since we wish to eliminate the numerical flux-linkage values Ψ_s, and the saturation current i_s, and leave only the rated phase current values and inductances. Also we note at this stage of derivation that if the rated current were to reduce to such a point so as to coincide with the saturation current in Eq. (16), the resulting area W' would be equal to the area found in Eq. (5) for the non-saturating SR machine, since all but the first term in the brackets would cancel out. Therefore the area calculation with Eq. (16) is the general case, whereas Eq. (5) is the special case as was noted earlier.

Considering **Figure 9** and Eq. (16), we draw yet two more corresponding diagrams: one for the flux-linkage quadrilateral as in **Figure 11** and another for the phase current profile as in **Figure 12**.

Figure 11 shows that, generally, the SR machine phase current commutation starts well before the fully aligned position [6], at point C instead of point B, since it would require an infinite DC bus voltage in order to bring the phase current to zero instantly so as to avoid the negative torque production which is undesirable in the motoring mode. Therefore the current will decay to point E before reaching zero value as before in **Figure 10**. The line AC in **Figure 11** can be expressed in terms of the inductances in Eqs. (9) and (11) and introducing the commutation factor c, being in the region of $0 < c \leq 1$, which is also shown in **Figure 11**:

$$AC = c \cdot AB = L_{sa} \cdot i_r + c\Psi_s - L_{uu} \cdot i_r; (V \cdot s) \tag{17}$$

since the unaligned unsaturated inductance is not affected by the commutation factor c, as in **Figure 11**.

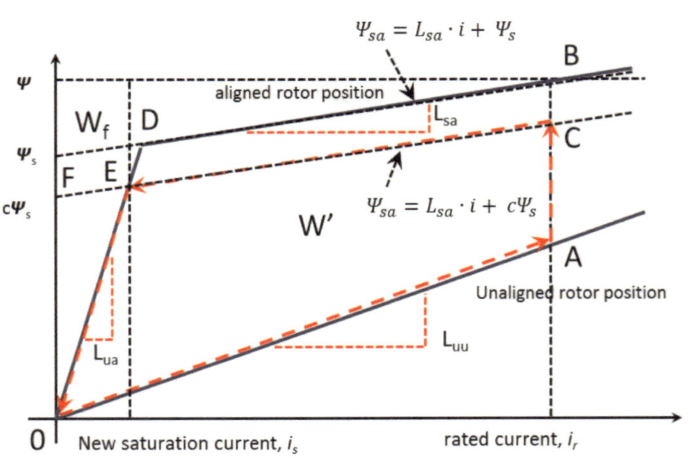

Figure 11.
Linearized flux-linkage curves for the 18/12 SRM2 and the reduced phase current locus (dashed red).

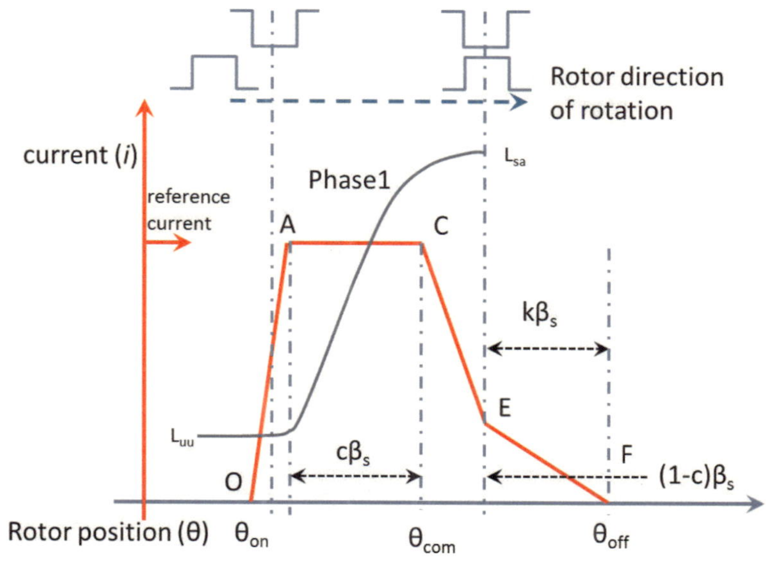

Figure 12.
Phase current profile in the phase windings resulting from the assumed linearized current profile in ***Figure 11****.*

The units of volt-seconds of the flux-linkage difference AC in **Figure 11** now comes very handy since if we know the angular speed of the SRM2 machine ω (radians/second) and the stator pole arc angle, β_s (radians), both as specified in **Table 1**, then Eq. (17) can be expressed in terms of the root-mean-square voltage $V_{rms(PWM)}$ applied to the phase winding from A to C, in order to maintain the phase current flat-topped, as in **Figure 11**:

$$AC = c \cdot AB = L_{sa} \cdot i_r + c\Psi_s - L_{uu} \cdot i_r = V_{rms(PWM)} \frac{c \cdot \beta s}{\omega}; (V \cdot s) \tag{18}$$

the last term in Eq. (18) indicating that the units of $(V \cdot s)$ result if the rotor rotates by the angle $c \cdot \beta_s$ when rotating at the angular speed ω, while simultaneously the voltage $V_{rms(PWM)}$ is applied across the phase winding. The $V_{rms(PWM)}$ is the voltage needed to counteract the build-up of the back-electromotive force, *back-emf*, being generated in the phase windings at the specific speed.

Rearranging Eq. (18) in order to obtain the saturation flux-linkage constant in **Figure 11** gives

$$\Psi_s = V_{rms(PWM)} \frac{\beta s}{\omega} - (L_{sa} - L_{uu}) \cdot \frac{i_r}{c} \tag{19}$$

Since the equality previously considered in Eq. (10) has changed as a result of the early commutation, it is rewritten here for convenience:

$$L_{ua} \cdot i_s = L_{sa} \cdot i_s + c \cdot \Psi_s \tag{20}$$

since the two linearized curves intersect at point E in **Figure 11**.

To obtain the new saturated phase current value i_s, remembering to include the phase commutation factor c, Eq. (20) is rearranged:

$$i_s = \frac{c \cdot \Psi_s}{L_{ua} - L_{sa}}, \tag{21}$$

and by using the result from Eq. (19), the saturation current, Eq. (21), can be expressed as

$$i_s = \frac{V_{rms(PWM)} \frac{c \cdot \beta s}{\omega} - (L_{sa} - L_{uu}) i_r}{L_{ua} - L_{sa}} \tag{22}$$

Therefore, having obtained Eq. (19) and (22), we are now in the position to express the co-energy W' given in Eq. (16) in terms of the inductances, the voltage, the rated phase current, the angular speed, and the stator pole arc angle, by substitution and simplification into Eq. (16):

$$W' = \frac{1}{2}\left[(L_{sa} - L_{uu}) i_r^2 + (L_{ua} - L_{sa}) i_s i_r + c \cdot \Psi_s i_r - c \cdot \Psi_s i_s\right] =$$

$$= \frac{1}{2}\left[2 \cdot V_{rms(PWM)} \frac{c \cdot \beta s}{\omega} \cdot i_r + (L_{uu} - L_{sa}) i_r^2 - \frac{\left(V_{rms(PWM)} \frac{c \cdot \beta s}{\omega} + (L_{uu} - L_{sa}) \cdot i_r\right)^2}{L_{ua} - L_{sa}}\right] \tag{23}$$

Eq. (23) is the fundamental expression for the computation of the energy conversion W' for any SR machine, including both saturating and non-saturating cases, as well as for the trapezoidal phase current profile operation [6] and for the general quadrilateral operation of the phase currents as in **Figure 11** and for the motoring and the generating modes of the machine operation. By comparison, Eq. (23) indicates significantly different behavior of the general SR machine case compared to the special cases in Eqs. (5) and (7).

The questions remain as to how to estimate the commutation factor c and the *rms* voltage, since Eq. (23) requires these numeric values to be assumed in order to calculate accurately the energy conversion area W', whereas only the DC bus voltage and the rated current are given (or are assumed to be known in the design process) as in **Table 1**. Considering **Figure 12**, corresponding to **Figure 11**, where the c factor is shown graphically, we assume that—strictly for the rated motor speed—the commutation of the phase current profile starts at point C, which is well before the rotor pole is fully aligned with the stator pole, and this is so that the current can be brought down in the shortest time possible to point E so as to maximize the torque-generating region.

Assuming that, first of all, the phase current reaches point A with the full wave DC bus voltage when the rotor is at the unaligned position, but also with the early commutation assumption in mind, we determine the time taken for the rated current i_r to fall to the saturation value i_s, from Eq. (11) and **Figure 11**, which is approximately

$$\begin{aligned}\Psi_{sa}(@B) - \Psi_{sa}(@D) &= (0.0004948(H) \times 320\ (Amp) + 0.4192920(V \cdot s)) \\ &\quad - (0.0004948(H) \times 62.6\ (Amp) + 0.4192920(V \cdot s)) \\ &= 0.12736\ (V \cdot s)\end{aligned} \tag{24}$$

Moreover, since the steady angular speed and the DC bus voltage at the point of the commutation are known, the time required for the current to fall from point C to point E in **Figure 12** is

$$\frac{\Psi_{sa}(@B) - \Psi_{sa}(@D)}{V_{DC}} = \frac{0.12736\ (V \cdot s)}{500\ (V)} = 2.547 \times 10^{-4}\ (s) \tag{25}$$

since the full wave voltage is applied as in **Figure 3**.

Therefore, the angle through which the rotor traverses during this time period can be found, since the rated angular speed is known:

$$\theta_{CE} = \frac{\Psi_{sa}(@B) - \Psi_{sa}(@D)}{V_{DC}} \times \omega = \frac{0.12736\ (V \cdot s)}{500\ (V)} \times \frac{1200(rpm)}{60} \times 2\pi$$
$$= 0.032(radians) = 1.83\ (degrees) \quad (26)$$

Finally, from the definition of the commutation factor c in **Figure 12**, we find:

$$c = 1 - \frac{\theta_{CE}}{\beta s} = 1 - \frac{1.83\ (degrees)}{10.5\ (degrees)} \approx 0.8 \quad (27)$$

Also of interest is the *rms* voltage that is applied across the phase winding when the peak current is flat-topped as in **Figure 12**; this current shape is assumed to be present at the rated motor speed. Therefore the *rms* voltage from point A to point C in **Figure 12**, using Eq. (19), is

$$V_{rms(PWM)} = \left(\Psi_s + (L_{sa} - L_{uu}) \cdot \frac{i_r}{c}\right) \frac{\omega}{\beta s} \approx 100(\mathrm{V}) \quad (28)$$

being only an approximate value since the winding resistance voltage is assumed to be unknown at this stage of analysis which would be added to the *rms* voltage.

Having estimated the commutation factor and the *rms* voltage, Eq. (23) is used to find the energy conversion capability of the chosen SRM2 machine at the rated angular speed of 1200 rpm and the other variables taken from **Table 1** and **Figure 9**, for a single phase:

$$W' = \frac{1}{2}\left[2 \cdot V_{rms(PWM)} \frac{c \cdot \beta s}{\omega} \cdot i_r + (L_{uu} - L_{sa}) i_r^2 - \frac{\left(V_{rms(PWM)} \frac{c \cdot \beta s}{\omega} + (L_{uu} - L_{sa}) \cdot i_r\right)^2}{L_{ua} - L_{sa}}\right]$$
$$= \frac{1}{2}[74.67 + 72.95 - 17.74] \approx 65\ (\mathrm{Joules}) \quad (29)$$

The total torque at the rated speed of the three-phase, q, SRM2 machine is thus estimated as follows:

$$T = W' \times \frac{q \cdot N_r}{2\pi} = 65 \times \frac{3 \times 12}{2\pi} \approx 372\ (N \cdot m) \quad (30)$$

The analytically calculated rated torque value of the chosen SRM2 SR machine compares favorably with the published measured rated torque value of 400 Nm, being about 8% underestimated. This small discrepancy arises since our calculations in Eq. (29) are influenced by two main factors as a consequence of the simplifications made. First—and this has an overestimating effect—the excursion of the phase current into the decreasing inductance region after the full stator and rotor alignment, EF in **Figure 12**, is considered in Eq. (23) as a positive torque generation region, whereas in reality it is a negative torque generation region. Secondly—and this leads to the underestimating effect in Eq. (29)—generally there is the phase overlap present, as in **Figure 3**, where a fraction of a leading phase energy conversion is being compensated by a fraction of the trailing phase which, if accounted for, increases the instantaneous torque in Eq. (23). Therefore, by accounting for the SRM2 machine phase overlap ratio R at the rated speed, we are able to compensate for the underestimated torque generation value in Eq. (29) for the overlapping torques:

$$R = 1 + \frac{1}{\beta s}\left(\beta s - \left(\frac{360}{N_r} - \frac{360}{N_s}\right)\right) = 1.0476 \tag{31}$$

where the stator pole arc angle βs is in mechanical degrees and N_s and N_r are the number of the stator and the rotor poles, respectively. The answer of Eq. (31) can now be multiplied by the computed torque value of Eq. (30) which makes the discrepancy with respect to the published measured torque of less than about 3% underestimated, which must be considered as relatively very accurate for the analytical method-based design process.

Other factors influencing the accuracy of the analytical torque calculations include the uncertainties in the estimated commutation factor c as well as the actual *rms* voltage, both not reported in [4]. However, the published maximum computed torque value in [4] was in the region of 415 (N-m) at the assumed 1200 (rpm) rated speed, as in **Table 1**. The reasons for the discrepancy between the reported measured and the reported computed rated torque values in [4] were not explained.

Finally, the electromechanical power at the rated speed for the SRM2 design can be estimated as

$$P_{em} = T \times \omega = W' \times \frac{q \cdot N_r}{2\pi} \times \omega \times R \approx 390\ (N \cdot m) \times \frac{1200\ (rpm)}{60} \times 2\pi = 49\ (\text{kW}) \tag{32}$$

which is an underestimate of about 2% compared to the published power levels at the rated speed [4]. However, the maximum power of the SRM2 design beyond the rated speed was quoted as 54 (kW).

Although Eq. (23) (or Eq. (29)) is rather complex, it represents the most suitable form of Eq. (16), since the numerical evaluation of all flux-linkage values, term by term, at each desired phase current level is no longer required. Even more importantly, Eq. (23) is very convenient to use as the operating parameters that are readily available to machine designers are typically expressed in terms of the inductance estimates, obtained either from finite element analysis or the analytic aligned and unaligned flux-linkage maps. Furthermore, the theory of other electric machines usually relies on inductances, and for this reason, Eq. (23) can be more accessible and intuitive for the control engineers.

The required operating parameters to be used in Eq. (23) for the fundamental energy conversion estimation of the SR machine are collected from **Figure 9** and Eqs. (27) and (28) and are summarized in **Table 2**.

Eq. (29) is further exploited to gain insights into the torque control effectiveness of the saturable SR machines via the rated phase current regulation. The operating parameters from **Table 2** were substituted into Eqs. (29) and (30), and the rated phase current was varied from some minimum practical value to the full rated phase current value, as plotted in **Figure 13**.

Figure 13 reveals critically important results regarding saturable SR machine operation when, at the fixed constant speed, the torque is regulated with the phase current only. If the start of the phase current rise is to be fixed at a certain rotor angle, θ_{on} in **Figure 12**, and the current decay should start at the commutation angle, θ_{com}, the electromagnetic torque production for the SR machine will be weakly nonlinear and will follow the power law in phase current as in **Figure 13**. This is significant from the control point of view since—if the linear torque control is to result from the linear increase of the phase current—the current will have to be

Energy conversion variables to be used in Eq. (23)	
Required SR machine design parameters:	**50 kW SRM2**
Saturated aligned inductance, L_{sa} [H]	0.0004948
Unsaturated aligned inductance, L_{ua} [H]	0.0071879
Unsaturated unaligned inductance, L_{uu} [H]	0.0012072
Stator pole arc angle β_s [degrees]	10.5
Rated angular speed ω [rad/s]	125.7
Rated current, i_r [A]	320
rms voltage, $V_{rms(PWM)}$ [V]	100
DC bus voltage [V]	500
Commutation factor c	0.8

Table 2.
50 kW SRm2 machine parameters.

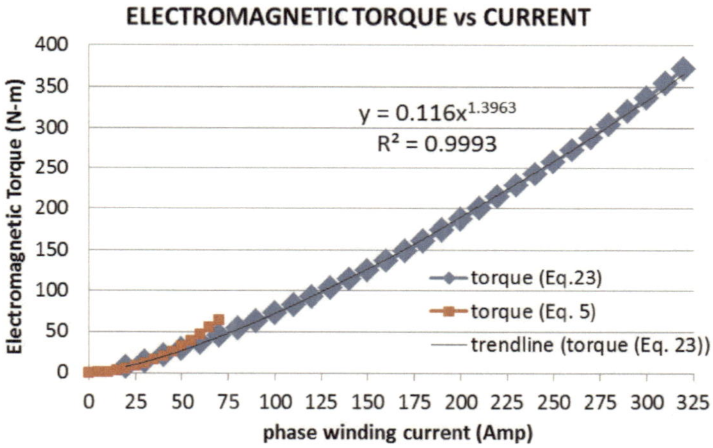

Figure 13.
The electromagnetic torque production as a function of the phase current regulation.

regulated simultaneously with the turn-on and the commutation angles, as defined in **Figure 12**. Therefore, for each required torque value at the fixed constant speed, an exhaustive search for the combination of the three control parameters will have to be performed, that is, the turn-on, the commutation, and the phase current magnitude. This type of search is effective only with the help of computer-based design routines; for example, see [8, 10].

Figure 13 also shows the torque obtained from Eq. (5) where the linear non-saturating SR machine is assumed for all rated current values. The linear theory suggests that the torque production follows the quadratic power law with the phase current, as is also evident from Eq. (5) itself. The torque values in the region below the saturation current are similar using both methods; however, once the current is increased beyond the saturation level, the linear theory overestimates the torque value and becomes invalid. To conclude, the general torque of the saturable SR machine can be computed relatively quickly and accurately with Eq. (23), even if the rated phase current value is reduced below the saturation current value.

Further discussion regarding Eq. (23) is as follows. It should be understood that the commutation factor c, as computable with Eq. (27), is not a constant but can be chosen or optimized, depending on the commutation timing as in **Figure 12**.

Likewise, the *rms* voltage is not a constant, as computable with Eq. (28), but is dependent on the commutation factor and the rated speed. The terms of Eq. (23) can be examined in turn in order to gain further insights into the operation and energy conversion process of the general SR machine.

The first term in Eq. (23), the "voltage" term,

$$2 \cdot V_{rms(PWM)} \frac{c \cdot \beta s}{\omega} \cdot i_r \tag{33}$$

indicates that the energy conversion part of this term is dominated by the rated current and the commutation factor; however, even with zero commutation factor, the machine will produce some useful torque, since the full wave DC bus voltage is still utilized, while with the *c* factor equal unity, the *rms* voltage-based energy component will be fully realized. Moreover, the *rms* voltage will tend to increase the energy conversion component, itself being a function of the operating speed at which it occurs. The *rms* voltage is created by the power electronic switches as shown in **Figure 3**.

The second term in Eq. (23), the "inductance" term,

$$(L_{uu} - L_{sa}) i_r^2, \tag{34}$$

contributes significantly to the energy conversion process and is dominated by the square of the rated current in the phase winding. Therefore, it is desirable to maximize this value in order to operate at the highest possible torque, yet giving proper consideration to the cooling requirements of the SR machine, as at very high phase current values, the cooling of the windings becomes very challenging. It therefore appears that in order to arrive at a torque dense SR machine design, as the main goal, it is important to drive as large a current through the motor windings as is practicable. Furthermore, the inductance difference $L_{uu} - L_{sa}$ should be made as large as possible. This can be achieved in practice if the saturated aligned inductance L_{sa} is minimized with respect to the unsaturated unaligned inductance L_{uu}.

The third term in Eq. (23), the "saturation quality" penalty term,

$$\frac{\left(V_{rms(PWM)} \frac{c \cdot \beta s}{\omega} + (L_{uu} - L_{sa}) \cdot i_r\right)^2}{L_{ua} - L_{sa}}, \tag{35}$$

is a relatively large negative quantity and is dominated by the denominator defined by the inductance difference. Therefore, to reduce this "penalty," it would be desirable to maximize the unsaturated aligned inductance L_{ua}, while the saturated aligned inductance L_{sa} should be kept as low as possible in relation to the unsaturated aligned inductance. This effect can only be achieved with the highest-quality electric steels, having very high saturation flux density values, which must therefore be used to make the core of the SR stator and rotor [7].

2.2 Saturable switched reluctance machine energy conversion ratio estimation

Since the SR machine requires a DC bus capacitor if the DC supply voltage is to be kept undisturbed by the returning magnetic field energy from the phase windings, as discussed and shown in **Figure 2**, the estimation of the magnitude of such returning energy is an important consideration when designing the h-bridge power electronic converter for the SR machine. As was shown in **Figure 10**, which for convenience is replotted here as **Figure 14**, the magnetic field energy W_f first increases as the phase current rises in the phase windings up to point *B*, at which point it is returned as the phase current is commutated to decay to zero.

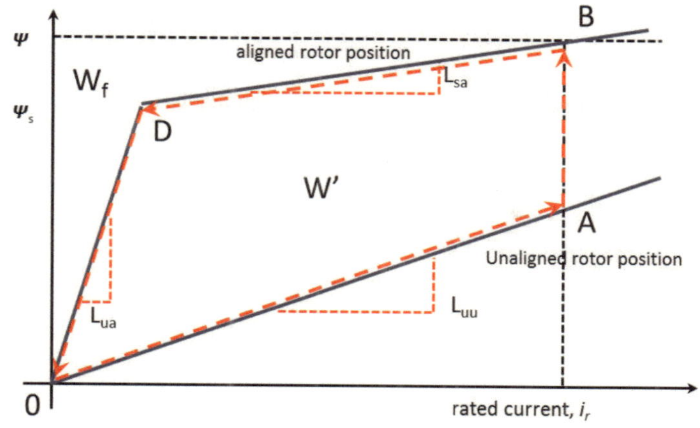

Figure 14.
Linearized flux-linkage curves for the 18/12 SRM2 and the definition of the magnetic field energy W_f.

The amount of energy (in joules) being created as a result of the phase current (and the magnetic field in the windings) build-up is represented by the area bounded by the entire aligned flux-linkage curve, the horizontal line at point B at which the maximum flux-linkage is being established by the current profile as shown in **Figure 14** and by the flux-linkage axis thus being a quadrilateral.

Knowing the magnitude of the stored magnetic field energy W_f in the phase winding at the rated speed of the SR machine enables the energy conversion ratio [6] to be determined as

$$\textit{energy conversion ratio} = \frac{W'}{W' + W_f} \tag{36}$$

In order to estimate the energy conversion ratio, it is necessary to estimate the stored magnetic field energy W_f along the lines of Eq. (23) and **Figures 9** and **10** with the help of Varignon parallelogram calculations. Detailed derivation is not shown here to save space, but readers are encouraged to verify the following result:

$$W_f = \frac{1}{2}\left[L_{sa}{i_r}^2 + \frac{\left(V_{rms(PWM)}\frac{c\cdot\beta s}{\omega} + (L_{uu} - L_{sa})\cdot i_r\right)^2}{L_{ua} - L_{sa}}\right] \tag{37}$$

Eq. (37), being somewhat less complex than Eq. (23), can be interpreted as follows: maximizing the rated current and the saturated aligned inductance, L_{sa}, will greatly increase the stored magnetic field energy. Furthermore, minimizing the difference between the inductances in the identified "saturation penalty" term in Eq. (37) will help to minimize the stored magnetic field energy, which is a desirable effect and can be achieved by selecting the electric steel with high saturation flux density values for the construction of the SR machine.

Substituting the known operating parameters of the SRM2 design from **Table 2**, the estimated field energy is found as

$$W_f = \frac{1}{2}\left[50.67 + \frac{0.1354}{0.0067}\right] \approx 35.5\ (\textit{joules}) \tag{38}$$

Using Eq. (37) for the operating parameters from **Table 2**, with the phase current magnitude varied from some lower practical limit to the full rated current, the energy conversion ratios were obtained with Eq. (36) as shown in **Figure 15**.

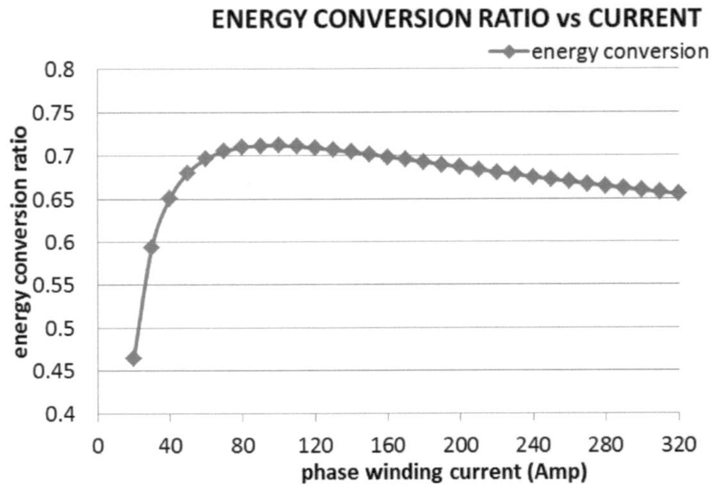

Figure 15.
Energy conversion ratio curve for the 18/12 SRM2, for the rated speed of 1200 (rpm).

From **Figure 15** it is seen that the amount of the energy converted by the SRM2 design is rather small at the low phase current values, but it improves quickly as the phase current increases, in effect encompassing more of the co-energy area W' as in **Figure 14**. However, there is a definite optimum phase current value above which the energy conversion ratio starts to decrease again, and this phase current value is well below the rated current. The reason as to why the energy conversion ratio is decreasing as the phase current is increased is due to the uneven incremental co-energy areas $\Delta W'$ being encompassed by the phase current profile with respect to the incrementally increased stored magnetic field energy ΔW_f being created, thus having the diminishing effectiveness as in **Figure 15** and given by Eq. (36), although this effect is not too severe for the most useful range of the currents.

2.3 Saturable switched reluctance machine average rated current and voltage estimation

Having successfully obtained the rated torque values for the selected SRM2 machine design using Eq. (29), we are now in a position to extend the analytical treatment for the estimation of the average rated phase current as well as the voltage. These quantities are very important during the preliminary SR machine design process in order to estimate the demand from the available DC bus supply, such as a battery of an electric vehicle. Furthermore, the knowledge of the rated current is required for the estimation of the heat dissipation requirements for the designed SR machine.

Consider **Figure 12**, which for convenience is redrawn here as **Figure 16**.

Since by using Eqs. (24) to (28) we were able to evaluate the commutation factor c and the *rms* voltage, similar reasoning will be applied to find the rest of the information needed to estimate the rated average phase current. The aim is to use the Varignon parallelogram principles as in **Figure 10** to compute the area of the quadrilateral $OFEC$ in **Figure 16** as well as the triangular areas θOA and OCA.

To find the time it takes for the phase current to rise to the rated value, using Eq. (25),

$$t_{\theta \to O} = \frac{\Psi_{uu}(@A) - 0}{V_{DC}} = \frac{0.3863\ (V \cdot s)}{500\ (V)} = 7.73 \times 10^{-4}\ (\text{s}) \tag{39}$$

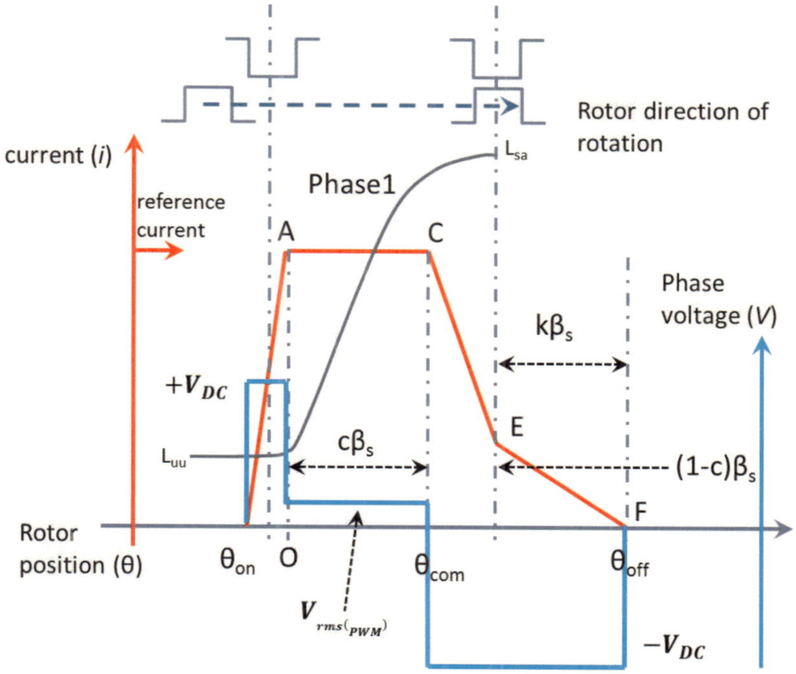

Figure 16.
Phase current profile in the phase windings as a function of the voltages and rotor position angles.

For the rotor to traverse from point *A* to point *C*, as in **Figure 16**,

$$t_{A\to C} = \frac{\Psi_{sa}(@D) - \Psi_{uu}(@A)}{V_{rms(PWM)}} = \frac{0.1074\ (V\cdot s)}{100\ (V)} = 1.074\times 10^{-3}(\mathrm{s}) \tag{40}$$

where the saturated aligned flux-linkage at *D* is

$$\Psi_{sa} = L_{sa}\cdot i_r + c\cdot \Psi_s = 0.0004948 i_r + 0.8\times 0.4192920 \tag{41}$$

Next we estimate the time needed for the phase current to decay from point *E* to point *F*, the $k\cdot\beta s$ factor, at which it is zero as in **Figure 16**:

$$t_{E\to F} = \frac{\Psi_{ua}(@E)}{V_{DC}} = \frac{0.3701\ (V\cdot s)}{500\ (V)} = 7.40\times 10^{-4}(\mathrm{s}) \tag{42}$$

where the saturating current needed to evaluate the unsaturated aligned flux-linkage in Eq. (42) was found from Eq. (22) using *c = 0.8* as before:

$$i_s = \frac{V_{rms(PWM)}\frac{c\cdot\beta s}{\omega} - (L_{sa} - L_{uu})i_r}{L_{ua} - L_{sa}} \approx 51.5\ (\mathrm{A}) \tag{43}$$

Finally, the time from *C* to *E* was estimated earlier using Eq. (25) as

$$t_{C\to E} = 2.547\times 10^{-4}(\mathrm{s}) \tag{44}$$

Thus the area bound by the quadrilateral *OFEC* can be found, as in Eq. (14):

$$\begin{aligned} \mathrm{A_{OFEC}} &= \frac{1}{2}\left[(\mathrm{t_{A\to C}} + \mathrm{t_{C\to E}})\times \mathrm{i_r} + (\mathrm{t_{C\to E}} + \mathrm{t_{E\to F}})\times \mathrm{i_s}\right] = \\ &= 0.2381\ (A\cdot \mathrm{s}) \end{aligned} \tag{45}$$

The triangular area OAC and θOA are easily found as

$$A_{OAC} = \frac{1}{2}[t_{A\to C} \times i_r] = 0.17184\ (\mathrm{A}\cdot\mathrm{s}) \tag{46}$$

and

$$A_{\theta OA} = \frac{1}{2}[t_{\theta\to O} \times i_r] = 0.12368\ (\mathrm{A}\cdot\mathrm{s}) \tag{47}$$

Now the areas are added and divided by the total rotor traverse time from θ_{on} to θ_{off} in **Figure 16**, to obtain the average phase current:

$$I_{AV} = \frac{0.53362(\mathrm{A}\cdot\mathrm{s})}{0.0028417(\mathrm{s})} \approx 188\ (\mathrm{A}) \tag{48}$$

However, by revisiting the argument regarding the overlapping phase operation of the selected SRM2 machine at the rated speed, we make use of the overlap ratio R defined in Eq. (31) in order to obtain a more conservative estimate of the average rated *rms* phase current expected to be demanded from the battery (see **Table 1**):

$$I_{BAT} = I_{AV} \times R = \frac{0.53362(\mathrm{A}\cdot\mathrm{s})}{0.0028417(\mathrm{s})} \times 1.045 \approx 196\ (\mathrm{A}) \tag{49}$$

Likewise, all of the information needed to compute the average phase voltage from the known instantaneous voltage waveform of **Figure 16** is readily available. The computed value found for the single phase at the rated speed is

$$V_{AV} \approx 350\ (\mathrm{V}) \tag{50}$$

Therefore, at the rated speed, the SRM2 machine will be demanding the electric power:

$$P_{EL} = I_{BAT} \times V_{AV} = 196\ (\mathrm{A}) \times 350\ (\mathrm{V}) = 68(\mathrm{kW}) \tag{51}$$

The result in Eq. (51) does not need to include the power factor at the rated speed since the phase currents and voltages are averaged over the full electrical cycle in Eqs. (49) and (50). Finally, knowing the demanded electric power and the generated electromagnetic power, the efficiency of the SR machine at the rated speed can be readily found which for this particular case is:

$$Efficiency = \frac{49\ (\mathrm{kW})}{68(\mathrm{kW})} \approx 72(\%) \tag{52}$$

which compares favorably with the published efficiency value of around 80% at the rated speed and torque [4].

The above estimated electric power demand compares well with the electromagnetic power production as found in Eq. (32). Therefore, the relatively accurate SR machine design process is accomplished.

2.4 Saturable switched reluctance machine converter volt-ampere requirement estimation

As was shown in **Figure 2**, the asymmetric h-bridge SR machine power electronic converter requires two active devices (transistor switches) and two passive devices (the power diodes per single phase) in order to realize the most flexible SR machine operation in a motoring mode or a generating mode [7]. Consequently, we can define the metric of the volt-ampere product that these power electronic devices have to be rated at in order to deliver the required phase current at the required DC bus voltage.

For a three-phase, $q = 3$, SR machine operated from a fixed supply DC bus voltage, V, and off the asymmetric half-bridge power electronic converter delivering the phase current, i, the maximum volt-ampere rating of the power electronic converter is defined as [6]

$$S = 2q \cdot Vi = 2 \times 3 \times 500(\text{V}) \times 350(\text{A}) = 1050(\text{kVA}) \tag{53}$$

Eq. (53) indicates that the stated maximum of the current will have to be delivered by the transistor switches having the stated voltage rating. The quantity obtainable with Eq. (53) is in units of kVA, and in this case, it is permissible to form a ratio of two quantities not having the same units [6] as long as both units are explicitly retained, and thus the converter volt-ampere value in Eq. (53) is divided by the produced electromagnetic power, as computed with Eq. (32), to obtain a measure of economic utilization of the converter in terms of kVA/kW:

$$S/P = \frac{4\pi \cdot Vi}{W'N_r} \ (\text{kVA/kW}) \tag{54}$$

where the volt-ampere quantity S is a constant and the electromagnetic power is computed from Eq. (32) with the current taken as a parameter. The ratio is then computed with Eq. (54) for the range of phase current values i and is plotted in **Figure 17** for the fixed constant rated speed of 1200 (rpm) for the SRM2 machine.

Figure 17 shows the h-bridge converter volt-ampere requirement for the rated angular speed of 1200 (rpm) with the phase current as a parameter. The curve in **Figure 17** indicates that approximately 25 (kVA/kW) rating is required at the power electronic converter end of the SR machine drive in order to deliver the rated torque, as found in the previous section using Eqs. (29) and (30). As the phase current magnitude is reduced in order to reduce the torque level, while the angular speed is kept constant, the *S/P* ratio tends to increase slowly first, but as the current

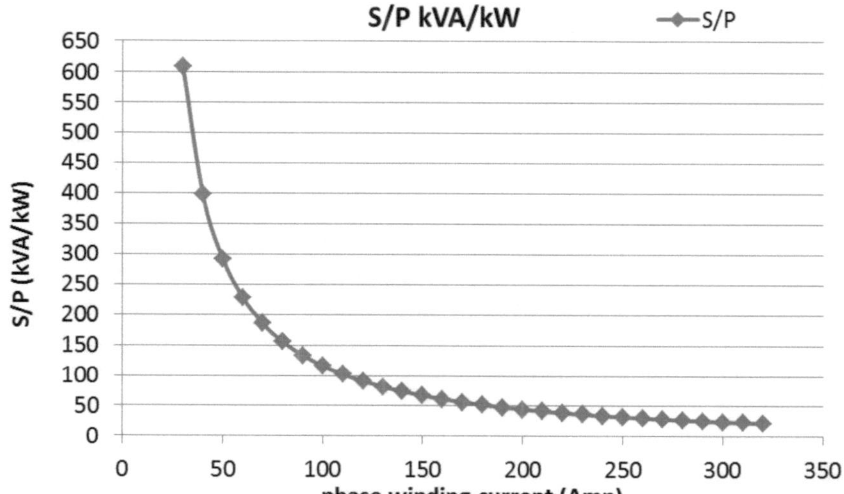

Figure 17.
Converter volt-ampere requirement curve for the 18/12 SRM2, for the rated speed of 1200 (rpm).

is reduced to a very low value, the ratio becomes very large. This trend only suggests the operating penalty when the demanded torque is low; however, the maximum rated torque operation of the saturable SR machine makes very clear economic sense compared to the non-saturating SR machine in terms of the size of the power electronic converter for the given amount of rated torque production [6]. Therefore, the saturable SR machine can be constructed economically since the power electronic transistors and diodes increase in price as their volt-ampere rating (kVA) is increased.

2.5 Saturable switched reluctance machine generator mode energy conversion estimation

Since SR machines are able to operate in the motoring as well as the generating mode, the energy conversion performance of an SR generator will also be quite accurately predicted with the Varignon parallelogram theorem as was the case for the motoring mode given by Eq. (23).

Consider **Figure 10** where for the SR machine-generating mode the locus of the phase current is reversed. Therefore the phase current will form the quadrilateral $V_0V_3V_2V_1V_0$ since the phase current will start flowing just before the rotor pole is aligned with the stator pole. The co-energy area W' for the generator encompassed by the phase current locus will be given in terms of the vector coordinates along the lines of Eq. (15), but with the vector sequence reversed, thus:

$$W' = \frac{1}{2}[(i_r - 0)(\Psi_{uu} - \Psi_{ua}) - (i_r - i_s)(\Psi_{sa} - 0)] \text{ (Joules)} \tag{55}$$

Having the co-energy relation for the generator, we then make use of Eq. (19) and Eq. (22) to express the generator co-energy in Eq. (55) in terms of the inductances, rated phase current, and *rms* voltage:

$$W' = \frac{1}{2}\left[(L_{uu} - L_{sa})i_r^2 + (L_{sa} - L_{ua})i_s i_r + c \cdot \Psi_s i_s - c \cdot \Psi_s i_r\right] = \\ = \frac{1}{2}\left[-2 \cdot V_{rms(PWM)} \frac{c \cdot \beta s}{\omega} \cdot i_r - (L_{uu} - L_{sa})i_r^2 + \frac{\left(V_{rms(PWM)} \frac{c \cdot \beta s}{\omega} + (L_{uu} - L_{sa}) \cdot i_r\right)^2}{L_{ua} - L_{sa}}\right] \tag{56}$$

Eq. (56) is the fundamental expression for the computation of the energy conversion W' for any SR generator, both saturating and non-saturating cases, as well as for the trapezoidal phase current profile operation [6] and for the general quadrilateral operation of the phase currents as in **Figure 11**.

The *rms* voltage and the commutation factor for the generating mode are both found from Eq. (27) and Eq. (28), and the generated average phase current will be found as in Section 2.3. Such generated DC current will be possible only if the external torque is applied to the SR generator, and the magnitude of such torque requirement as given in Eq. (56) and Eq. (30) will be a negative quantity, indicating that the external torque input is required.

3. Conclusions

Having successfully found the rated torque and power production capability of the analyzed SR machine design, we are now in the position to summarize the

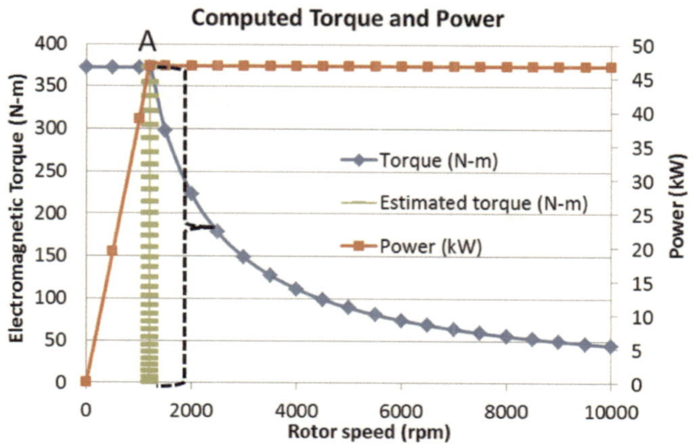

Figure 18.
Estimated speed-torque and speed-power characteristics for the analyzed SR machine design.

advantages of the proposed analytic method based on Varignon parallelogram and vector-based computations.

The estimated rated torque and power of a general SR machine are the two most important quantities a machine designer is seeking at early stages of the machine design process. Once these two estimates are known for the required rated speed, the speed-torque and power-torque characteristics for the SR machine can be constructed as in **Figure 18**.

The particular characteristics in **Figure 18** are assumed to obey the constant-torque and constant-power operation of the SR machine with respect to the speed of rotation. Such specific characteristics are most desirable from the accurate torque and power control point of view and are used for electric vehicle propulsion applications or the general industrial motor drive applications.

The convenient analytical form of Eq. (23) enables the SR machine design engineer to estimate relatively accurately the electromagnetic torque of the machine at the assumed required rated speed, at point A in **Figure 18**, which in turn is dictated by the particular practical application. Furthermore, this electromagnetic torque is estimated with the known (or application dictated) quantities such as DC bus voltage, rated current, and the geometric parameters of the assumed SR machine design. More importantly, the SR machine design engineer is able to parametrize Eq. (23) in terms of the phase current in order to estimate the range of the rated torque with respect to the change in the phase current magnitude as indicated by the bracket in **Figure 18**. At this stage of the SR machine design process, the engineer has enough information to construct the constant-torque and constant-power characteristic curves, as in **Figure 18**, with the help of Eq. (32). Such a characteristic map can be enhanced with further operating quantities of the average phase winding current and average voltage. The other performance metrics for the designed SR machine to be estimated with the help of the proposed analytical method are the volt-ampere requirements of the power electronic converter for the SR machine being designed. This entire information regarding the designed SR machine enables the decision-making process if such SR machine is to be further analyzed in more detail, this time with advanced computer-based optimization routines [8].

Author details

Aleksas Stuikys[1*] and Jan Sykulski[2]

1 RETORQ Motors Ltd., London, UK

2 Electronics and Computer Science, University of Southampton, Southampton, UK

*Address all correspondence to: a.stuikys.ac@gmail.com

References

[1] Nasar SA. D.C.-switched reluctance motor. Proceedings of the Institution of Electrical Engineers. 1969;**116**(6):1048

[2] Widmer JD, Martin R, Kimiabeigi M. Electric vehicle traction motors without rare earth magnets. Sustainable Materials and Technologies. 2015;**3**:7-13

[3] Krishnan R. Switched Reluctance Motor Drives: Modeling, Simulation, Analysis, Design, and Applications. Boca Raton, FL: CRC Press; 2001

[4] Takeno M, Chiba A, Hoshi N, Ogasawara S, Takemoto M, Rahman MA. Test results and torque improvement of the 50-kW switched reluctance motor designed for hybrid electric vehicles. IEEE Transactions on Industry Applications. 2012;**48**(4): 1327-1334

[5] Meeker D. Finite Element Method Magnetics (FEMM), FEMM Reference Manual. Available rom: http://www.femm.info/wiki/Documentation/ [Accessed: 5 September 2017]

[6] Miller TJE. Converter volt-ampere requirements of the switched reluctance motor drive. IEEE Transactions on Industry Applications. 1985;**IA-21**(5): 1136-1144

[7] Miller TJE. Switched Reluctance Motors and Their Control. Hillsboro, OH/Oxford: Magna Physics Pub.; Clarendon Press; 1993

[8] Stuikys A, Sykulski J. Rapid multi-objective design optimization of switched reluctance motors exploiting magnetic flux tubes. IET Science Measurement & Technology. 2017

[9] Coxeter HSM, Greitzer SL. Geometry Revisited, 10. print. Mathematical Association of America: Washington; 2005

[10] Stuikys A, Sykulski JK. An efficient design optimization framework for nonlinear switched reluctance machines. IEEE Transactions on Industry Applications. May 2017;**53**(3): 1985-1993

Chapter 13

Numerical and Experimental Analysis of Vibrations in a Three-Phase Linear Switched Reluctance Actuator

José António da Costa Salvado,
Maria do Rosário Alves Calado and
António Eduardo Vitória do Espírito-Santo

Abstract

This chapter focuses on the analysis and characterization of the vibrations produced by switched reluctance actuators. The emphasis stands on the linear configuration of this type of machine. The complexity of the mechanical system and the materials is used to define the modal frequencies. Moreover, the power controller topology, the excitation regimes, and the switching frequency used for the actuator operation can excite the natural modes and put restrictions on its usage. The analysis considers both numerical and experimental methodologies. The numerical technique relies on the finite element method (FEM) using the 3D model of the actuator to find its natural frequencies up to ~1.3 kHz. The experimental characterization counts on the operational modal responses and the acoustic noise emitted. We identify the regions of interest to measure the local accelerations and collect data for post-processing and record the audible noise emitted for signal analysis. The popular discrete Fourier transform and the joint wavelet-Fourier analysis are used for signal analysis. The reliability and the suitability of this approach are verified comparing both the numerical and the experimental outcomes and support the identification of the switching frequencies with high potential to excite the natural modes under the regular operation of the machine and to choose the proper control strategy.

Keywords: linear switched reluctance drives, force ripple, vibrations, acoustic noise, numerical analysis, operational modal responses, hybrid Spectrum

1. Introduction

Despite their simplicity, robustness, and reliability, switched reluctance drives face some limitations on their acceptance and usage due to the vibrations produced, and the acoustic noise emitted under operation. These are significant weaknesses and concerns and have been subject of research since the early 90s [1–5]. Several studies have been published focusing either on the origin of the vibrations or proposing methodologies to mitigate its effects by using the right control strategy, or on avoiding its presence at the design stage.

The principle of operation of switched reluctance drives, producing either rotational or linear movements, is based on the variation of the magnetic circuit reluctance, according to the relative position and the alignment of its fixed and moving parts [6]. The electromagnetic force ripple produced is the primary cause of vibrations. However, while in the rotational configuration, that is, the switched reluctance motor (SRM), the vibrations reveal stationary and periodic characteristics, for the linear switched reluctance actuator (LSRA), they tend to be nonstationary and confined to specific regions. That happens mostly due to the finite dimensions of the actuator parts and is related to wave propagation and reflection phenomena, as well as the geometry of the actuator and the properties of the materials.

This chapter addresses the modal analysis of the vibrations through the finite element method (FEM) and its evaluation and validation through experimental methods using the operational responses of the LSRA. The methodology is similar to operational modal analysis (OMA) and was considered due to the characteristics related to the operation of the actuator and its physical dimensions [7].

The LSRA considered in this research is a three-phase longitudinal magnetic flux actuator designed, built, and validated for precision positioning applications. **Figure 1** shows a general overview and the cross-section of the actuator. The LSRA dimensions in SI units are length: 2000 mm; width: 500 mm, height: 250 mm (includes the leveling feet), and weight: 47 kg. The actuator is built mainly with aluminum 6060-O (AlMgSi 0.5F25) profile frames forming the base assembly, two longitudinal aluminum 6060-O bars, and two ball guide rails made of chrome-plated steel, one per each side. Two aluminum plates on top of the structure are solidary bonded and constitute and constitute the moving platform. The platform has six steel ball runner blocks that fit into the ball rail guides and allow the longitudinal movement.

The parts belonging to the magnetic circuit, namely the poles of the primary and the teeth of the secondary, and the cores of the phase coils are made of carbon steel alloy S235 JR according to EN 10025 (ST37–2 under DIN 17100). Each coil is constituted by 1092 turns of AWG20 copper wire with a resistance of ~10 Ω. **Figure 2a** shows a detailed view of the primary circuit, and **Figure 2b** shows a front view of one position with alignment in phase *b*. The magnetic circuit for each phase does not have shared cores or paths and therefore can be considered independent and the flux linkage constant.

According to the shapes of the poles and teeth, its alignment position, and the excitation current profiles, the attraction and thrust forces developed at the LSRA air gap can reach, respectively, 750 and 150 N (for $i_{as} = 4$ A) [8]. Consequently, the operation of the LSRA is sensitive to ripple, namely for frequency values able to

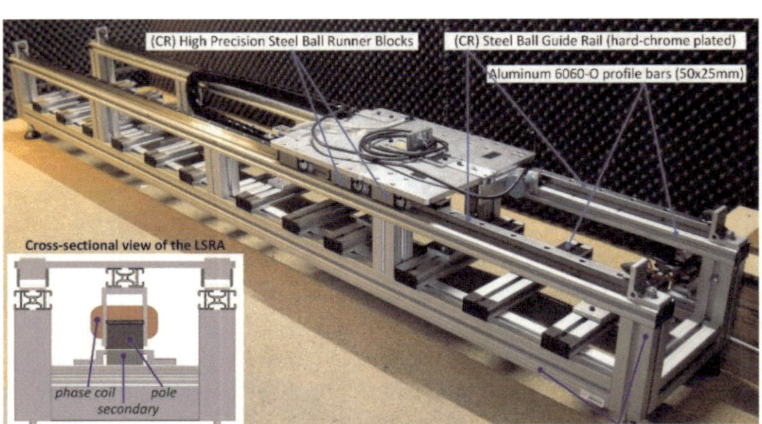

Figure 1.
Overview of the LSRA structure and details of its cross-section.

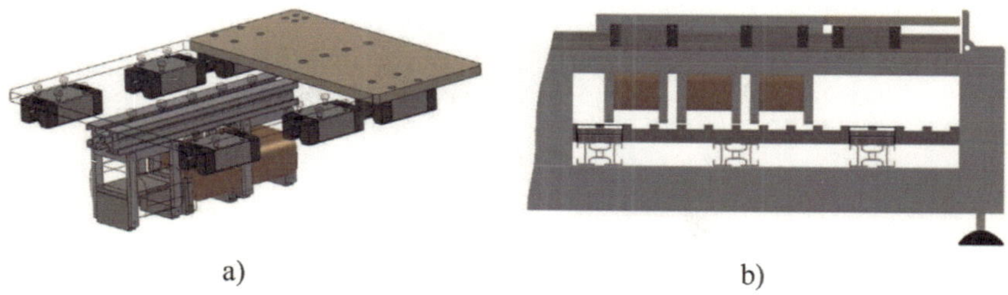

Figure 2.
Detailed view of (a) the subassembly of the primary and (b) example of front view position of the translator with full alignment in phase b.

excite its natural modes. Moreover, the interactions between the electromagnetic motoring and the regenerative forces can also generate vibrations.

This chapter is organized as follows. Section 2 describes the operation basics of the LSRA, the power converter topology, the forces developed, and its relations with the origins of vibrations. The methods for dynamic analysis of mechanical systems are mentioned in section 3. Section 4 proposes the test setup and the configuration of instrumentation to perform the experimental tests for operational modal analysis. The analysis tools developed are also presented. Section 5 presents the results obtained, which are analyzed and discussed. Finally, the conclusion in section 6 summarizes the main achievements and contributions of this research work.

2. Operation of the LSRA

The principle of operation of switched reluctance machines is based on the variation of the magnetic reluctance according to the relative alignment positions of the poles and teeth of the primary and secondary, respectively, as shown in **Figure 3**.

The phase inductance, L_{as}, varies with the position, x_m, due to the longitudinal displacement along x, and reaches its maximum value with full alignment, being minimum at the unaligned position. By energizing phase a at the unaligned position (when the magnetic reluctance is maximum) with a coil current profile i_{as}, the work done by the developed electromagnetic forces tends to move the translator toward the aligned position where the inductance is maximum (minimum reluctance). The real current profile, i_{as}, tends to approach the ideal profile and is modeled by the switching signals, T_{a1} and T_{a2}, which depend on the switching technique, the power converter topology, and the control strategy adopted.

Due to the switching process, an induced electromagnetic force (emf) appears at the terminals of the phase coil. The induced emf can be expressed as

$$v_{La} = i_{as} \cdot \frac{\partial L_{as}(x_m, i_{as})}{\partial x} \cdot \frac{\partial x}{\partial t}. \tag{1}$$

The switching frequency, the duty cycle, and the control strategy used at the power converter define the phase current waveform. If the magnetic circuit has uniform geometry and the magnetic field is homogeneous, and neglecting the system losses, then, the linkage magnetic flux, λ_{as}, is proportional to the inductance and the current, according to expression

$$\lambda_{as}(x_m, i_{as}) = i_{as} \cdot L_{as}(x_m, i_{as}). \tag{2}$$

The magnetic energy stored in the coupling field depends on the inductance and the linkage flux. For linear and homogeneous systems with null initial linkage flux,

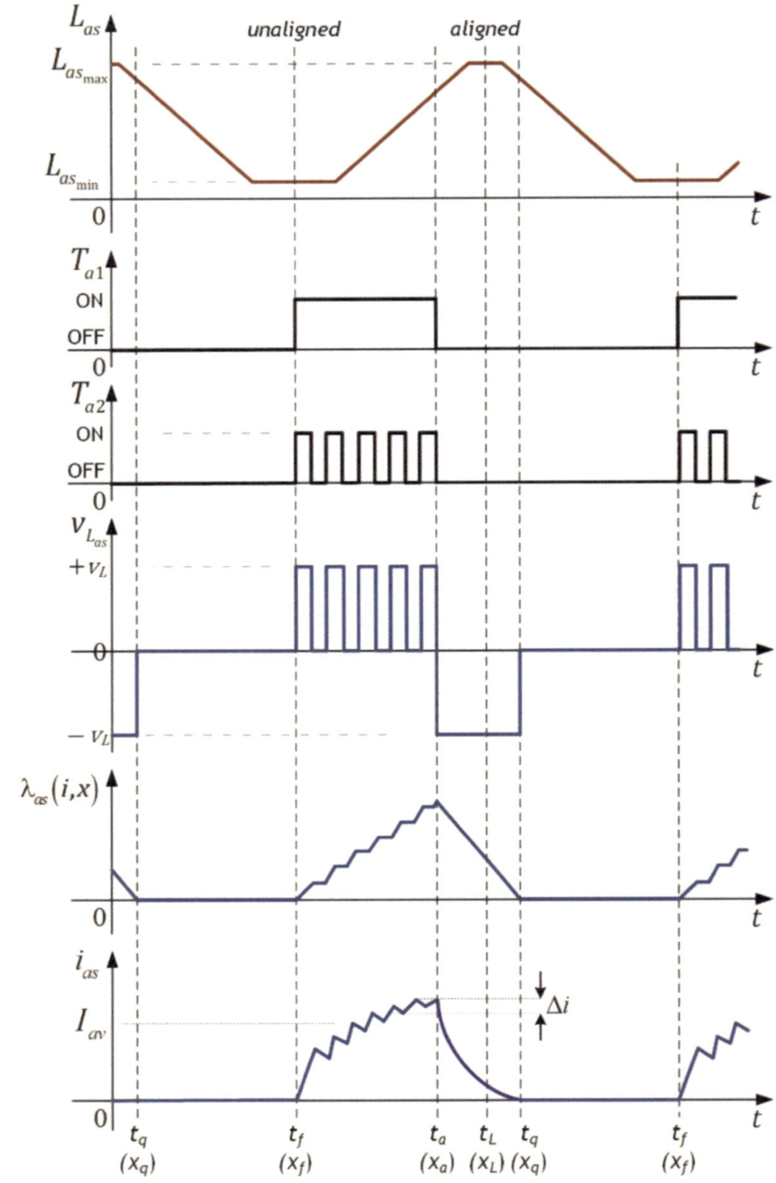

Figure 3.
Operation of phase a: variation of inductance with the position, time evolution of the switching control signals, time evolution of the voltage across the coil terminals, flux linkage and waveform of the phase current.

the electromagnetic energy associated with the field, W_{em}, is related to the variation of the magnetic flux linkage and also the current, and is expressed by

$$W_{em}(\lambda_{as}, x_m) = \int_{\lambda} i_{as}(\lambda_{as}, x_m) \cdot d\lambda_{as} = \frac{\lambda_{as}^2}{2 \cdot L_{as}(x_m, i_{as})}. \tag{3}$$

If the system can be considered homogeneous, and the flux linkage is uniform, the energy associated with the field in Eq. (3) is independent of the integration path. This means that the system is linear and conservative, and both, the mechanical and the electrical systems interact, and therefore, the variation of energy and the system losses are uniform. Following these conditions, the variation of energy can be expressed in terms of partial derivatives with respect to its independent state variables as

$$dW_{em}(\lambda_{as}, x_m) = \frac{\partial W_{em}}{\partial \lambda_{as}} \cdot d\lambda_{as} + \frac{\partial W_{em}}{\partial x} \cdot dx = i_{as} \cdot d\lambda_{as} - f_{em} \cdot dx, \tag{4}$$

where, f_{em} is the electromagnetic motoring force.

If the coil excitation current is almost constant near its average value, $i_{as} \approx I_{av}$, then, using Eqs. (3) and (4), the attraction and motoring (traction) electromagnetic forces produced at the air gap, respectively, are given by

$$f_{ea} = \left. \frac{\partial W_{em}(\lambda_{as}, x_m)}{\partial z} \right|_{\lambda_{as}=C^{te}} = \frac{i_{as}^2}{2} \cdot \frac{dL_{as}(x_m, i_{as})}{dz} \tag{5}$$

and

$$f_{em} = \left. \frac{\partial W_{em}(\lambda_{as}, x_m)}{\partial x} \right|_{\lambda_{as}=C^{te}} = -\frac{i_{as}^2}{2} \cdot \frac{dL_{as}(x_m, i_{as})}{dx}. \tag{6}$$

Likewise, a regenerative force which tends to oppose movement is developed at the air gap and expressed as

$$f_{er} = -f_{em} = \frac{i_{as}^2}{2} \cdot \frac{dL_{as}(x_m, i_{as})}{dx}. \tag{7}$$

The switching frequency, the topology of the converter, and the control strategy cause variations in the coil excitation currents [9]. The controller used to activate the LSRA that supports the research described here is a three-phase half-bridge asymmetric power converter topology depicted in **Figure 4**, operating with a soft chopping control strategy, whose control signals, voltages, and currents relative to the operation of phase *a* are depicted in **Figure 3**.

These current variations foster ripple in the electromagnetic motoring and regenerative forces, whose interactions induce mechanical vibrations that can excite the natural modes of the LSRA. Therefore, knowledge of the natural structural frequencies of the actuator and its mode shapes is paramount.

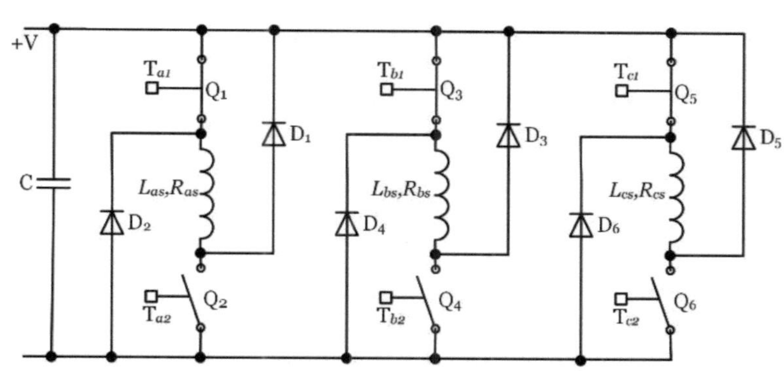

Figure 4.
Electrical diagram representing the basis of operation of a three-phase asymmetric half-bridge converter in soft chopping mode.

3. Modal analysis via the finite element method

The vibrations and the acoustic noise problems in machinery depend on both the forces applied and structural aspects, namely, the system characteristics, the properties of the materials, and the boundary conditions. The diversity of sizes and shapes of its parts and their interconnections often leads to high-complexity mechanical models and multiple degrees of freedom (MDOF) systems.

Following the classical theories for structural dynamics analysis, systems are commonly described as a set of interconnected masses acting through a network of springs and dampers, with viscous damping, and submitted to a steady-state sinusoidal excitation by the action of the external forces, where the number of independent coordinates establishes the number of degrees of freedom.

3.1 Dynamic analysis for multiple degrees of freedom systems

The Lagrange's equation is an alternative approach to Newton's laws formulation that uses the equilibrium of energies in the system to describe its motion. Considering the kinetic and potential energies stored in the system, W_{kin} and W_{pot}, respectively, the condition of equilibrium is given by,

$$\frac{d}{dt}\cdot\frac{\partial(W_{pot}-W_{kin})}{\partial\dot{x}_i}-\frac{\partial(W_{pot}-W_{kin})}{\partial x_i}=0 \qquad i=1,2,\ldots,n, \tag{8}$$

where x_i and $\dot{x}_i$ represent, respectively, the generalized displacements and velocities at generalized coordinates i. The total kinetic energy stored in the springs due to compression and displacement Δx with respect to an initial position considers all coordinates i associated with the system [10]. For n-DOF systems, the Lagrange's equation is

$$\frac{d}{dt}\cdot\left(\frac{\partial W_{kin}}{\partial\dot{x}_i}\right)-\frac{\partial W_{kin}}{\partial x_i}+\frac{\partial W_{pot}}{\partial x_i}+\frac{\partial D_R}{\partial\dot{x}_i}=Q_i \qquad i=1,2,\ldots,n. \tag{9}$$

Here, D_R is the Rayleigh's dissipation function representing the energy losses at the dumpers, and Q_i refers to the generalized non-conservative forces. For each element of order i, the quadratic forms of the kinetic and potential energies and the Rayleigh's dissipation function can be expressed in matrix notation as

$$W_{kin}=\frac{1}{2}\cdot\dot{\mathbf{x}}^T\cdot\mathbf{M}\cdot\dot{\mathbf{x}}, \tag{10}$$

$$W_{pot}=\frac{1}{2}\cdot\mathbf{x}^T\cdot\mathbf{K}\cdot\mathbf{x}, \tag{11}$$

$$D_R=\frac{1}{2}\cdot\dot{\mathbf{x}}^T\cdot\mathbf{C}\cdot\dot{\mathbf{x}}. \tag{12}$$

Here, **M**, **C**, and **K** are square and symmetric matrices that represent the inertia (mass), and the damping and stiffness coefficients, respectively; $\mathbf{x}$ and $\dot{\mathbf{x}}$ are the vectors of generalized displacements and velocities. The off-diagonal coefficients of matrices **M**, **C**, and **K** are the coupling terms that depend on the configuration of the system and its dimensions depend on the number of degrees of freedom.

Considering small-amplitude displacements with reference to the equilibrium points, that is, $x_i \approx x_{ei}$, the Lagrange's equation can be linearized and expressed using matrix notation in the form

$$\mathbf{M} \cdot \ddot{\mathbf{x}}_i + \mathbf{C} \cdot \dot{\mathbf{x}}_i + \mathbf{K} \cdot \mathbf{x}_i = \mathbf{Q}_i. \tag{13}$$

Finding the solutions of Eq. (13) for MDOF systems with several parts, of several dimensions and shapes and interconnections, often requires considerable computational time and powerful numerical tools. Therefore, a common approach consists of performing the modal analysis using the finite element method (FEM).

3.2 Modal analysis via the finite element method

The free vibration result for undamped systems is labeled as mode and corresponds to a standing wave where all points move at one frequency, that is, the natural frequency, with a particular phase relation. Consequently, for any mode of order r, the eigenvalue, ξ_r, corresponds to the undamped natural frequencies and the eigenvectors, $\boldsymbol{\psi}_r$, contain information on the mode shapes associated with these frequencies. As modes have a weighted orthogonality property [10] (pp. 309–310) and, the mass, stiffness, and damping matrices are in general Hermitian, it is possible to perform matrix diagonalizations and obtain their eigenvectors, therefore simplifying the problem. Hence, in real systems, one can consider proportional damping as a linear combination of the mass and stiffness matrices and perform the decoupling of MDOF systems into modal coordinates [11]. Thus, it is possible to represent the motion by rewriting Eq. (13) and obtain

$$\mathbf{M}_r \cdot \ddot{\mathbf{x}}_i + \mathbf{C}_r \cdot \dot{\mathbf{x}}_i + \mathbf{K}_r \cdot \mathbf{x}_i = \boldsymbol{\psi}_r^T \cdot \mathbf{Q}_i. \tag{14}$$

where

$$\mathbf{M}_r = \boldsymbol{\psi}^T \cdot \mathbf{M} \cdot \boldsymbol{\psi}, \tag{15}$$

$$\mathbf{K}_r = \boldsymbol{\psi}^T \cdot \mathbf{K} \cdot \boldsymbol{\psi}, \tag{16}$$

$$\mathbf{C}_r = \alpha \cdot \mathbf{M}_r + \beta \cdot \mathbf{K}_r = \boldsymbol{\psi}^T \cdot \mathbf{C} \cdot \boldsymbol{\psi}. \tag{17}$$

Each row in Eq. (14) corresponds to a single-degree of freedom (SDOF) system whose equation of motion considers its modal parameters: mass, damping, and stiffness. The undamped natural frequency and the damping coefficient for mode r, respectively, are given by

$$\omega_r = \sqrt{\frac{k_r}{m_r}}, \tag{18}$$

$$c_r = -2 \cdot \sigma_r \cdot m_r, \tag{19}$$

where k_r and m_r are the stiffness constant and mass, respectively, and c_r is the damping coefficient.

The use of Lagrange's equation to MDOF systems is possible by selecting a set of independent coordinates and applying successive differentiation along each coordinate. Hence, the analysis via FEM corresponds to decomposition into many building blocks to find the approximate solutions of the differential equations that describe the model. This process corresponds to a system discretization technique. Moreover, it allows the collection of the complete system responses as linear combinations of the set of responses from numerous single degree of freedom systems.

3.3 Results from FEM simulations

Most parts of the LSRA in **Figure 1** are uniform plates, and uniform bars mounted either transversely or longitudinally. However, the multiplicity of parts, their shapes and configurations and its interconnections, the diverse materials used, and the geometry increase the complexity of the system. That leads to the use of FEM simulation tools as an efficient approach for analysis of the natural modes and the vibration frequencies of the LSRA structure.

Based on the LSRA dimensions and symmetry characteristics, five scenarios are considered according to the position of the translator, which correspond to configurations P1–P5, as shown in **Figure 5**. For the requisites of experimental modal excitation analysis, five positions are established to apply excitation from external forces: E1–E5.

Based on these configurations and the materials used in the real actuator, we used SolidWorks to design the 3D model for each LSRA configuration (P1–P5) and built-in simulation tools to perform the FEM-based modal analysis study. The global contacts are considered ideal and set as bonded with no clearance and no penetration among parts. To optimize the simulation results, we use:

- curvature-based mesh, Jacobian, tetrahedral elements (four points);
- fine-density mesh, with elements of size among 1.2 (min) and 12 mm (max); and
- growth ratio of 1.6 per element, and at least eight elements to describe circles.

These simulation arrangement presets lead to a high-quality mesh, whose detailed view is presented in **Figure 6**. With this mesh configuration, the total number of elements is ~1,221,800 of which 28% have an aspect ratio less than 3, and 8% have an aspect ratio greater than 10. The number of nodes is around 2,226,200, hence achieving ~6,678,600 degrees of freedom for 3D coordinates.

To complete the modal analysis study, we selected the FFEplus solver embedded in SolidWorks software simulation tools and performed the frequency analysis for each scenario in the range up to 1.3 kHz. The FFEplus solver adopts an iterative integration method to find the solutions and evaluates the error for each iteration; when the latter is small enough, the result is assumed stable. When compared to the direct method, this approach leads to gains in accuracy but with higher computational costs. For example, for each configuration model, the iterative process requires around 30 hours for mesh construction and analysis, on a PC with Intel

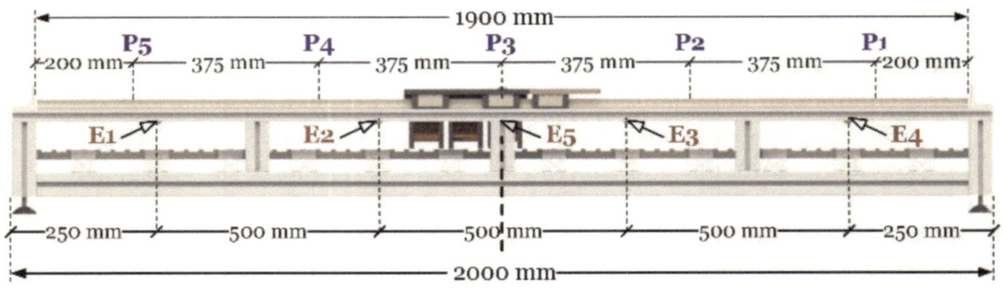

Figure 5.
Detailed view of the configurations considered for experimental tests: positions of the translator (Px) and points to apply external excitation forces (ex).

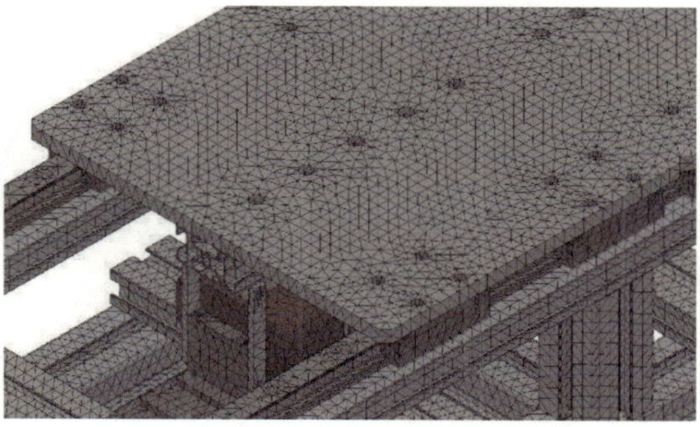

Figure 6.
Detailed view of the mesh and tetrahedral elements for FEM analysis.

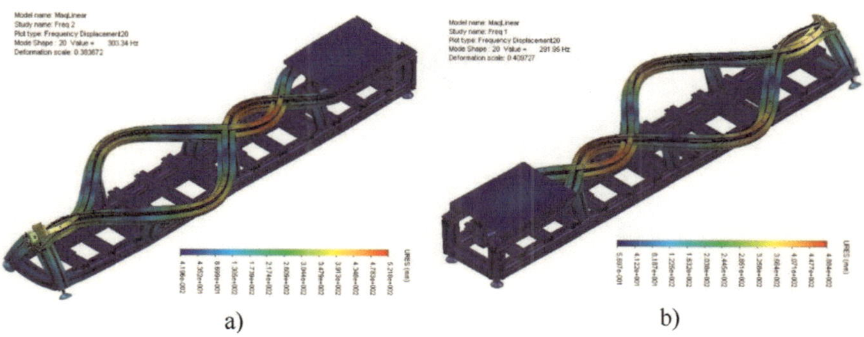

Figure 7.
Modal shapes for mode #20: (a) at position P1 (fn = 303 Hz) and (b) at position P5 (fn = 292 Hz).

Core 2 Duo E7500 CPU @ 2.93 GHz, 6 GB RAM DDR3 @ 1.333 Mbps (36 ns) and Windows 7 Pro 64-bit operating system.

The modal analysis results for each 3D model reveal a set of 80 natural modes from which the first six frequencies correspond to rigid body vibration modes, as expected. Accordingly, modes of order 7 correspond to the first useful natural modes, in a set of 74 significant modes whose complete list and the corresponding analysis are made in [12]. Here, we only present a brief summary of the results obtained and show a few examples of the modal shapes.

For all configurations considered and for same-order modes, the vibration frequencies are close to each other. On the other hand, for sequential modes at one configuration, the differences in frequency are small. The average modal frequency interval is ~15 Hz with minimum differences as small as 1 or 2 Hz (e.g., for modes 41 and 42 or modes 57 and 58 at P1). Moreover, for adjacent modes, the modal frequencies can have very close values meaning that a given vibration frequency can be excited for different modes in different actuator configurations.

Because each mode corresponds to a dominant vibration with displacement at specific parts or subassemblies, it is imperative to look also at the modal shapes. **Figure 7a** and **b** illustrates the mode shapes associated with mode #20 for positions P1 and P5, respectively, whose frequencies are ~303 Hz and ~292 Hz.

The mode shapes at positions P1 and P5 are symmetric and reveal displacements at the top bars associated with elastic waves in phase opposition. It must be noted that the movements shown at these images are not at the correct scale, serving merely to illustrate the most significant displacements for each mode. **Figure 8a** and **b** shows the counterpart and symmetric modal shapes for P3 corresponding to modes #13 (~226 Hz) and #14 (~230 Hz), correspondingly. Although with different-order modes, the modal shapes are similar, and frequency values are close.

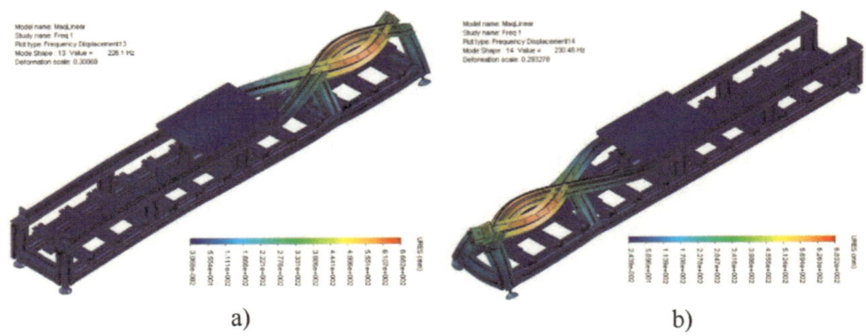

Figure 8.
Modal shapes at position P3: (a) mode #13 (fn = 226 Hz) and (b) mode #14 (fn = 230 Hz).

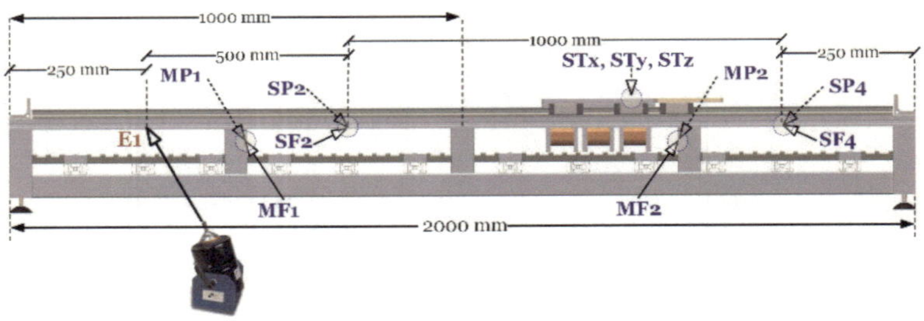

Figure 9.
Detailed view of the configurations and instrumentation considered for the experimental modal excitation tests.

At position P3, the length of the bar's segments is identical on both sides, but the center of mass is not aligned with the symmetry axis. That influences the symmetric characteristic of the vibration modes, namely the shapes of its deformation waves. Hence, while the mode shapes of first-order modes (modes 13 and 14) exhibit symmetry, the second-order modes show anti-symmetric mode shapes. These characteristics apply to other modes at P3.

3.4 Experimental modal excitation

The modal frequencies were evaluated through experimental modal excitation. For that purpose, we used an electrodynamic shaker to apply the external forces at excitation points E1–E5, and several accelerometers are placed along with the structure of the actuator, as shown in **Figure 9**.

The accelerometers are installed over a layer of bee wax with ~0.5 mm thickness at the areas with significant displacement: two at the top longitudinal bar at the front side (positions SF*n*); two at the opposite face (positions SP*n*); and four accelerometers at middle vertical bar; two at the front side and other two at the rear (positions MF*n* and MP*n*, respectively). All these accelerometers have one-axis sensitivity (ADXL001-70Z). Moreover, one three-axis accelerometer (ADXL326Z) is mounted on top of the translator plate, to measure the vibrations along the X-, Y-, and Z-axis (STx, STy, and STz, respectively).

4. Operational modal responses

The modal frequencies whose mode shapes denote vertical displacement are of interest for analysis, due to their potential for excitation during the regular operation of the LSRA. To evaluate the operational modal responses, we measure the local accelerations at the structure. To achieve this objective, we establish a test setup and define the experimental procedures. To collect data of the local responses, we designed and developed a modular data acquisition system and developed specific analysis tools.

4.1 Experimental setup and instrumentation

The experimental setup is formed by the actuator, the half-bridge power converter, power supply units, regular equipment for test and measurement (oscilloscope, multimeter, and function generator), a host PC and a modular, and distributed system for data acquisition (MDSAM) to measure the local responses and collect data from sensors [13, 14]. The instrumentation set is composed by accelerometers placed at selected positions of the LSRA structure. Due to the symmetry characteristics of the actuator, only half of the structure needs to be instrumented for data collection. **Figure 10** shows the locations of 10 one-axis ratiometric accelerometers with analog output response (voltage) and ± 70 g sensitivity (ADXL001-70Z), placed at the left side of the structure with the translator at configuration P2.

The accelerometers are also installed over bee wax with ~ 0.5 mm thickness with the following distribution criteria: four at the front side on the top longitudinal bar (positions SFn); four at top longitudinal bar on the opposite face (positions SPn); and two accelerometers mounted at the bar that constitutes the secondary of the actuator (SS1 and SS2). The three-axis accelerometer (ADXL326Z) is mounted on top of the translator plate, to measure the vibrations along the axes X, Y, and Z (STx, STy, and STz, respectively). The phase current is evaluated by using current sensors and oscilloscope probes to monitor its waveforms; an ampere meter is also used to measure the average current values. The accelerometers connect to the MDSAM through the intelligent and interface modules based on 16-bit low-power microcontrollers (MSP430F5419 or MSP430F5438), and anyone of the 13 output terminals from the accelerometer modules corresponds to one measurement channel of the MDSAM. The intelligent sensor modules communicate with the host PC through a USB 2.0 bus, and the MDSAM is fully controlled through the host PC via MATLAB environment.

To conduct the operational modal tests, the translator is fixed and locked at configuration P2, in a position that corresponds to 50% alignment between poles and teeth, as depicted in **Figure 11a**. To produce the electromagnetic excitation forces, we consider exclusively the activation of phase a, whose coil current is fed by the asymmetric half-bridge converter operating at soft chopping switching strategy. The soft chopping operation mode is achieved by setting ON the control signal T_{a1}, while T_{a2} is adjusted to a given switching frequency—f_{Ta2}—and a duty cycle of $d = 85$ % (see **Figures 3** and **4**). This configuration allows an average phase

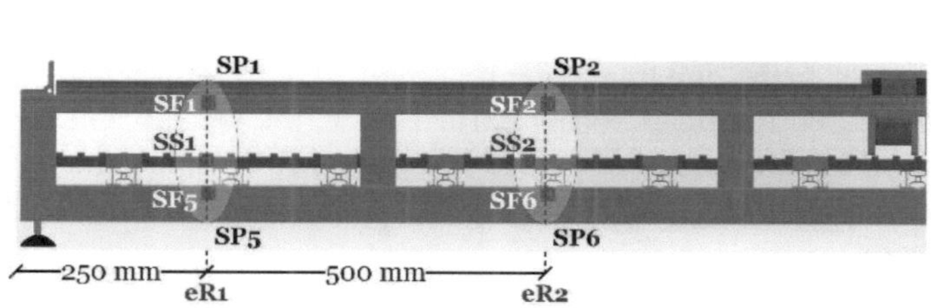

Figure 10.
Placement of accelerometers at the left half of the LSRA structure for local response measurements with translator positioned at P2.

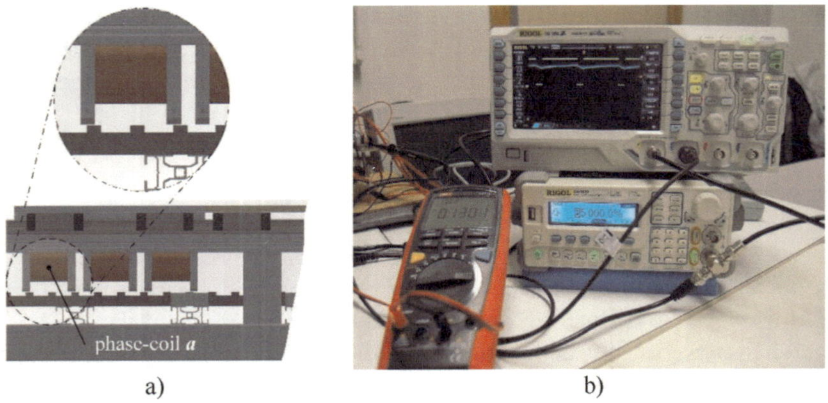

Figure 11.
Detailed view of (a) the magnetic circuit with 50% alignment, and (b) test bench setup to measure the current in phase a ($I_{av} = 1.3\ A$ @ $d = 85\%$).

current of 1.3 A through the phase coil when a 30 V voltage is used for phase activation, as shown in **Figure 11b**.

The 50% alignment and this phase current profile generate an electromechanical traction force of approximately 30 N, whose work tends to move the translator toward the full alignment position. However, as the translator is locked, and due to the nature of the forces generated, the excitation forces act similarly as hammer impacts with periodic steady state [7].

4.2 Analysis of stationary and nonstationary signals

The data collected through the MDSAM are organized according to the type of sensor and with correspondence to the point of measurement. The data obtained from one measurement set are stored in MAT-files (*.mat) for further processing. After reading, extracting data from the selected MAT-file, and identifying the type of response associated to the signal (phase current or local acceleration), the user establishes the parameters for analysis and defines the desired representation, either the time response or the spectrum. For that purpose, we developed a set of MATLAB scripts and functions for MDSAM control, data acquisition, and storage and for signal analysis of the local responses. The signal analysis tools are also based on MATLAB and use the windowed discrete Fourier transform (DFT), whose coefficients can be calculated through Eq. (20). The options of window function include Rectangular, Hamming, Hann, Bartlett, and Flat-Top, selected according to their performance and suitability for this application [15]. The DFT is a popular solution that performs quite well for stationary signals and therefore is used in a vast number of applications.

However, in LSRA, the vibrations are in most cases localized and characterized by nonstationary signals, due to an eventual mixture with other vibration frequencies associated with other neighboring parts. Hence, the localization and nonstationary characteristics of the vibrations in LSRA demand innovative approaches and specialized tools for signal analysis, rather than merely the Fourier analysis methods.

An alternative for signal analysis is the discrete wavelet transform (DWT) which has proper localization in the frequency domain and whose basis functions are orthogonal and scalable in the time domain. The most prevalent wavelet constructions include the Daubechies family of orthogonal wavelets with n vanishing moments (DBn) and the Cohen-Daubechies-Feauveau (CDF) family of biorthogonal wavelets. The DWT can be implemented through banks of finite impulse response (FIR) filters, thus allowing signal analysis with multiresolution [ref] as shown in **Figure 12**.

The left-hand side of **Figure 12** represents a multiresolution scheme that uses low-pass, $\tilde{h}$, and high-pass, $\tilde{g}$, analysis filters to obtain the DWT coefficients. Each stage l of decomposition consists of obtaining the approximation, $a_{l,\,n}$, and the detail coefficients, $d_{l,\,n}$, resulting from the low-pass and high-pass filtering of the signal, and applying a decimation factor of 2 for in-place computation. The first frequency band of the signal is organized in octaves where the lower band contains a smooth representation of the signal; the following bands contain the details with scaled resolution. Successively applying the analysis filters to the detailed coefficients in each level leads to a wavelet packet (WP) decomposition scheme as depicted at the right-hand side in **Figure 12**, also known as wavelet packet analysis (WPA). Using this scheme and for l levels of analysis, the initial bandwidth is divided into 2^l regularly distributed bands. Hence, both the DWT and WPA schemes are suited for the analysis of nonstationary signals with adjustable resolution.

4.3 Software tools for signal analysis and representation

Specific and dedicated software tools were developed by the authors in MATLAB using built-in and custom functions to perform the analysis of the results obtained from experimental modal excitation and from the operational modal responses.

The flowchart of **Figure 13** shows the sequence of actions for the analysis of the local responses whose data are stored in *.mat files. Through a dialog box, the user chooses the file, and the relevant data are extracted, namely the arrays containing the data of each channel. Then, each array is processed according to the type of data and the desired analysis following a control order: "A" for audio; "V" or "T" for accelerations.

The analysis of signals using the Fourier method and its spectral representation is obtained through the discrete Fourier transform (DFT), expressed as,

$$X_w[q] = \sum_{n=0}^{N} \{x[n] \cdot w[n]\} \cdot e^{-j\frac{2\pi nq}{N}}, \quad 0 \le q \le N-1 \tag{20}$$

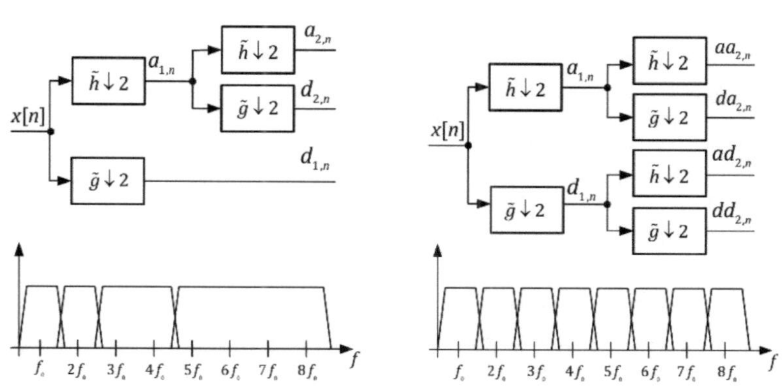

Figure 12.
Diagrams for two-level multiresolution signal analysis via DWT (left) and through wavelet packet decomposition (right).

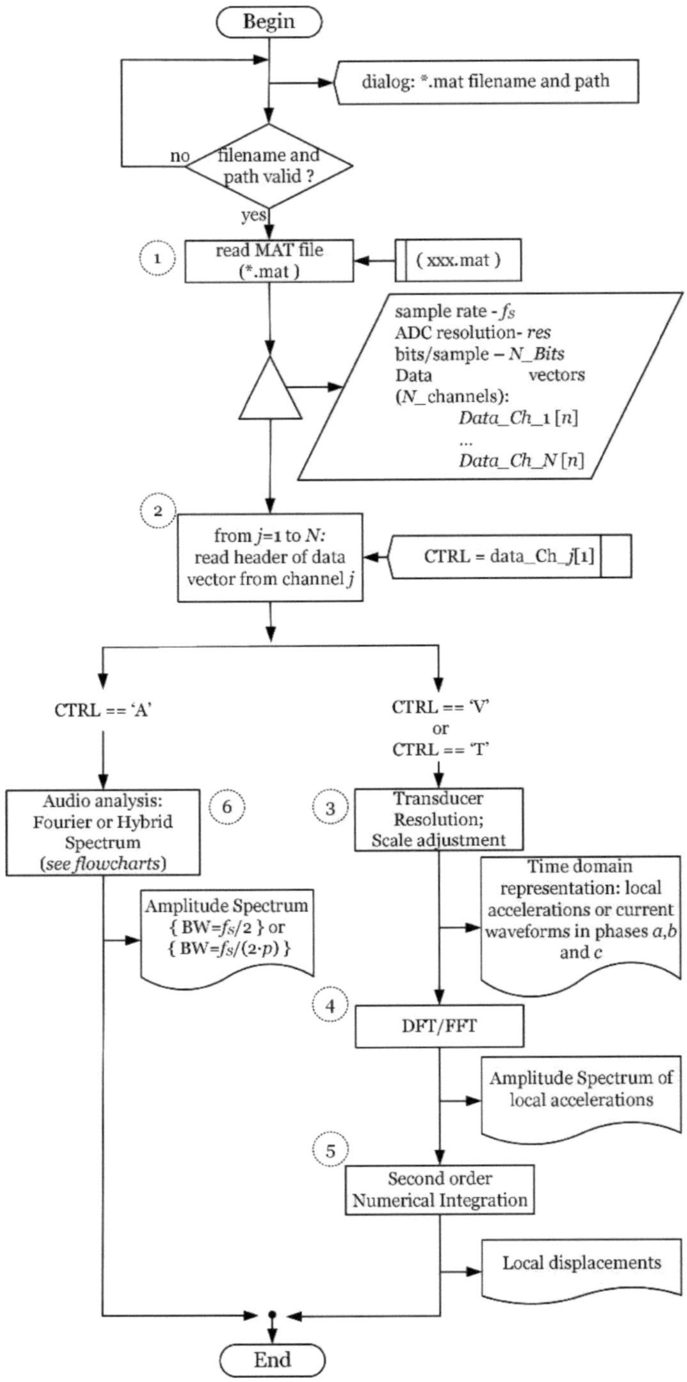

Figure 13.
Description of the tool developed for signal analysis.

where $x[n]$ is the discrete-time signal and $w[n]$ is the finite length window function of size N and defined within the interval $0 \leq n \leq N-1$. The same is possible via the fast Fourier transform (FFT). The flowchart of **Figure 14** illustrates the sequence of actions corresponding to the DFT or FFT and the amplitude spectrum of the responses, either for the accelerations, phase currents, or audio signals. This sequence is activated at stage 6 at the flowchart of **Figure 13**.

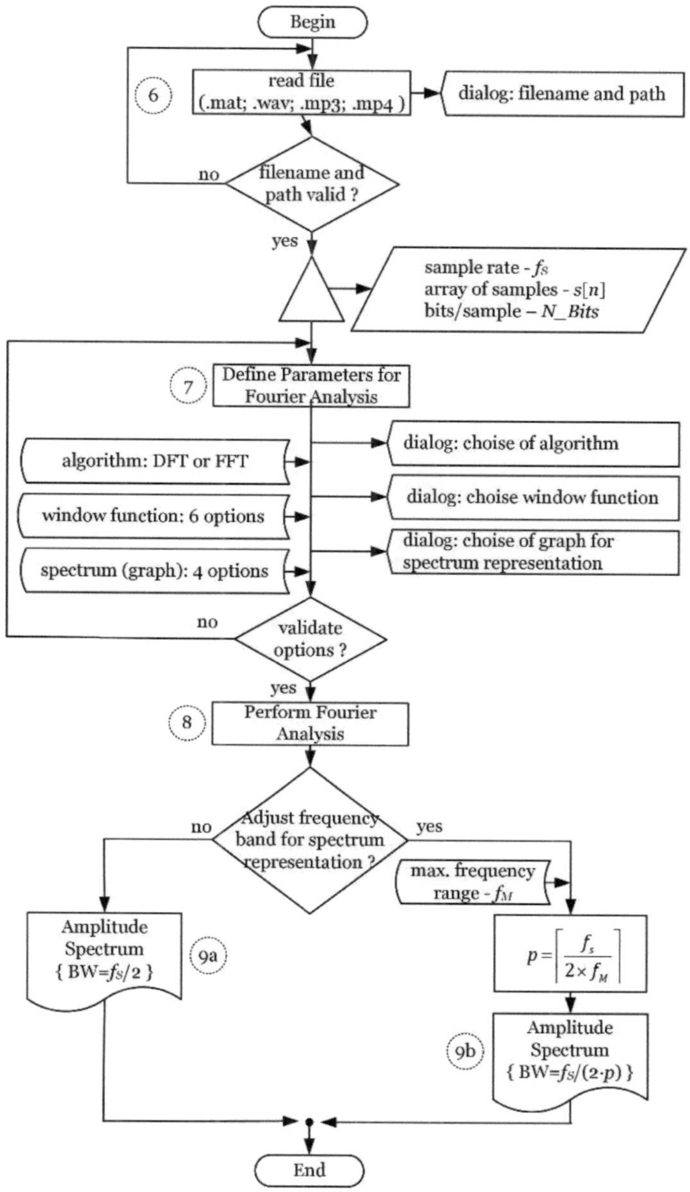

Figure 14.
Flowchart illustrating the operation of the Fourier analysis tool.

The window function is chosen among six options: rectangular, Hann, Bartlett, Hamming, Bartlett-Hann, and Flat-Top. With exception to the rectangular window, which is suited for impact test analysis, the remaining functions are suited for periodic excitation. By default, the Hann window is used. The options for spectrum representation permit four graphic types: amplitude vs. frequency using linear scales; a variant with amplitude scale in dB; and the same options but using semi-logarithmic scales. In any case, the frequency range is adjustable. By default, it uses the signal bandwidth. However, due to the nonstationary characteristics of the vibrations associated with the operation of the LSRA, an alternative analysis method is available. This technique, which combines wavelet and Fourier analyses to represent the hybrid spectrum, is described in the flowchart of **Figure 15**. After selecting the file and data extraction, the user establishes the parameters for WPA, namely

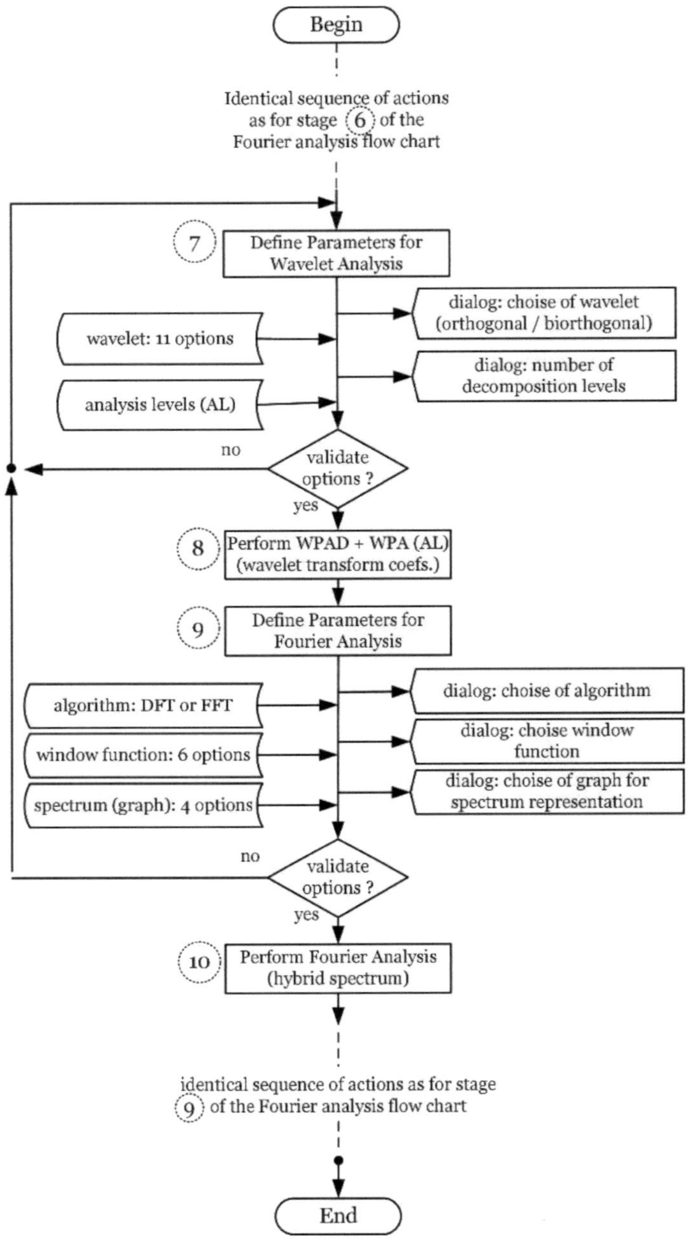

Figure 15.
Sequence of action involved in combined wavelet-Fourier analysis.

the wavelet function to use (orthogonal DBn or biorthogonal CDF families), and the number of analysis levels—ALs.

If these parameters are valid, a one-level wavelet packet analysis and denoising (WPAD) is performed, using the Shannon entropy associated with the signal. The signal is then reconstructed and submitted to WPA using the wavelet function specified with AL decomposition levels [16]. As a consequence, the wavelet transform coefficients are organized in frequency sub-bands, as shown in **Figure 12**. The bandwidth in each sub-band is established by dividing the Nyquist bandwidth by a factor expressed as a power of two, such that $r = 2^{AL}$. Then, the wavelet coefficients in sub-bands are submitted to Fourier analysis to obtain the hybrid wavelet-Fourier spectrum [17].

The Normalized Cross-Correlation (NCC) is a traditional technique to evaluate the resemblances between two signals, expressed as

$$\|r_{sy}[l]\| = \frac{r_{sy}[l]}{\sqrt{r_{ss}[0] \cdot r_{yy}[0]}}, \tag{21}$$

for discrete-time signals $s[n]$ and $y[n]$. Here, $r_{ss}[0]$ and $r_{yy}[0]$ represent the auto-correlation of these signals with null lag, and $r_{sy}[l]$ the cross-correlation coefficients with lag l, defined as

$$r_{sy}[l] = \sum_{n=-\infty}^{\infty} s[n] \cdot y[n-l], \quad l = 0, \pm 1, \pm 2, \ldots, \tag{22}$$

The NCC supports the evaluation of spectral similarities of the audio signals emitted for two different configurations of the LSRA, considering the amplitude spectrum coefficients and a given frequency lag.

Another method to estimate the spectral similitudes is the Cross-Power Spectrum Density (CPSD) which corresponds to the DFT of the cross-correlation (Eq. (22)). The CPSD is expressed as

$$S_{sy}[l] = \frac{1}{N}\sum_{n=0}^{N-1} r_{sy}[n] \cdot e^{-j\frac{2\pi nl}{N}} = R_{sy}[l] \cdot e^{-j\phi_{sy}[l]}, \tag{23}$$

where $R_{sy}[l]$ and $\phi_{sy}[l]$ are, respectively, the amplitude cross-spectrum and the phase cross-spectrum. The CPSD is estimated via the Welch periodogram [18].

5. Experimental characterization and results of local responses

To characterize the vibrations of the LSRA through experimental methods, we established a test setup to conduct a set of tests for experimental modal excitation ad for operational modal analysis. For these tests and according to the instrumentation configuration, we collected data of the local accelerations and recorded the audio emitted. For that purpose, during the tests, the actuator remained inside of a semi-anechoic chamber [19].

5.1 Results from experimental modal excitation

For any excitation point of **Figure 9**, a set of tests were performed for each configuration, the local accelerations are measured, and the audio recorded is analyzed to obtain its spectral components. The overabundance of results leads to a quite exhaustive analysis; therefore, only a few examples are presented here. **Figures 16** and **17** show the amplitude spectrum of the audio signals for configurations P1 and P2, respectively, both obtained for a steady-state sinusoidal force of 4.5 N applied at E1 (see **Figure 9**), and linear frequency variation from 10 Hz to 1.3 kHz.

Both Figures reveal prominent amplitude peaks at 65 Hz, which correspond to the first mode for P1 and P5 (dual configurations). The nearest peaks at the right-hand side have a frequency near 75 Hz and correspond to the first mode in positions P2 and P4. The 180- and 201-Hz frequency components are related to modes 12 and 13, respectively, and the frequencies of 282 and 290 Hz are connected to mode 19.

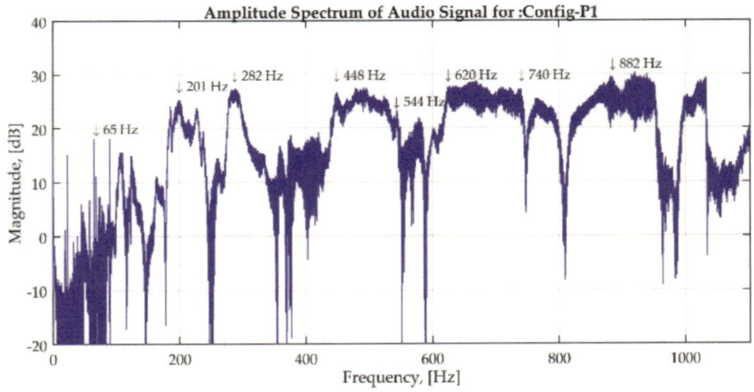

Figure 16.
Spectral components relative to modal frequencies excited at P1.

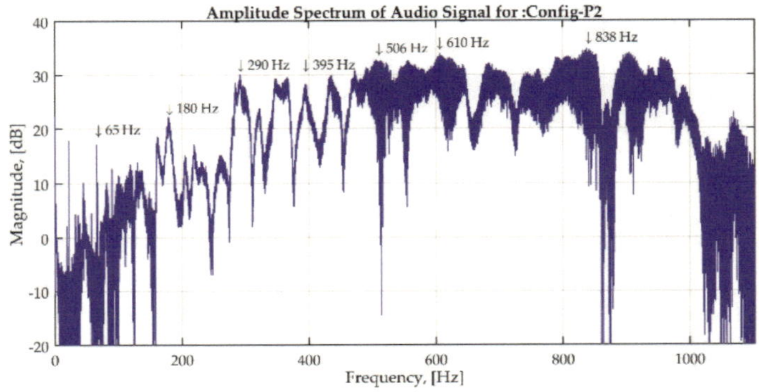

Figure 17.
Spectral components relative to modal frequencies excited at P2.

The range of frequencies around 610/620 Hz corresponds to the modal frequencies of modes 40–44.

Moreover, the highlighted frequencies, 838 and 882 Hz, are linked to vibrations almost exclusively at the primary circuit's parts. The deviations among the experimental results and the values obtained via FEM are due to non-ideal contacts and tolerances present at the real actuator. Still, the test data results are coherent and close to those obtained by numerical methods.

5.2 Characterization for operational modal excitation

For the experimental evaluation of the local vibrations produced with the LSRA in operation, we consider the test setup referred in subsection 4.1, and configuration P2. The switching frequency f_{Ta2} is set to 465 Hz and duty cycle to 85%, obtaining an average current of 1.3 A and a traction force of ~30 N [20]. To evaluate the vibrations due to operation using these conditions, we take modes #30-P2 (444 Hz) and #33-P2 (487 Hz) as references. On one side, the value of the switching frequency is in between the values of the two natural frequencies obtained from FEM and corresponding to these two modes. On the other side, the modal shapes associated with these modes reveal transverse displacements along the vertical direction, as shown in **Figures 18** and **19**, respectively. The modes with deflections through the vertical direction are relevant for this analysis. Mode #67-P2 (976 Hz) is also pertinent for analysis as its frequency corresponds to the first

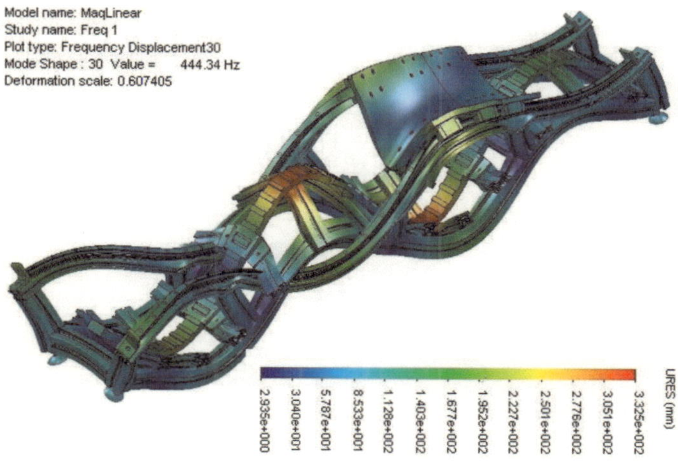

Figure 18.
Details of the modal shape related to mode #30 for configuration P2.

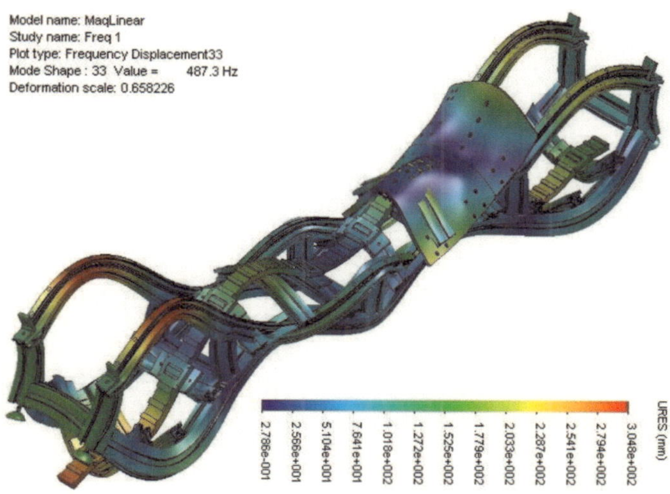

Figure 19.
Details of the modal shape related to mode #33 for configuration P2.

harmonic of f_{Ta2}, and its modal shape denotes movement almost exclusively at the secondary of the actuator, as shown in **Figure 20**.

The peaks of deformation in the modal shapes for configuration P2 are almost coincident with the positions considered for measuring the local responses (see **Figure 10**). Among these points, we refer mainly to SF1, SP1, SS1, SS2, and STz (see **Figures 18–20**). The characteristics above mentioned highlight the potential of excitation of these modes under operation, thus turning these test conditions into a good example and representative case for analysis.

Figures 21 and **22** show the time representation of the signal controlling switch, T_{a2}, measured with an oscilloscope probe, and the current through phase a, i_{as}, measured via a ratiometric current sensor (voltage output), respectively.

Observing **Figure 22**, we verify that the waveform of the AC component of the phase current (almost triangular) and two frequencies are visible at the amplitude spectrum: one at 464 Hz and the other one at 933 Hz (first harmonic). These frequency components are as expected and coherent with the switching frequency

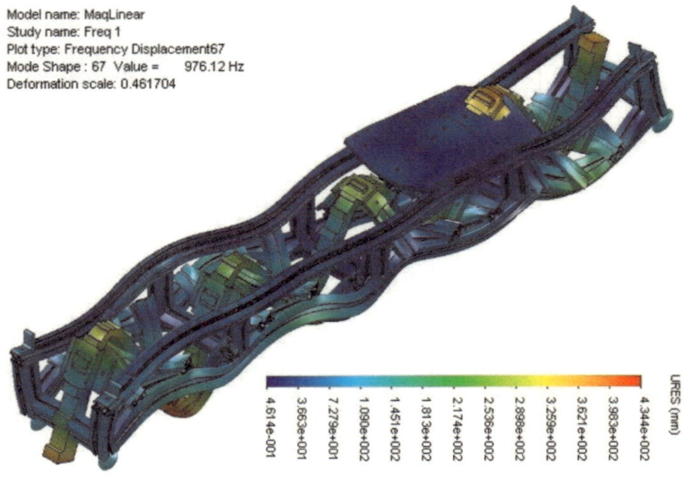

Figure 20.
Details of the modal shape related to mode #67 for configuration P2.

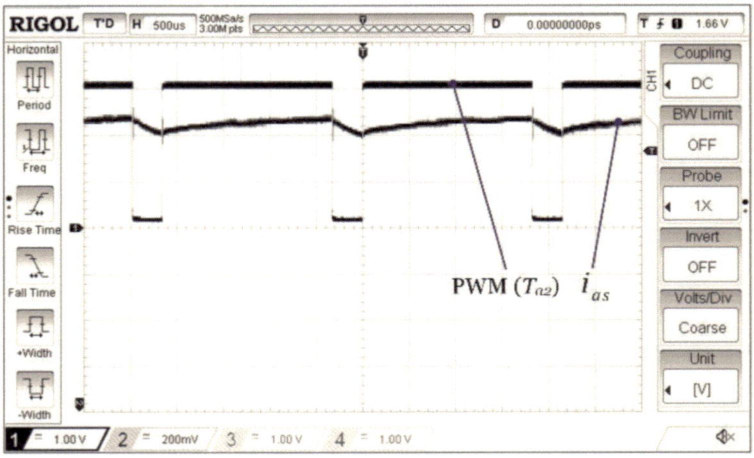

Figure 21.
Waveforms of the current through phase a *and of control signal* T_{a2} *(PWM).*

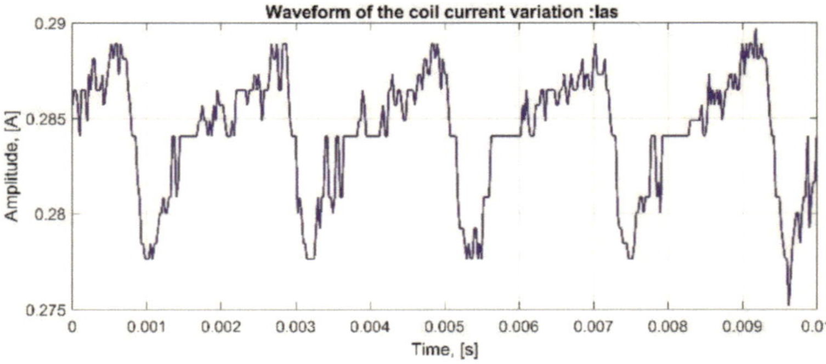

Figure 22.
Waveform related to the AC component of the current in phase a.

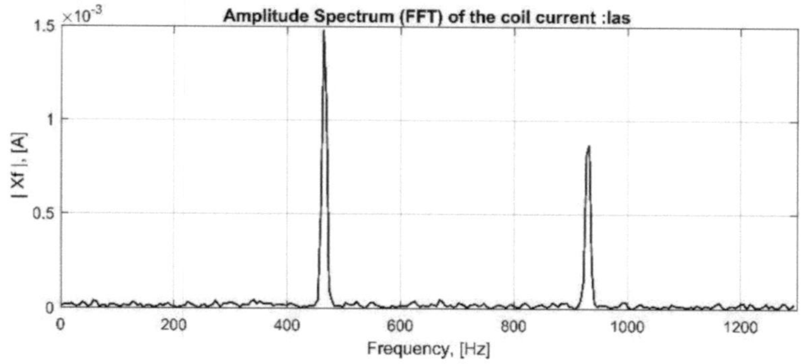

Figure 23.
Spectral components from the analysis of the current in phase a.

used to control the excitation, and the amplitude of the harmonic is coarsely half of that of the primary frequency.

Figure 23 presents the harmonic analysis of the phase current using the windowed fast Fourier transform (FFT) using the Hann window. The resulting amplitude spectrum considers the band up to 1300 Hz (frequencies of interest). From Eqs. (5)–(7), we can expect ripple in the forces generated in the actuator, which influences these two frequency components. Therefore, for a 5% error margin, the ripple in the electromagnetic forces and their interactions have the potential to excite the natural modes considered as reference.

The local vibrations caused by the action of the attraction and thrust forces are evaluated through the response of the accelerometers placed at the structure. We consider mostly the vibrations with vertical displacement at the platforms associated with the primary, and at the bar forming the secondary of the actuator. The frequency analysis of the local accelerations is also performed through the FFT using the Hann window. We start by probing the local responses measured at the points in **Figure 10** that are located close to the primary and the secondary of the actuator.

The accelerometer at the top of the platforms measures the responses along the three axes: X (longitudinal), Y (transverse), and Z (vertical). Among these, the acceleration along the Z-axis is of singular interest for analysis. The time evolution of its signal is shown in **Figure 24** and resembles the waveform of the phase current in **Figure 22**. This is related to the phase coils of the primary, which are assembled just below the top platforms. The ripple is mainly related to the attraction force and is straightly perceived with peak values of acceleration at +0.6 and $-1\ \mathrm{ms}^{-2}$.

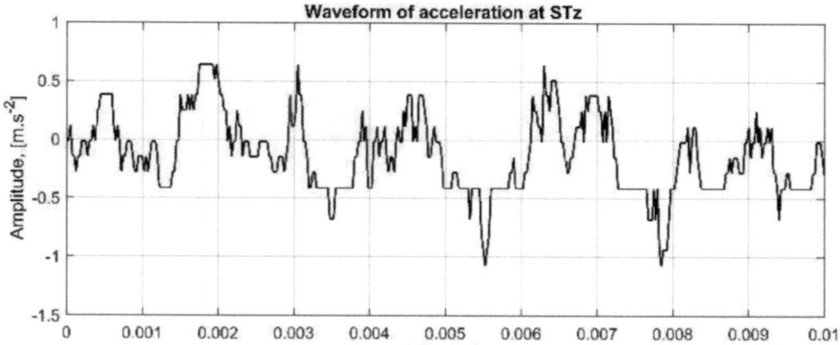

Figure 24.
Time evolution of the local accelerations along the Z-axis at STz.

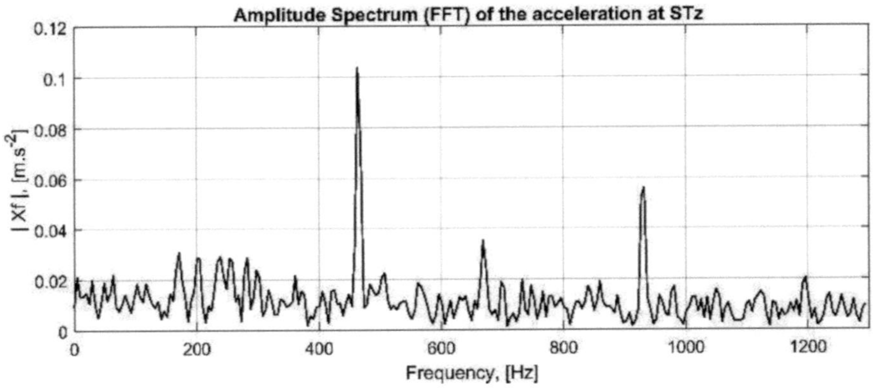

Figure 25.
Spectral components associated with the acceleration at STz.

Figure 25 shows the amplitude spectrum of the acceleration measured through STz with spectral components equivalent to the current. A less significant component around 669 Hz also appears which is related to modes considered exclusively associated with vibrations at the primary, as its frequency is almost constant regardless of the position of the translator.

Figure 26 presents the time evolution of the acceleration measured at SS1 (bar of the secondary), showing peak values reaching ± 20 ms^{-2}, which are higher than those obtained for STz. Two main reasons explain this: (1) the proximity of SS1 to the air gap where the forces are generated and (2) the higher value of stiffness associated to the material of the secondary bar (steel vs. aluminum).

The spectrum of accelerations at SS1 shown in **Figure 27** reveals frequency components with higher values than those obtained in the previous case, which is confirmed by its amplitude spectrum.

Besides the spectral components at 464 and 928 Hz, which are familiar to the spectrums of SS1 and STz, it exposes frequency components at 869 Hz (#56-P2 ~858 Hz), 1050 Hz (#72-P2 ~1055 Hz), and 1143 Hz (#76-P2 ~1141 Hz). All these frequencies correspond to modal shapes of configuration P2 that present displacements along the bar of the secondary.

The homologous spectrum at SS2, shown in **Figure 28**, presents similar frequency components as in SS1, namely the switching frequency, its first harmonic, and component around 1050 Hz. However, in this case, the amplitude values are lower, which is explained by the proximity to the air gap. Besides, two new

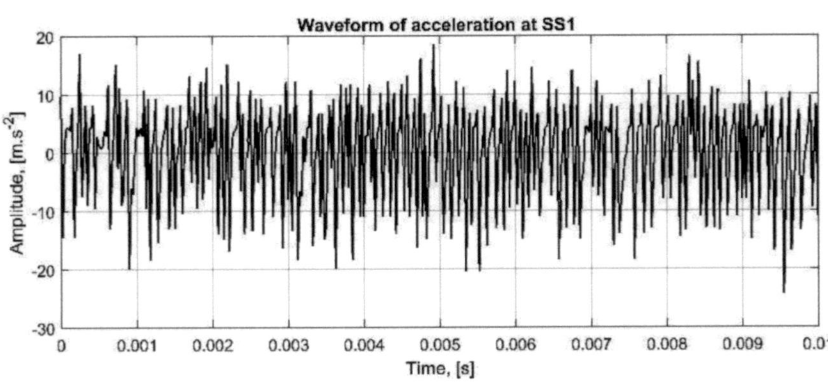

Figure 26.
Time-domain evolution of acceleration along the Z-axis at SS1.

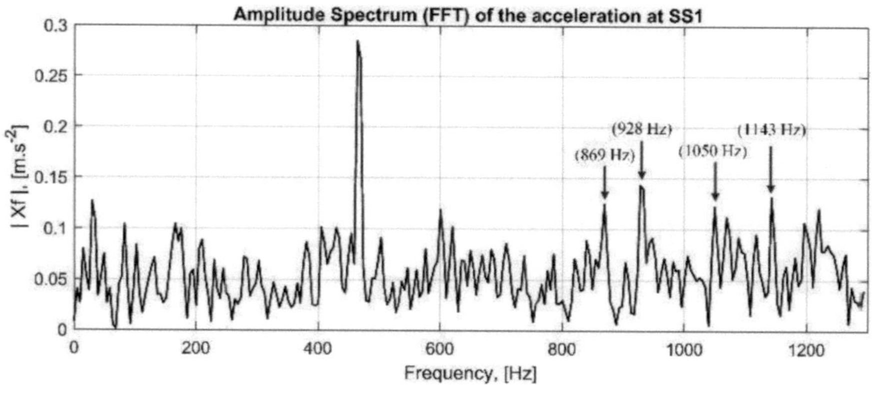

Figure 27.
Spectral components associated with the vertical acceleration at SS1.

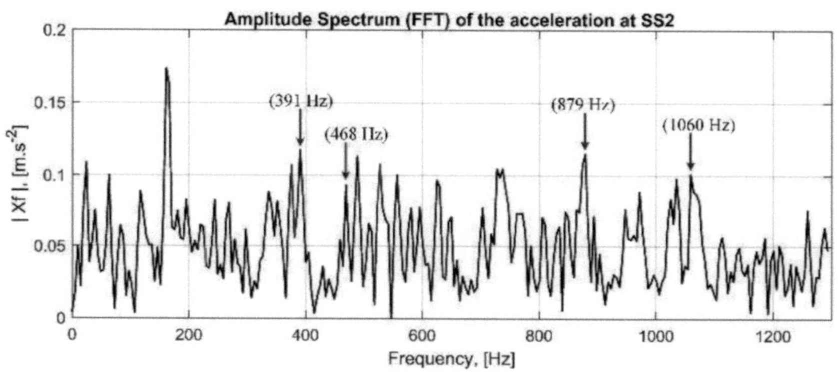

Figure 28.
Spectral components associated with vertical accelerations at SS2.

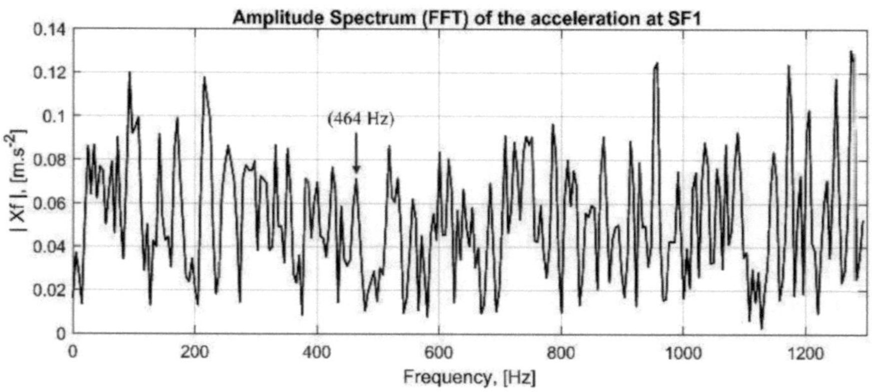

Figure 29.
Spectral components associated with vertical accelerations at SF1.

components are exposed for 161 and 391 Hz, both related to configuration P3, where the first is more significant.

Finally, the amplitude spectrum of the local vibrations measured at SF1 and SF2 reveal a more complex harmonic content, as presented in **Figures 29** and **30**.

We also notice in both cases that the switching frequency is not the most relevant component and that the number of modes excited with a single frequency is significant. For SF1, the frequency components with an amplitude equal to or

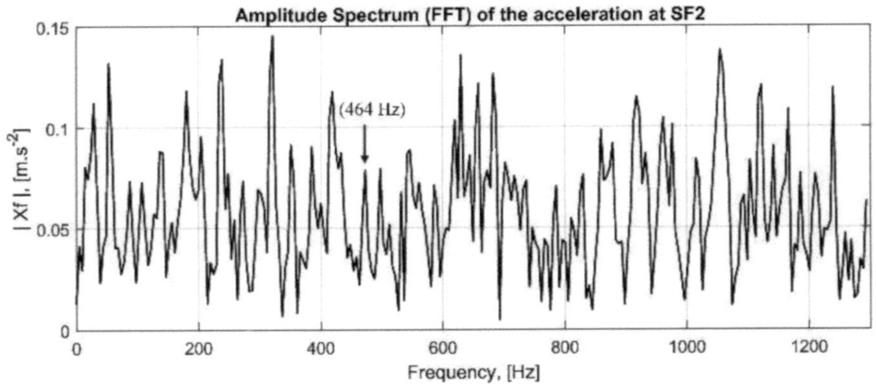

Figure 30.
Spectral components associated with vertical accelerations at SF2.

greater than 0.1 ms^{-2} are (in Hz): 93, 215, 957, 1172, and 1274. For SF2, the components with amplitude equal to or greater than 0.12 ms^{-2} are (in Hz): 53, 239, 322, 630, 660, 684, 918, 1055, 1123, and 1240. Two reasons explain these phenomena: first, the number of degrees of freedom for movement at the top longitudinal bars is greater, due to the absence of other parts assembled transversely; second, the influence of elastic wave propagation, or transmission, with origin in vibrations from other parts. These features demonstrate the potential to excite multiple modal frequencies with a single switching frequency during the power converter operation.

6. Spectral analysis of the audio emitted

The spectral signature that characterizes the LSRA vibrations can also be identified by analysis of the audio noise emitted during the experimental tests [21]. The audio is recorded in "wave" file format (*.wav) with a sample rate of 22,050 sa/s represented with 16 bits per sample. The tests consist of coupling an electrodynamic shaker to structure and apply transverse forces at the excitation positions defined in **Figure 9**. The intensity of the forces is fixed while its frequency varies linearly from 10 to 1300 Hz. These forces (sinusoidal) excite essentially the frequencies whose mode shapes represent transverse deformations along the horizontal direction. The symmetry characteristics suggest that the pairs E1-P2, E2-P2, E3-P4, and E4-P4 can be considered as representatives of the natural vibrations of the actuator.

Figures 31 and **32** represent the NCC of the amplitude spectrums of the audio signal considering the similarities of the following excitation-configuration pairs: E1-P2, E2-P2, E3- P4, and E4-P4.

Concerning the symmetric excitation pairs, the similarities are higher than 95% for frequency deviations up to 6 Hz, 90% for $\Delta f \in [-27, +10]$ Hz, and 80% for $\Delta f \in [-41, +20]$. For the anti-symmetric pairs, and using similar values of the frequency lag, the maximum likeness is 84.5% for the normalized cross-correlation considering E1-P2 vs. E4-P4 and 75.2% for E2-P2 vs. E3 P4. The NCC reduces to 70% for frequency deviations near ± 40 Hz. These values of NCC confirm the quasi-symmetric characteristic of the actuator.

Figures 33 and **34** show the spectral resemblances of the audio at configurations P2 and P4, respectively, obtained through the CPSD and considering two excitation points for each configuration. Each graph denotes a total of 28 spectral components, which are considered the main vibration frequencies for each configuration.

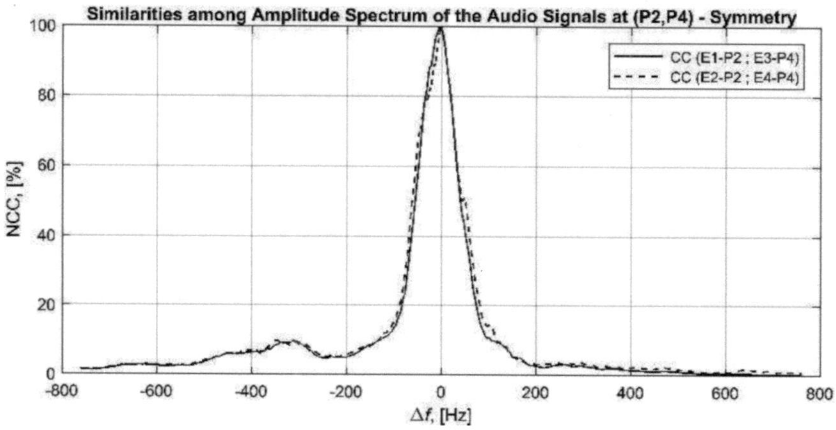

Figure 31.
Normalized cross-correlation relative to the symmetric pairs.

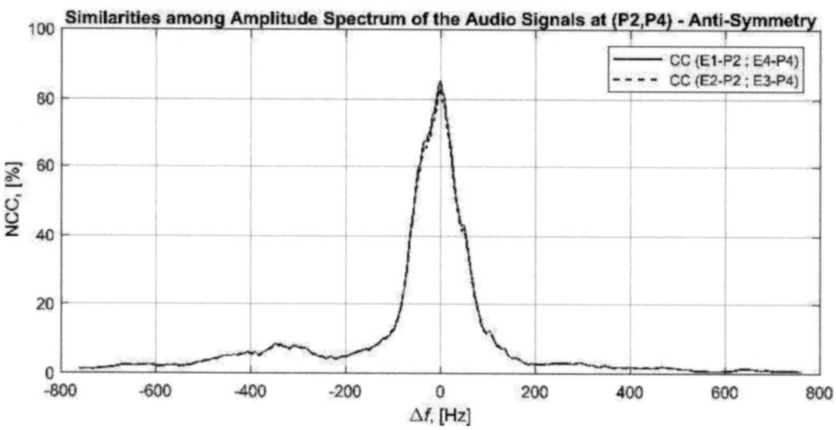

Figure 32.
Normalized cross-correlation relative to the anti-symmetric pairs.

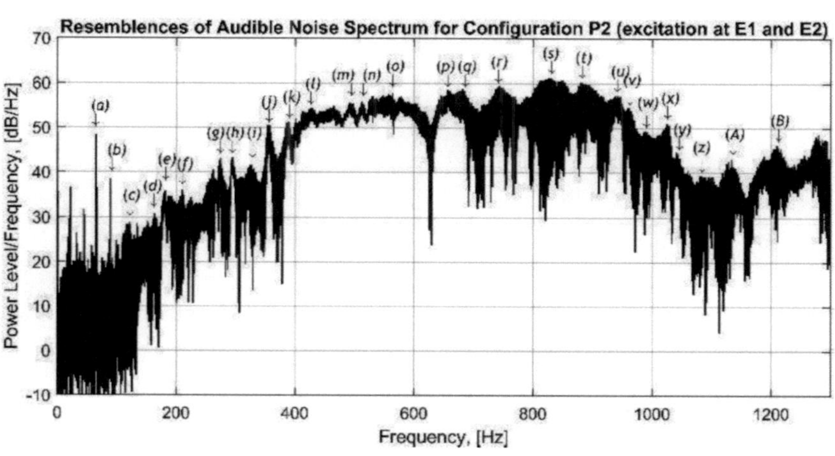

Figure 33.
Similarities among frequency components at P2 (CPSD and FFT).

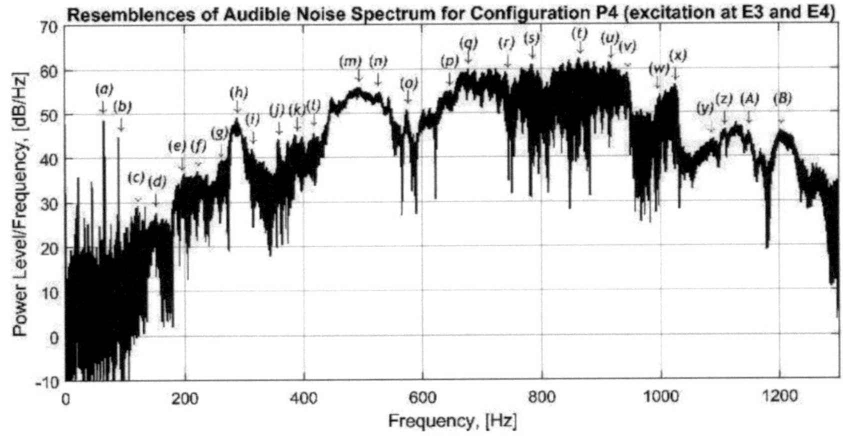

Figure 34.
Similarities among frequency components at P4 (CPSD and FFT).

Figures 35 and **36** show the hybrid spectrum of the audio signals at E1-P2 and E2-P2, respectively, obtained as described in the flowchart of **Figure 15**, using biorthogonal 4.4 wavelets for WPA with three levels of analysis. The resulting DWT coefficients are organized in sub-bands with 1.38 kHz bandwidth, corresponding to the frequencies of interest (~1.3 kHz).

The similarities between the amplitude hybrid-spectrum for E1-P2 and for E2-P2 are noticeable in **Figures 35** and **36**. Moreover, with these graphs, a total of 40 frequency components are identified, including the 28 components previously recognized and referred for similar configurations. Hence, 12 new frequencies referenced by (+) are perceptible.

Table 1 presents a list of all frequency components recognized through the CPSD and the error values with respect to the results from FEM analysis.

The relative average frequency deviation in **Table 1** is 2.67%. Except for the first three spectral components, the relative frequency deviation is, in general, less than 5%. It is important to note that some spectral components have close frequencies, which present difficulties in its identification. That is the case, for example, of component (*h*) which can be associated either to mode #19 or to mode #20. The same applies to frequency component (*p*) with respect to modes #45 and #46, and to component (*t*) and modes #59 and #60, respectively. In observations, (P)

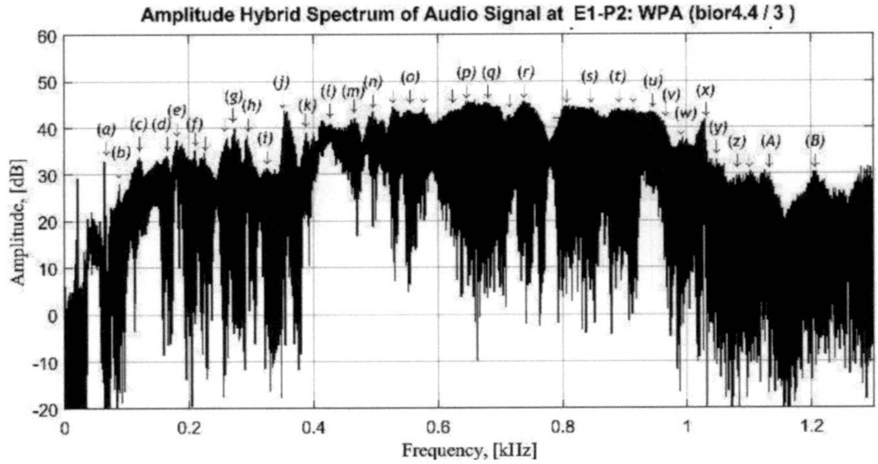

Figure 35.
Frequency components' similarities at E1-P2 (wavelet-Fourier).

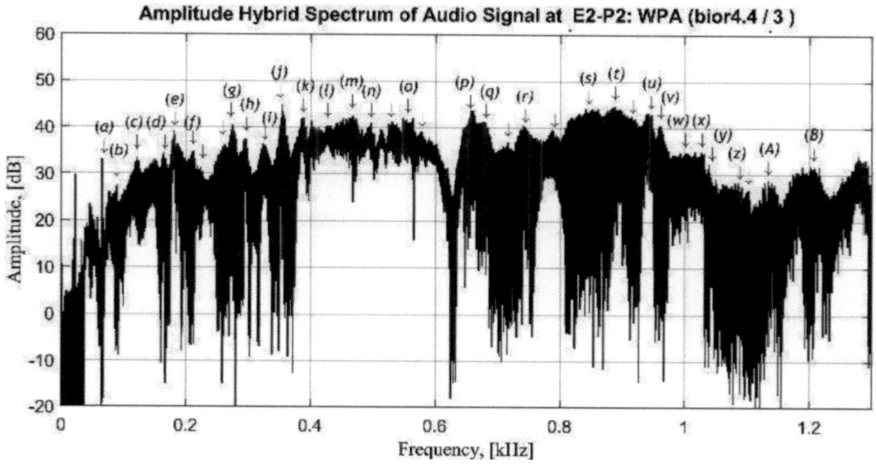

Figure 36.
Frequency components' similarities at E2-P2 (wavelet-Fourier).

Freq. Comp.	$f_{n\ FEM}$ [Hz]	$f_{n\ E1}$ [Hz]	$f_{n\ E2}$ [Hz]	$\Delta f_{n\ av}$ [%]	#mode (FEM)	*Obs.*
(*a*)	76	65	65	14.47	7	
(*b*)		89	89			#7-F3
(*c*)	133	121	121	9.02	9	
(*d*)	177	164	164	7.34	10	
(*e*)	180	181	180	0.28	11	
(*f*)	201	210	210	4.28	13	
(+)	234	224	224	4.27	15	
(+)	260	262	262	0.77	16	
(*g*)	266	274	273	2.82	17	
(*h*)	284	295	296	4.05	19	#20
(*i*)	325	327	323	0.62	22	
(*j*)	370	354	354	4.32	25	
(*k*)	387	389	386	0.39	26	
(+)	394	400	400	1.52	27	(P)
(+)	413	415	413	0.24	28	(P)
(*l*)	420	429	427	1.91	29	
(*m*)	460	467	467	1.52	32	
(*n*)	496	496	497	0.10	34	
(+)	528	528	530	0.19	35	
(*o*)	552	555	554	0.45	38	
(+)	568	578	578	1.76	39	
(+)	639	648	NI	1.41	44	
(*p*)	665	660	657	0.98	45	#45
(*q*)	681	684	678	0.44	47	
(+)	722	726	717	0.62	49	

Freq. Comp.	$f_{n\ \mathrm{FEM}}$ [Hz]	$f_{n\ \mathrm{E1}}$ [Hz]	$f_{n\ \mathrm{E2}}$ [Hz]	$\Delta f_{n\ av}$ [%]	#mode (FEM)	*Obs.*
(*r*)	751	739	740	1.54	51	
(+)	790	805	787	1.14	54	*torsion*
(*s*)	855	850	849	0.64	55	(P)
(*t*)	885	890	888	0.45	60	#59
(+)	908	913	914	0.61	62	#63
(*u*)	933	943	942	1.02	65	
(*v*)	960	963	961	0.21	66	*hidden*
(*w*)	983	994	1004	1.63	69	
(*x*)	1032	1028	1027	0.44	71	
(*y*)	1055	1048	1039	0.96	72	
(*z*)	1085	1085	1086	0.04	73	
(+)	1096	1102	1103	0.59	75	
(*A*)	1153	1145	1134	1.17	77	
(*B*)	1196	1208	1202	0.75	79	#78

Table 1.
List of vibration frequencies identified by the wavelet-Fourier analysis.

denotes modes relative to vibration frequencies associated with deformations exclusively located to the primary circuit parts.

These results show that the wavelet-Fourier approach allows a gain of 42.86% for frequency components identification with respect to the CPSD. Moreover, with the wavelet-Fourier analysis, both the magnitudes and the relevance of each component become more evident, as shown by comparing **Figures 33** and **35** and taking components (j), (k), and (m) as reference.

The hybrid spectrum representations for configuration pairs E3-P4 and E4-P4 reveal similar results like the ones obtained for P2. These features demonstrate the effectiveness of this approach for the identification and characterization of the vibrations in the LSRA. The wavelet-Fourier method can also be used for audio signals collected during the tests of operational modal excitation of the actuator, as shown in **Figure 37**, using the 4.4 biorthogonal wavelets with three levels of analysis.

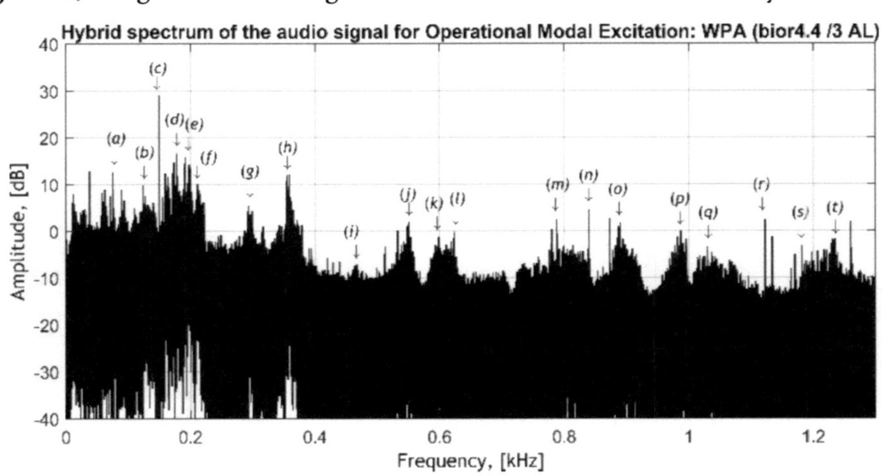

Figure 37.
Frequency components identified by wavelet-Fourier analysis of the audio emitted during the experimental tests.

Among the components identified, the following relations with natural modes are possible: (*f*) relates to mode #14, (*n*) links to both modes #55 or #56, (*p*) is related to modes #67 or #68, and (*r*) corresponds to mode #74. These results are related to those obtained by FEM analysis whose modal shapes denote transverse movement along the vertical direction, and those with displacements almost exclusively located at magnetic circuit parts.

7. Conclusion

This chapter describes a research work aimed to evaluate and characterize the vibrations in a three-phase LSRA. The actuator is described in terms of its structure and according to its modes of operation.

To analyze and evaluate the structural vibrations, we built the 3D mechanical model and performed numerical simulations using the finite element method and iterative methods to find the solution of the system dynamics. The natural frequencies obtained and the corresponding mode-shapes support the characterization of the vibrations at the actuator for both, external forces applied or forces developed during the operation of the actuator. The numerical evaluation of the natural frequencies considers five configurations of the actuator, according to the position of its platforms. For each configuration, a total of 74 modal frequencies are considered in the range up to nearly 1.3 kHz. Besides the frequency values, the mode shapes, associated with each mode, are also crucial for analysis as they evidence the parts with relevant deformations. Owing to the complexity of the LSRA structure and its symmetry characteristics, in general, the vibration frequencies, and the mode shapes are related for configurations corresponding to dual positions.

An experimental setup for the evaluation of the vibrations is also proposed, including the test bench configuration and the instrumentation set used to permit a collection of data results, showing the most relevant modal frequencies associated with the LSRA operation. The data acquisition system and the analysis tools were developed specifically for this research.

In addition to experimental modal excitation, operation modal excitation tests were also performed to identify the potential of excitation of the natural modes, and the critical values for the switching frequency. The operational responses produced, either the phase current or the local acceleration, were measured and evaluated. The spectral components related to these quantities at the selected points were analyzed and discussed regarding the results from FEM analysis. Besides Fourier analysis, a tool performing a combined wavelet-Fourier analysis of signals through a wavelet packet decomposition was also developed, proposed, and evaluated with quality improvements to the identification of the frequency components.

The results of the hybrid spectrum representation demonstrate the reliability and suitability of the wavelet-Fourier approach. The Normalized Cross-Correlation and the spectrum coherence using the Cross-Power Spectrum Density were also used. Moreover, the spectrum of the audio signals emitted was collected and analyzed, serving to confirm the data obtained by the simulation and supporting the experimental characterization of the actuator.

The results obtained are consistent and agree with the analytical estimations obtained through FEM simulations and confirm the experimental results of modal excitation obtained by experimental methods.

The data collected allow establishing a framework to fully characterize the linear actuator, regarding the vibration modes due to its regular operation, and to identify the vibration modes that can be excited for different positions of the actuator's moving parts. The information gathered serves also to improve the actuator's control by avoiding excitation regimes with frequencies near its natural vibration frequencies. That also permits to modify the structural configuration of the actuator aiming for low vibrations, either by choosing the right materials for the critical parts or by adopting suitable mechanical project strategies to redesign the structure of the actuator to avoid sensitive vibration frequencies.

Acknowledgements

The authors wish to thank Instituto de Telecomunicações for funding this work, with the investigation scholarship UID/EEA/50008/2019.

Nomenclature

$c_r,\ m_r,\ k_r$	dumping, mass and stiffness coefficients for mode r
$\mathbf{C},\ \mathbf{K},\ \mathbf{M}$	matrices of generic dumping, mass, and stiffness coefficients
$\mathbf{C}_r,\ \mathbf{K}_r,\ \mathbf{M}_r$	matrices of dumping, mass, and stiffness for mode r
D_R	Rayleigh dissipation function
$f_{ea},\ f_{em},\ f_{er}$	electromagnetic attraction, motoring, and regenerative forces
$i_{as},\ L_{as}$	coil excitation current and inductance of phase a
$Q_i,\ \mathbf{Q}_i$	generalized non-conservative forces (scalar and matrix forms)
$r_{sy}[l]$	cross-correlation of discrete-time signals $s[n],\ \ y[n]$ with lag l
$R_{sy}(l)$	the cross-spectrum amplitude of discrete-time signals $s[n],\ \ y[n]$
S_{sy}	cross-power spectrum density of discrete-time signals $s[n],\ \ y[n]$
$S_w[k]$	discrete Fourier transform of $s[n]$ for discrete window $w[n]$
v_{La}	electromotive force induced at the coil of phase a
$x_i, \dot{x}_i, \ddot{x}_i$	displacement, velocity, and acceleration at generic coordinate i
$\mathbf{x}_i, \dot{\mathbf{x}}_i, \ddot{\mathbf{x}}_i$	vectors of displacement, velocity, and acceleration at coordinate i
x_m	longitudinal displacement (position)
W_{em}	electromagnetic energy stored in the coupling field
W_{kin}	kinetic energy stored in the mechanical system
W_{pot}	potential energy stored in the mechanical system
$\phi_{sy}(l)$	phase cross-spectrum of signals $s[n],\ \ y[n]$ with lag l
λ_{as}	flux linkage in phase a (for values of i_{as} and L_{as})
$\xi_r, \boldsymbol{\Psi}_r$	eigenvalue and eigenvector relative to natural mode of order r
ω_r	undamped angular natural frequency of mode r

Author details

José António da Costa Salvado[1*], Maria do Rosário Alves Calado[2] and António Eduardo Vitória do Espírito-Santo[2]

1 School of Technology, Polytechnic Institute of Castelo Branco, Castelo Branco, Portugal

2 Faculty of Engineering, Instituto de Telecomunicações and University of Beira Interior, Covilhã, Portugal

*Address all correspondence to: josesalvado@ipcb.pt

References

[1] Castano SM, Bilgin B, Fairall E, Emadi A. Acoustic noise analysis of a high-speed high-power switched reluctance machine: Frame effects. IEEE Transactions on Energy Conversion. 2016;**31**(1):69-77

[2] Furqani J, Kawa M, Kiyota K, Chiba A. Current waveform for noise reduction of switched reluctance motor in magnetically saturated condition. In: 2016 IEEE Energy Conversion Congress and Exposition (ECCE). 2016;**54**(1): 213-222

[3] Bayless J, Kurihara N, Sugimoto H, Chiba A. Acoustic noise reduction of switched reluctance motor with reduced RMS current and enhanced efficiency. IEEE Transactions on Energy Conversion. 2016;**31**(2):627-636. Available from: http://ieeexplore.ieee.org/document/7349203/

[4] Ma C, Qu L, Mitra R, Pramod P, Islam R. Vibration and torque ripple reduction of switched reluctance motors through current profile optimization. In: 2016 IEEE Applied Power Electronics Conference and Exposition. Long Beach, CA, USA: IEEE; 2016. pp. 3279-3285. Available from: http://ieeexplore.ieee.org/document/7468336/

[5] Mollet Y, Gyselinck J, Sarrazin M, Van Der Auweraer H. Experimental noise and vibration analysis of switched reluctance machines comparison of soft and hard chopping in transient conditions. In: 2015 International Conference on Renewable Energy Research and Applications. Palermo: IEEE; 2015. pp. 420-425. Available from: http://ieeexplore.ieee.org/document/7418448/

[6] Krishnan R. Switched Reluctance Motor Drives: Modeling, Simulation, Analysis, Design, and Applications. Industrial Electronics. Boca Raton, FL: CRC Press; 2001

[7] Mohanty P, Rixen DJ. Identifying mode shapes and modal frequencies by operational modal analysis in the presence of harmonic excitation. Experimental Mechanics. 2005;**45**(3):8

[8] Espírito-Santo AE, Calado MRA, Cabrita CMP. On the influence of the pole and teeth shapes on the performance of linear switched reluctance actuator. COMPEL—The International Journal for Computation and Mathematics in Electrical and Electronic Engineering. 2011;**30**(2): 412-430. Available from: http://dx.doi.org/10.1108/03321641111101005

[9] Barnes M, Pollock C. Power electronic converters for switched reluctance drives. IEEE Transactions on Power Electronics. 1998;**13**(6):1100-1111. DOI: 10.1109/63.728337

[10] Meirovitch L. Fundamentals of Vibrations. Boston: McGraw-Hill; 2001

[11] Schwarz B, Richardson M. Proportional damping from experimental data. The Conference Proceedings of the Society for Experimental Mechanics Series. 2014; **45**(7):179-186

[12] Salvado J, Calado M do R, Santo AE, Guerman A. Numerical modal analysis of vibrations in a three-phase linear switched reluctance actuator. Modelling and Simulation in EngineeringModel. 2017;**2017**:1-18. Available from: https://www.hindawi.com/journals/mse/2017/3258376/

[13] Salvado J, Espírito-Santo A, Calado M. An intelligent sensor array distributed system for vibration analysis and acoustic noise characterization of a linear switched reluctance actuator. Sensors (Basel). 2012;**12**(6):7614-7633. Available from: http://www.pubmedcentral.nih.gov/articlerender.fcgi?artid=

3435993&tool=pmcentrez&rendertype=abstract

[14] Salvado J, Espirito Santo A, Calado MRA. Improved distributed system for analysis of vibrations in linear switched reluctance actuators. In: ICIT 2012, IEEE International Conference on Industrial Technology. Athens: IEEE; 2012. pp. 991-996. Available from: http://ieeexplore.ieee.org/xpl/articleDetails.jsp?arnumber=6210068&newsearch=true&searchWithin=%22First Name%22:Jose&searchWithin=%22Last Name%22:Salvado

[15] Wickramarachi P, Corporation DP, Jose S. Effects of windowing on the spectral content of a signal. Sound and Vibration. 2003;**37**(1):10-11. (ISSN 15410161)

[16] Bianchi D, Mayrhofer E, Gröschl M, Betz G, Vernes A. Wavelet packet transform for detection of single events in acoustic emission signals. Mechanical Systems and Signal Processing. Elsevier; 2015;64-65:441-451. DOI: 10.1016/j.ymssp.2015.04.014

[17] Tarasiuk T. Hybrid wavelet-Fourier spectrum analysis. IEEE Transactions on Power Delivery. 2004;**19**(3):957-964. Available from: http://ieeexplore.ieee.org/lpdocs/epic03/wrapper.htm?arnumber=1308314

[18] Akçay H. Spectral estimation in frequency-domain by subspace techniques. Signal Processing. Elsevier. 2014;**101**:204-217. DOI: 10.1016/j.sigpro.2014.02.015

[19] Salvado JA, Calado MR, Santo AE. A simple yet effective semi-anechoic chamber for the acoustic characterization of an LSRA. IEEE Latin America Transactions. 2013;**11**(5):1190-1200

[20] Salvado JA, Calado MR, Santo AE. Evaluation of the operational modal responses in linear switched reluctance actuators. In: 2018 IEEE International Conference on Environment and Electrical Engineering, 2018 IEEE Industrial and Commercial Power Systems Europe (EEEIC/I&CPS Europe). Palermo, Italy: IEEE; 2018. pp. 1-6

[21] Salvado JA, Calado MR, Santo AE. Wavelet-Fourier analysis of audible signals to characterize the vibrations in LSRA. In: 2018 XIII International Conference on Electrical Machines. Alexandroupoli, Greece: IEEE; 2018. pp. 1738-1744

Chapter 14

Control and Dynamic Simulation of Linear Switched Reluctance Generators for Direct Drive Conversion Systems

Rui Pedro Gouveia Mendes, Maria do Rosário Alves Calado and Sílvio José Pinto Simões Mariano

Abstract

This chapter addresses the dynamic simulation and control of linear switched reluctance generators for direct drive conversion systems. The electromechanical energy conversion principles of linear switched reluctance machines are briefly explained. A detailed mathematical model is developed for linear switched reluctance generators. The different types of control strategies adopted for switched reluctance generators are referred. The hysteresis controller is applied to control the conversion system with constant damping load. The proposed control strategy also includes a DC/DC isolated converter to control the system DC bus voltage by adjusting the energy flow between the conversion system and the resistive load. The mathematical model is applied to simulate the behavior of a tubular linear switched reluctance generator as power take-off system in an ocean wave point absorber device. To accomplish this task, the dynamic equations of the point absorber are presented and integrated with the linear switched reluctance generator dynamic model. In the simulation process, the system is driven by a regular ocean wave and operates with constant damping load. The system performance is evaluated for different load values, and the simulation results are presented for the optimal damping load case scenario.

Keywords: tubular linear generator, switched reluctance machine, direct drive conversion, ocean wave energy

1. Introduction

The switched reluctance machine (SRM) is an electromechanical structure with salient poles defined by a fixed and a movable part as the main constitutive elements. This machine is characterized by the absence of permanent magnets and by the electric phase coils only being present in just one part (usually the fixed part). The part where the phase coils are located is designed as primary and the other part, which works as passive element, as secondary. The SRMs can be rotating or linear. Since these two configurations are homologous, they share the same operation principles with the exception of the electromagnetic force direction [1]. In SRMs, a force is developed when the magnetic structure tends to minimize the magnetic energy by displacing the movable part to achieve a configuration with minimum reluctance. This type of machine is characterized by a nonlinear electromagnetic behavior and can work as actuator or generator when the electric phase is energized during the period of increasing or decreasing inductance, respectively [2]. The SRM may have a flat structure [3] or a tubular one [4]. The tubular topology may overcome some disadvantages of the flat machine. The former has no edge effects due to the closed geometry. Also, the radial attraction forces are canceled due to its axisymmetric configuration [5].

This type of machine is suitable for low-speed applications and is a robust and low-cost solution since no permanent magnets are required for its operation. Also, the energy losses from Joule effect occurs in just one part and can, therefore, be more easily dissipated [6]. These characteristics make the SRM a strong candidate for generator in direct drive conversion systems. It has already been adopted as linear generator for ocean wave point absorbers [3]. Due to their unique advantages, the application of linear switched reluctance generators (LSRG) in direct drive conversion systems is of significant importance. Thus, design tools are necessary to develop and evaluate this type of machines, and suitable control techniques must be applied to ensure its reliable operation.

The aim of this chapter is to present the mathematical model to assist the design and assessment of LSRG. Also, control techniques are proposed for low-velocity operation. The mathematical formulation is applied to simulate the performance of a tubular linear switched generator (TLSRG) operating as a power take-off system in an ocean wave point absorber converter.

2. Electromagnetic conversion principles

In SRM, the magnetic circuit of each electric phase may be characterized with different values of magnetic reluctance for distinct relative positions of the movable part, in respect to the static part. When a magnetic field is established in this circuit, an electromagnetic force is developed to displace the movable part to the position with minimum magnetic reluctance in order to minimize the energy in the system. The referred position, designed as alignment position z_a, is characterized by the structural configuration where the saliencies of the movable part are aligned with the saliencies of the static part. In this situation, the self-inductance of the electric phase may achieve its maximum value. The nonalignment is verified for all the remaining relative positions, between the movable and static parts. For these

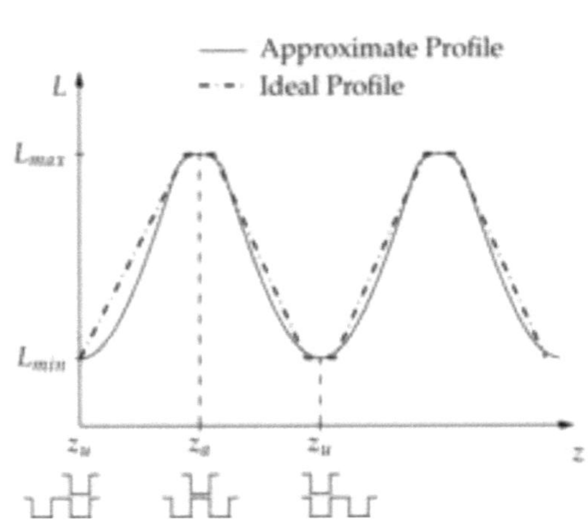

Figure 1.
Typical inductance profile for SRMs.

configurations, the magnetic circuit of the electric phase is characterized by superior magnetic reluctance achieving its maximum value at the unaligned position z_u. This position is defined by the minimum value of inductance for the electric phase.

The typical linear inductance profile for SRMs is displayed in **Figure 1**. The inductance is given as function of the relative position between the movable and fixed parts. As consequence of the structural pattern inherent to these machines, when the movable part is moved in a single direction, the inductance is characterized by a periodic and symmetrical profile resulting from successive movements, which are approaching (the rate of change of the inductance L is positive) and leaving (the rate of change of the inductance L is negative) from the alignment position.

By energizing the electric phases in the approximation region or separation region, the SRM may operate as an actuator or a generator, respectively. In the first case, magnetic flux established by the electric phase will tend to minimize its energy developing an electromechanical force to achieve structural alignment. As a result, a linear force will act on the machine's movable part. In the presence of an external load, which can overcome the referred electromagnetic force, the generator is displaced from the alignment position, increasing the reluctance of the active phase, which will reduce the respective magnetic flux. As a consequence, a back-electromotive force is developed, seeking to increase the electric current to restore the magnetic flux. During this procedure, mechanical energy is extracted from the movable part of the generator and converted to electrical energy [7].

The electric phase equation is given as:

$$\begin{aligned} u_k &= R_a i + \frac{\mathrm{d}\lambda}{\mathrm{d}t} \\ &= R_a i + L(i, z_k)\frac{\mathrm{d}i}{\mathrm{d}t} + i\frac{\mathrm{d}z_k}{\mathrm{d}t}\frac{\mathrm{d}L(i, z_k)}{\mathrm{d}z_k} \end{aligned} \tag{1}$$

where for phase k, u_k is the voltage across the electric phase, R_a is the internal electric resistance, λ is the flux linkage, i is the electric current, and z_k is the electric position. In the last part of Eq. (1), emf is the back-electromotive force developed by the phase:

$$emf = i\frac{\mathrm{d}z_k}{\mathrm{d}t}\frac{\mathrm{d}L(i, z_k)}{\mathrm{d}z_k} = iv\frac{\mathrm{d}L(i, z_k)}{\mathrm{d}z_k} \tag{2}$$

with v as the velocity of the movable part of the linear SRM.

These conversion processes may only occur with appropriate power electronic converter to control the energy flow. With proper switch commands, the conversion periods may be established, and electric phase current intensity may be regulated by the converter [8].

In [9, 10], several electronic converters are identified that are suitable for electric generation with SRMs. The asymmetric H-bridge converter is the typical choice. This converter topology can be found, for one phase, in **Figure 2**. It is a practical and simple solution, but it is characterized with a variable output voltage due to the self-excitation process. However, this drawback may be minimized with an external voltage source [11]. This converter has the less apparent power required to operate and is classified as an economical solution [10].

With the converter illustrated in **Figure 2**, the conversion cycle may be defined by the excitation period, the generation period, and the free-wheeling period, an intermediate stage between the first two. The circuit configurations for the different periods are illustrated in **Figure 3**. Assuming that the capacitor is already at its nominal operating voltage, the excitation period is initiated when the switches S_1

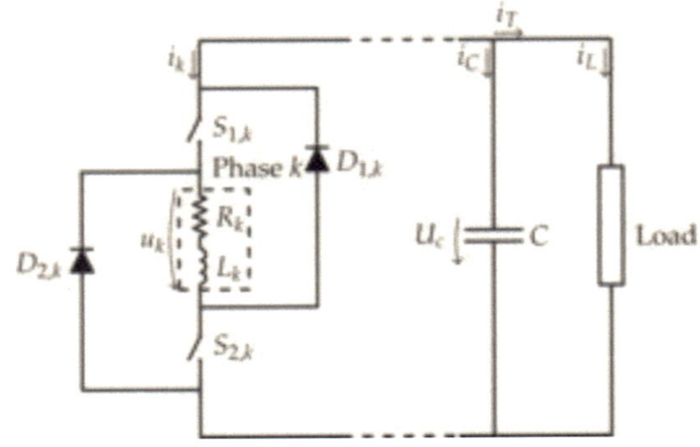

Figure 2.
H-bridge asymmetric electronic power converter for one electric phase.

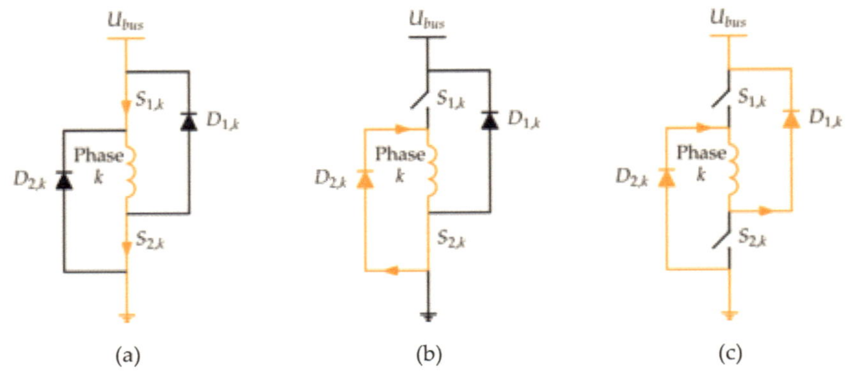

Figure 3.
Circuit configuration and electric current path for (a) excitation period, (b) generation period, and (c) free-wheeling period (adapted from [12]).

and S_2 are closed. Usually, this occurs when the phase is near the position that corresponds to its maximum inductance.

When the current intensity reaches a certain value, the switches are opened and the generation period begins. During this period, the back-electromotive force increases the electric current as a consequence of the magnetic flux reduction. Thus, the electric current is maintained through the diodes $D_{1,k}$ and $D_{2,k}$, delivering the generated electric energy to the capacitor and to the load. The transition between these two conversion periods is characterized by hard commutation because the voltage is inverted. In the free-wheeling period, just one switch is closed to provide a zero voltage across the generator electric phase. Only the back-electromotive force acts on the phase electric current. This period can be implemented to achieve a soft commutation where, after the excitation period, the voltage is first annulled and, only then, inverted [12]. In each conversion cycle, the electric energy supplied by the converter during the excitation period is given by:

$$W_{exc} = \int_{t_{on}}^{t_{off}} ui \, \mathrm{d}t \tag{3}$$

where t_{on} is the start time of the excitation period and t_{off} is the respective finish time.

The electric energy returned by the phase to the converter is:

$$W_{out} = \int_{t_{off}}^{t_{ext}} ui \, dt \tag{4}$$

with t_{ext} as the time when the electric current is extinguished. The amount of electric energy generated by the electromechanical conversion process is computed as follows:

$$W_{gen} = W_{out} - W_{exc} \tag{5}$$

3. Dynamic model

To assess the dynamic characteristics of switched reluctance generator, a mathematical model must be formulated to describe the system behavior in transient state. The mathematical model can be solved with numerical methods and computational calculus in order to estimate the system response for the given operating conditions. The mathematical model of SRMs is obtained from Eq. (1) that describes the transient phenomena involved in the electromagnetic conversion of each electric phase. The solution for this equation may be obtained by applying time integration to the linkage magnetic flux [13] or to the electric current derivative with time [14]. For both methods, it is required to relate the machine electromagnetic characteristics with the phase electric current and the movable part position. Several approaches have been proposed in the literature to include the machine nonlinear nature in the mathematical model. In [15], look-up tables are used to model the electromagnetic characteristics of the SRM. These are obtained from a static analysis and are used as data base to build the look-up tables. Analytic functions are also used to represent the machine magnetic characteristics within the mathematical model. The use of analytical expressions simplifies the computation process related to the differentiation and integration of the electromagnetic entities [16]. This approach can be achieved by fitting an appropriate function to discrete data [17]. Fourier series expansion may also provide analytical expressions to represent nonlinear electromagnetic characteristics in the SRM mathematical model [18]. A simpler method, but less precise, comprises the use of piecewise functions to establish a linear [19] or nonlinear [20] relation with the independent variables. Artificial intelligence-based methods have already been adopted in the dynamical analysis of SRMs. These methods comprise the use of neural networks and fuzzy logic with real data to develop a mathematical representation of SRM electromagnetic characteristics [21].

All the referred approaches require the representative data of the electromagnetic characteristics, usually expressed as a function of the electric phase current intensity and movable part relative position. With look-up tables, these data are used directly. Methods based on analytic expressions or artificial intelligence models apply these data to develop appropriate mathematical expressions.

The required discrete data may be obtained through experimental measurements. This process provides realistic curves, but a physical model of the machine is needed for it. In electric machine design, where several structural possibilities must be assessed, it becomes unpractical to get the machine electromagnetic characteristics with experimental evaluation, as numerical or analytical methods are needed to perform this task. As an alternative to experimental tests, the finite element method

(FEM) is used for electromagnetic characterization of SRMs, providing results with great precision, but needing for a sophisticated mathematical implementation and large amount of computational resources to process the solution [22]. However, with the existence of FEM-based commercial software, it is possible to evaluate electromagnetic systems without a deep knowledge of electromagnetism. Thus, FEM is getting more adopted in the design of electric machines.

The mathematical model of switched reluctance generator results from the analysis of the associated power converter. Since all electric phases are defined by identic magnetic and electric circuits, they are represented by the same dynamic equations. Its analysis is generalized following the notation specified in **Figure 2**. The voltage across each electric phase k is:

$$u_k = R_{a_k} i_k + \frac{d\lambda_k}{dt} \tag{6}$$

where for each phase k, R_{a_k} is the internal electric resistance, i_k is the electric current, λ_k is the magnetic linkage flux, and z_k is the electric position. The magnetic linkage flux is given by:

$$\lambda_k = \sum_{j=1}^{q} \lambda_{kj}(z_j, i_j) \tag{7}$$

with $\lambda_{kj}(z_j, i_j)$ as the magnetic linkage flux in phase k due to current in phase j and q as the number of electric phases. Replacing Eq. (7) in Eq. (6):

$$u_k = R_{a_k} i_k + \lambda_k = \sum_{j=1}^{q} \frac{d\lambda_{kj}(z_j, i_j)}{dt} \tag{8}$$

Applying the chain rule to Eq. (8), one has:

$$\begin{aligned} u_k = {} & R_{a_k} i_k + \left(L_k(z_k, i_k) + i_k \frac{dL_k}{di_k}(z_k, i_k) \right) \frac{di_k}{dt} + v i_k \frac{dL_k}{dz_k}(z_k, i_k) \\ & + \sum_{\substack{j=1 \\ j \neq k}}^{q} \left[\left(M_{kj}(z_j, i_j) + i_j \frac{dM_{kj}}{di_j}(z_j, i_j) \right) \frac{di_j}{dt} + v i_j \frac{dM_{kj}}{dz_j}(z_j, i_j) \right] \end{aligned} \tag{9}$$

where L_k is the self-inductance of phase k and M_{kj} is the mutual inductance between phases k and j. The linear velocity of the movable part v is:

$$v = \frac{dz_k}{dt} = \frac{dz_j}{dt} \tag{10}$$

The electromotive force, which results from the change of magnetic flux in phase k, is:

$$emf_k = v \left[i_k \frac{dL_k}{dz_k}(z_k, i_k) + \sum_{\substack{j=1 \\ j \neq k}}^{q} i_j \frac{dM_{kj}}{dz_j}(z_j, i_j) \right] \tag{11}$$

During the conversion period, electric energy is exchanged between the capacitor of the converter and the electric phases of the generator. In the excitation period, the capacitor supplies energy to the phase and in the generation period receives energy from it. The voltage across capacitor U_c is related with its capacity and input current i_c by:

$$i_c = C\frac{dU_c}{dt} \tag{12}$$

The total net value of the electric energy flowing between the electric phases and the capacitor i_T is given by:

$$i_T = \sum_{j=1}^{q} i_k \tag{13}$$

The bus voltage U_{bus} of electronic power converter is imposed by capacitor voltage:

$$U_{bus} = U_c \tag{14}$$

In order to fully define the model of the switched reluctance generator system, the voltage across each electric phase must be known. Its value depends on the different circuit configurations for the electric energy flow in the phase. According to the possible combinations of switch states, there are three different configurations, as illustrated in **Figure 3**, each of them corresponding to a distinct period of electromagnetic conversion.

The voltage across each electric phase, according to the switch states, is:

$$u_k = \begin{cases} U_{bus} - 2U_s & S_{1,k} \text{ and } S_{2,k} \text{ closed} \\ -U_{bus} - 2U_D & S_{1,k} \text{ and } S_{2,k} \text{ open} \\ -U_D - 2U_s & S_{1,k} \text{ open and } S_{2,k} \text{ closed} \end{cases} \tag{15}$$

with U_s and U_D as the voltage drops across the electronic switch and diode, respectively.

The linear force exerted by the generator F_{gen} is:

$$F_{gen}(t) = \sum_{k=1}^{q} F_{em,k}(z_k, i_k) \tag{16}$$

where $F_{em,k}(z_k, i_k)$ is the individual component of force provided by electric phase k during the conversion cycle. The generator electric efficiency can be determined by the mean values of the generated electric power $\overline{P}_{gen}$ and extracted mechanical power $\overline{P}_{mec}$ as:

$$\eta_e = \frac{\overline{P}_{gen}}{\overline{P}_{mec}} \tag{17}$$

All the formulation presented so far was defined for a phase electric position falling within the two opposing nonalignment positions. As the movable part of the generator is displaced, the electric position describes a periodic profile which is identical in all electric phases of the SRM. However, these electric positions have distinct values for the same absolute position because the phases are shifted. The relation between the electric position of one phase z_k and the mechanical absolute position z_{mech} is defined by:

$$z_k = -S_t + \left(\frac{z_{mech} - k_{offset} - S_t}{2S_t} - \left\lfloor \frac{z_{mech} - k_{offset} - S_t}{2S_t} \right\rfloor \right) 2S_t \tag{18}$$

where k_{offset} is the shift distance between the same electric position of two consecutive phases and S_t is the distance between the aligned and unaligned positions of the phase.

The graphical representation of Eq. (18) can be found in **Figure 4**.

The mathematical model of the LSRG system is defined by a nonlinear problem of initial value and is schematized in **Figure 5**. To obtain the respective solution, it is necessary to perform the time integration of the differential equations that govern the system electromechanical behavior. Also, a controller must be included to provide the appropriate switching commands to the electronic power converter.

The bus voltage of the LSRG converter should be kept as constant as possible to allow for a proper self-excitation operation. Also, since high voltage levels are required to energize each phase in due time, it may be desirable to reduce the output voltage applied to the electric load. To achieve these requirements, an H-bridge isolated DC/DC converter is admitted in the conversion system to control the energy flow between the conversion system and the electric load, as schematized in **Figure 6**. With the ideal model of the H-bridge isolated DC/DC converter, it is possible to establish a relation between the mean values of the input and output voltages (U_1 and U_2) and of the input and output currents (I_1 and I_2).

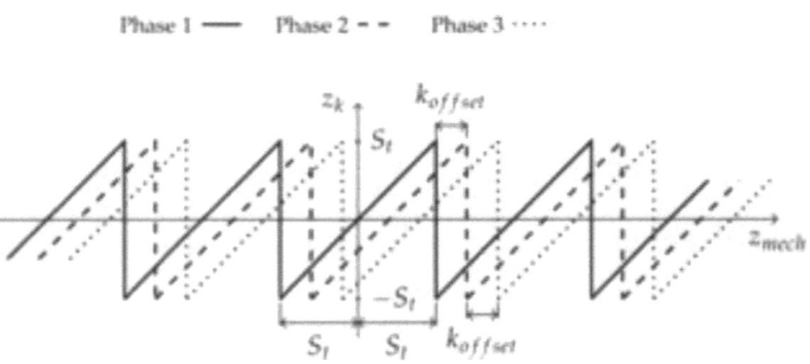

Figure 4.
Electric position as function of mechanical position.

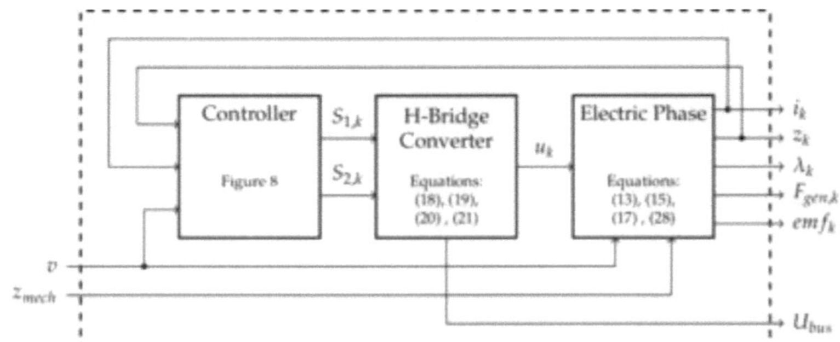

Figure 5.
Diagram of the mathematical model of the LSRG system.

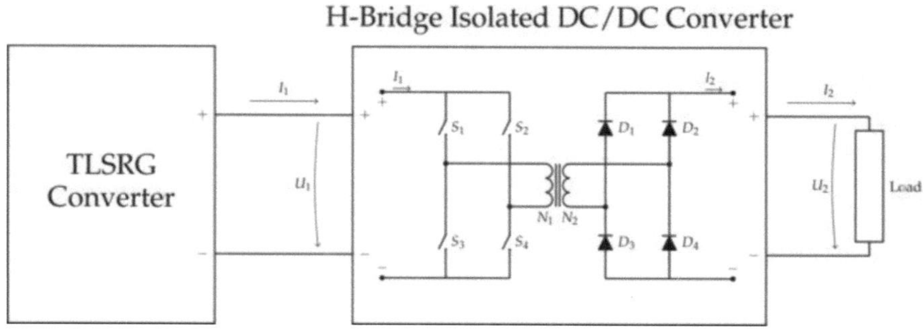

Figure 6.
TLSRG conversion system with a DC/DC conversion stage.

The duty cycle D is the control parameter to regulate the output voltage at constant chopping frequency. The relations between the referred electric entities are defined in [23] as:

$$\overline{U}_2 = 2\overline{U}_1 D\frac{N_2}{N_1} \tag{19}$$

$$\overline{I}_1 = 2\overline{I}_2 D\frac{N_2}{N_1} \tag{20}$$

where, respectively, $\overline{U}_1$ and $\overline{U}_2$ are the mean values of input and output voltages, $\overline{I}_1$ and $\overline{I}_2$ are the mean values of input and output electric currents, and N_1 and N_2 are the numbers of turns of the transformer primary and secondary coils.

4. Control

The SRM operation relies on the switching electric positions. Fixed values for these positions may cause system instability, especially when operating at variable velocity. As a consequence, the converter bus voltage may change considerably, depending on the system electric load [24]. Therefore, a closed loop control is required. As stated in [25], when operating as generator, the control of SRMs must be applied to preserve the converter output voltage, conditioned by the current flow in the electric phases and electric load.

The amount of energy extracted for conversion relies on the electromechanical force exerted by the generator which is also affected by the electric phase current. Thus, with a proper switching strategy, it is possible to control the electric current intensity to attain a desirable voltage level as well as to improve the system conversion efficiency. With the additional DC/DC converter, the bus voltage may also be kept near a nominal value by modifying the energy flow between the LSRG converter and the load. This method is proposed in [24] to achieve maximum energy conversion by controlling the voltage level according to the velocity of the generator.

At low operation velocities, the back-electromotive force is inferior to the bus voltage and the current is forced to decrease gradually. In this situation, the phase must be submitted to successive commutations to adjust the applied voltage and achieve the desirable current intensity [7]. This process can be accomplished with hysteresis band control [26]. This control is characterized with a variable switching frequency because it is conditioned by the rate of change of the electric phase current [27].

The control of the switched reluctance generator is the selection of appropriate parameter values that are responsible for its behavior. For velocities superior to the base velocity, it is only necessary to account the electric positions that define the conversion cycle. For lower velocities, it is also required to define a reference value for the phase current. With the mathematical model, it is possible to establish an optimal relation between the values of the control parameters and the physical entities that need to be controlled and include it in the control process through look-tables or fitted analytical expressions. Thus, the appropriate control parameters can be defined as function of the operation variables [28].

The proportional integral (PI) control may be applied to control the switched reluctance generator. It has been used in real-time optimal control of rotating generators, where the reference current and commutation angles are computed from the error of the output voltage [29]. PI control was adopted in [30] to adjust the phase current intensity and minimize the converter output voltage ripple in linear generators.

4.1 Hysteresis control

With proper switching commands, the current is maintained within the hysteresis band h_b , which is a range of values established around the reference current i_{ref}. The typical current profile obtained with this control is shown in **Figure 7(a)**. When the current falls below the inferior limit of the hysteresis band $-h_b/2$, the phase is energized to increase the current. If it is superior to the upper limit of the hysteresis band $h_b/2$, the switches are opened and the generation period begins until the lower limit is reached again [31]. The phase current error e_i is used to control the switches in agreement with the logic schematized in **Figure 7(b)**.

The conversion cycle is characterized by successive excitation and generation periods that occur while the phase electric position is between z_{on} and z_{off} and the movable part is in motion. For zero velocity, there is no electric generation and the switches remain opened. Thus, for each phase k, the hysteresis controller provides the electronic switching commands $S_{1,k}$ and $S_{2,k}$ as a function of the phase electric position z_k, movable part velocity v, current reference value i_{ref}, hysteresis band h_b, and electric positions z_{on} and z_{off}. In **Figure 8** the algorithm of the hysteresis controller is illustrated. For an optimal control, the parameters must be chosen to maximize the generated electric energy. With the mathematical model of the conversion system, these parameters may undergo an optimization process to find the best values for the best performance.

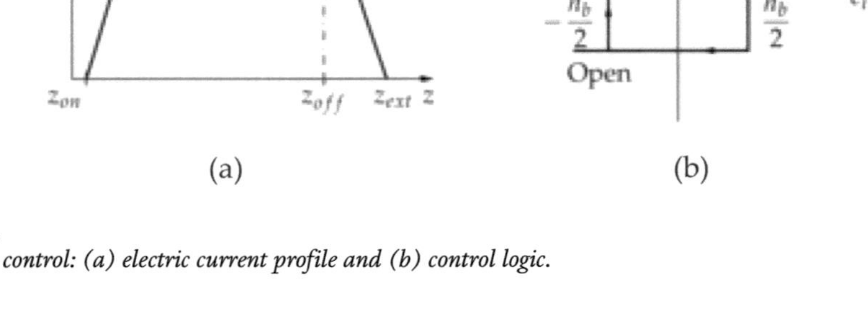

Figure 7.
Hysteresis control: (a) electric current profile and (b) control logic.

DOI: http://dx.doi.org/10.5772/intechopen.89291

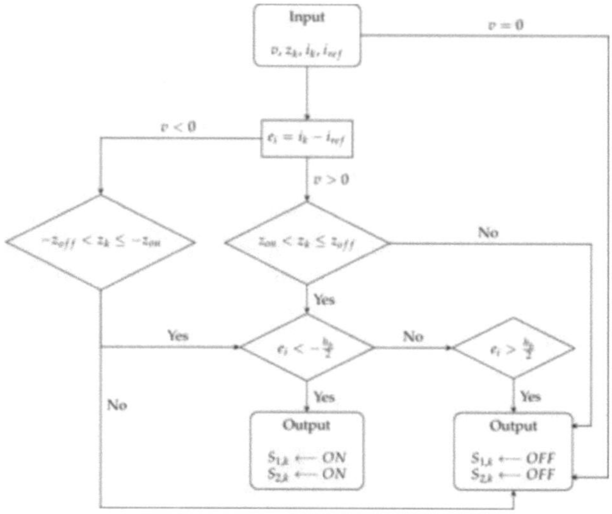

Figure 8.
Block diagram of algorithm proposed to apply the hysteresis control.

4.2 Proportional integral (PI) control

The H-bridge isolated DC/DC converter is used to maintain the system DC bus voltage level close to a reference value. In order to accomplish it, a PI control is used to properly compute the duty cycle value for the applied switching commands. The control variable s is given as function of error e between the variable to be controlled and the respective reference value:

$$s(t) = K_p e(t) + K_i \int_0^t e(\tau)\, d\tau \tag{21}$$

with K_p and K_i as the controller proportional and integral gains, respectively. The duty cycle value is given as:

$$D(t) = D_{init} + \Delta D(t) \tag{22}$$

where D_{init} is the initial value of the duty cycle and $\Delta D(t)$ is the incremental duty cycle provided by the PI controller:

$$\Delta D(t) = K_p e_u(t) + K_i \int_0^t e_u(\tau)\, d\tau \tag{23}$$

with e_u as the normalized error between DC bus voltage U_{bus} and the reference value U_{ref}:

$$e_u(t) = \frac{U_{ref} - U_{bus}(t)}{U_{ref}} \tag{24}$$

5. Practical case study scenario

The described formulation will be applied to evaluate the performance of a tubular linear SRM working as generator in an ocean wave point absorber device schematized in **Figure 9**.

The system comprises a floating body that drives the generator by action of the incoming ocean waves. The linear generator is a three-phase machine with coils located in the inner part. The outer part is rigidly coupled to a floating body that only allows a vertical motion. Mechanical springs are used to link the generator outer part to the reference system, which is fixed to the ocean bottom. The tubular LSRG is illustrated in **Figure 10**.

5.1 Point absorber mathematical model

The mathematical model of an ocean wave direct drive converter is also needed to fully define the conversion system and to compute the mechanical entities that are used as input in the generator dynamic model.

The dynamic behavior of the point absorber device can be found in [32] and is described by the expression:

$$m_b\ddot{z} = F_{exc}(t) + F_{rad}(t) + F_H(t) + F_V(t) + F_{gen}(t) \qquad (25)$$

where m_b is the combined mass of the floating body and generator movable part, $\ddot{z}$ is the bodies' acceleration, F_{exc} is the wave excitation force, F_{rad} is the radiation force, F_H is the hydrostatic force, F_v is the viscous damping force, and F_{gen} is the generator damping force.

The excitation force is given by:

$$F_{exc}(t) = \Re\left[S_a\tilde{F}_{exc}(\omega)e^{i(\omega t+\varphi)}\right] \qquad (26)$$

with S_a as the wave amplitude, $\tilde{F}_{exc}$ as the complex value of the excitation force per meter of wave amplitude, ω as the wave angular frequency, and the φ as wave phase.

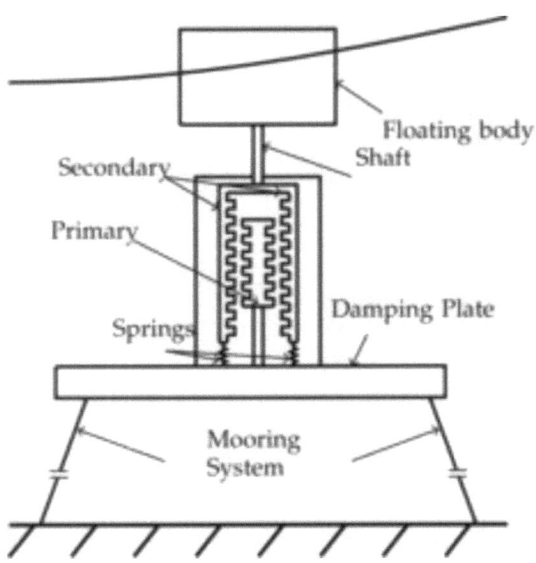

Figure 9.
Point absorber schematics.

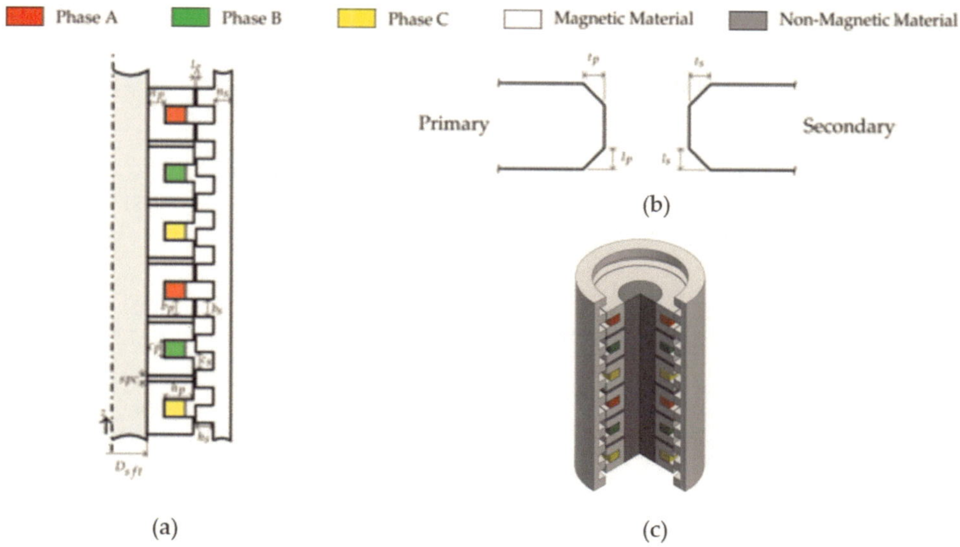

Figure 10.
Tubular LSRG (TLSRG) used as PTO: (a) longitudinal cross-sectional profile, (b) tooth shape, and (c) 3D illustration.

The radiation force is calculated as follows:

$$F_{rad}(t) = -m_{\infty}\ddot{z} - \int_{0}^{t} K_r(t-\tau)\dot{z}(\tau)\,\mathrm{d}\tau \tag{27}$$

In Eq. (27), m_{∞} is the added mass for infinite frequencies, K_r is the radiation velocity impulse response, and $\dot{z}$ is the floating body and generator movable part.

The hydrostatic force is:

$$F_H(t) = -\rho g A_w z \tag{28}$$

with ρ as the specific mass of seawater, g as the gravity acceleration, A_w as the cross-sectional area of the submerged part of the floating body, and z as its vertical position.

The floating body is subjected to a viscous drag force given by:

$$F_V(t) = -\frac{1}{2}\rho C_D A_D|\dot{z} - \dot{\eta}|(\dot{z} - \dot{\eta}) \tag{29}$$

where A_D is the cross-sectional area of the floating body, C_D is the viscous drag coefficient, and $\dot{\eta}$ is the vertical velocity of the water surface. The mathematical equations of TLSRG and the point absorber devices are combined to fully define the dynamic model for the direct drive ocean wave conversion system, as schematized in **Figure 11**.

5.2 System simulation/simulation results

A sinusoidal ocean wave with a period of 7.7 s and a height of 1.3 m was considered to drive the floating body of the point absorber. The floating body is cylinder with a diameter of 3.4 m and a vertical extension of 4.9 m. The TLSRG was designed with a mean damping force of 40 kN, which is the maximum load

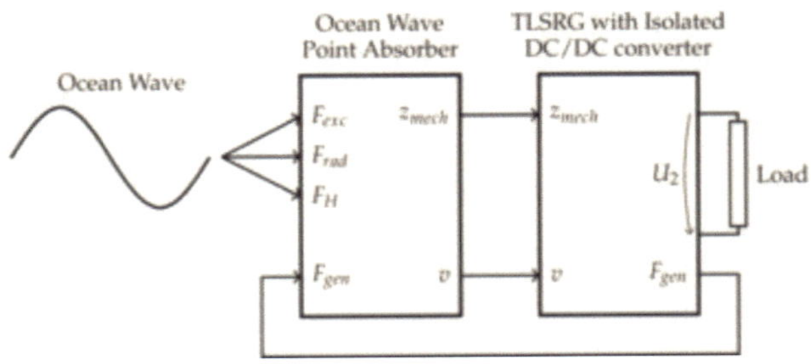

Figure 11.
Schematics of the ocean wave conversion system dynamic model.

l_g [mm]	N_t	D_{sft} [mm]	w_h [mm]	w_v [mm]	b_p [mm]	c_p [mm]	n_p [mm]	h_p [mm]	l_p [mm]	t_p [mm]	k_{offset} [mm]
5.1	230	499.5	0.1	1.7	170	152	222.4	46.7	13.9	9.3	107.3

Table 1.
Dimensional parameters of the TLSRG.

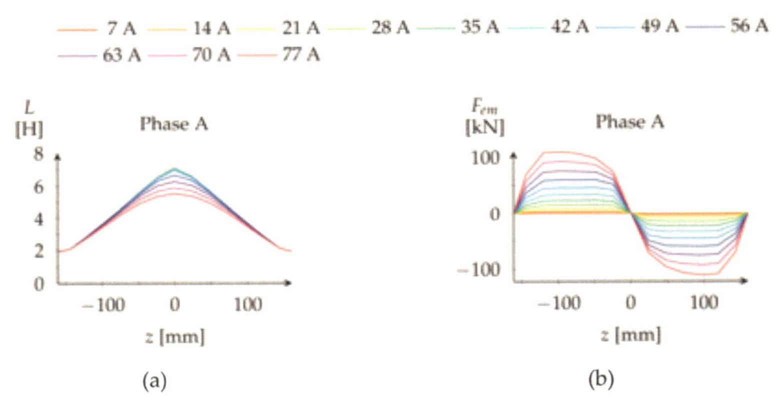

Figure 12.
Electromagnetic characteristic for the electric phase A of TLSRG: (a) inductance and (b) axial force.

expected for the ocean wave considered for simulation. The dimensions of the TLSRG are presented in **Table 1**. The electromagnetic characteristics of the machine, displayed in **Figure 12**, were obtained with MagNet®, a commercial software that uses the finite element method for electromagnetic analysis. The control parameters h_b, z_{on} and z_{off} were optimized for different combinations of velocity and phase current to maximize the mean value of the generated electric power. In this chapter, only the dynamic simulation of the system is important, so the applied optimization process is not presented. The optimal values are displayed in **Figure 13**.

These values were included in the mathematical model through 2D look-up tables and were computed, as function of the TLSRG movable part velocity and phase current, by linear interpolation. The capacitor in the converter was defined with a capacitance of 0.05 F.

A value of 41,430 kg was assumed for the mass of the oscillating body m_b, and the added mass for infinite frequencies m_∞ was quantified as 9951.1 kg. The specific mass of seawater ρ was set as 1025 kg/m^3, and for the viscous drag coefficient, a value of 0.88 was defined. The excitation force profile and impulse response function were computed with NEMOH and are displayed in **Figure 14**. For the H-bridge

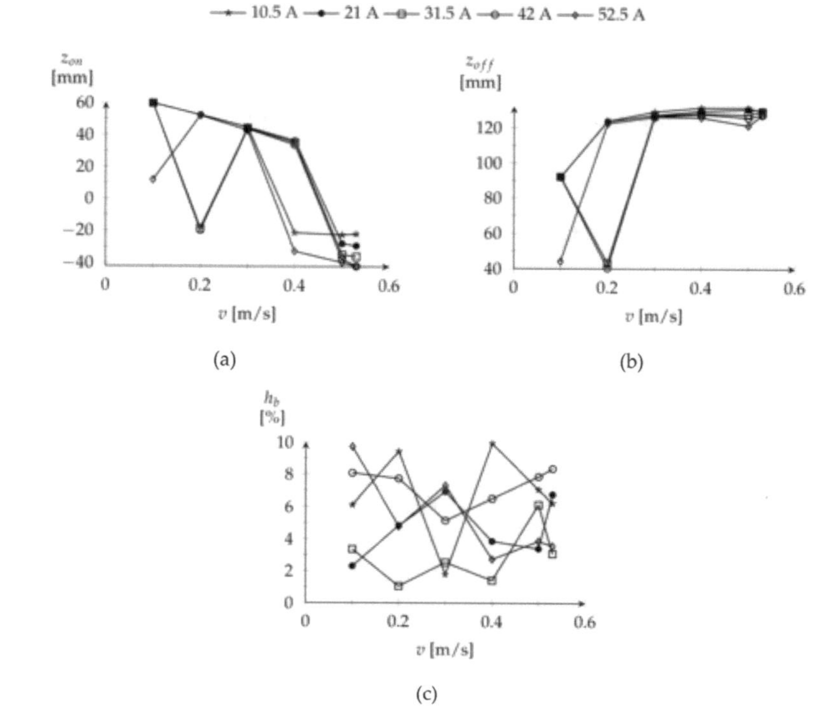

Figure 13.
Optimal values for (a) h_b, (b) z_{on}, and (c) z_{off}.

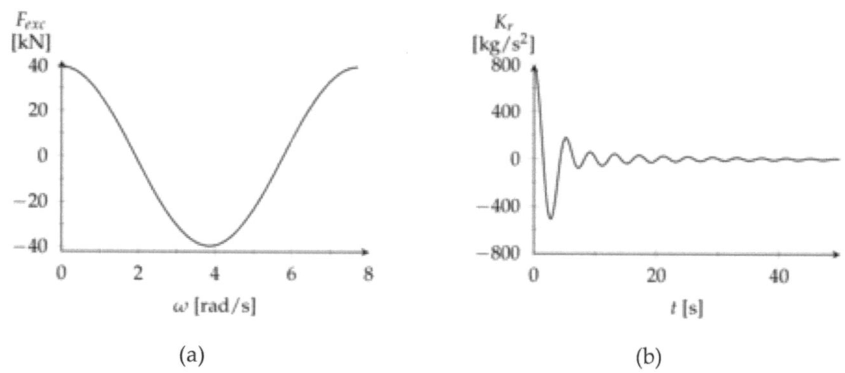

Figure 14.
Wave-body interaction loads: (a) excitation force profile and (b) impulse response function.

isolated DC/DC converter, the ratio $\frac{N_2}{N_1}$ was chosen as 10 to supply an output voltage of 400 V for a DC bus voltage of 4 kV and a duty cycle of 0.5. The PI controller was configured with a proportional gain K_p of 4089.3 and an integral gain K_I of 639.6. The system's mathematical model was implemented in Simulink® and was solved by the Dormand-Prince method, with a relative tolerance of 1×10^{-3} and a maximum step of 8×10^{-4} s. The system's mathematical model was simulated for distinct values of i_{ref} in order to evaluate the generator performance with different damping forces. The mean value of generated electrical power and electric conversion efficiency, as function of i_{ref}, can be found in **Figure 15**.

Attending to the simulation results, the greatest average of electric generated power was 7.6 kW for a reference current of 35 A. For this reference, current value also had the best electric efficiency of 43%. **Figure 16** shows the oscillating body absolute position and velocity obtained from the simulations with i_{ref} of 35 A.

The electric phase current and electromechanical force profiles are shown, respectively, in **Figures 17** and **18**. The TLSRG converter DC bus voltage level is displayed in **Figure 19**. With the application of the DC/DC isolated converter, the DC bus voltage was kept near its reference value with a maximum error of 1.05%.

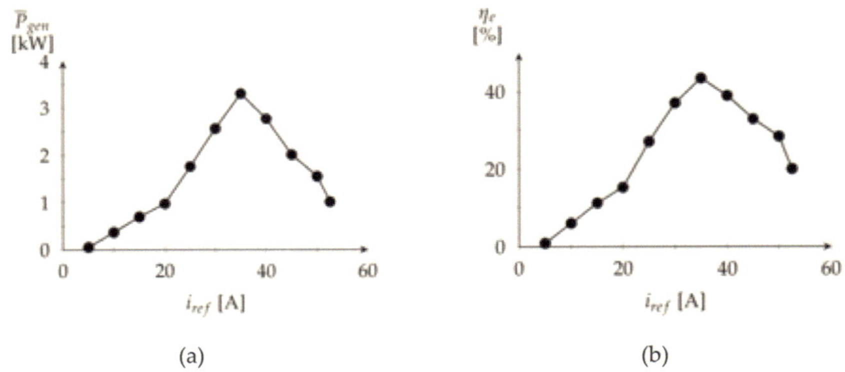

Figure 15.
TLSRG performance for different values of i_{ref}: (a) mean value of generated electric power and (b) electric conversion efficiency.

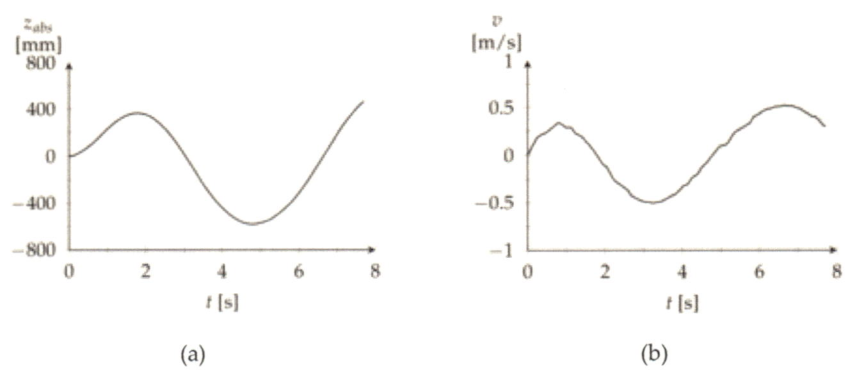

Figure 16.
Oscillating body profile for (a) absolute position and (b) linear velocity.

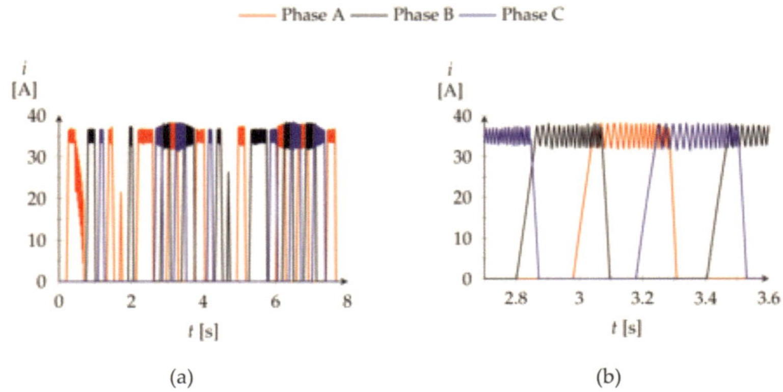

Figure 17.
Electric current profile for each electric phase: (a) normal view and (b) zoomed view.

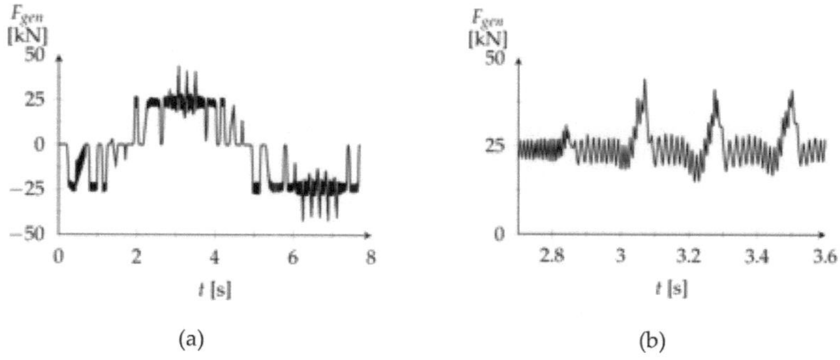

Figure 18.
Electromagnetic linear force profile: (a) normal view and (b) zoomed view.

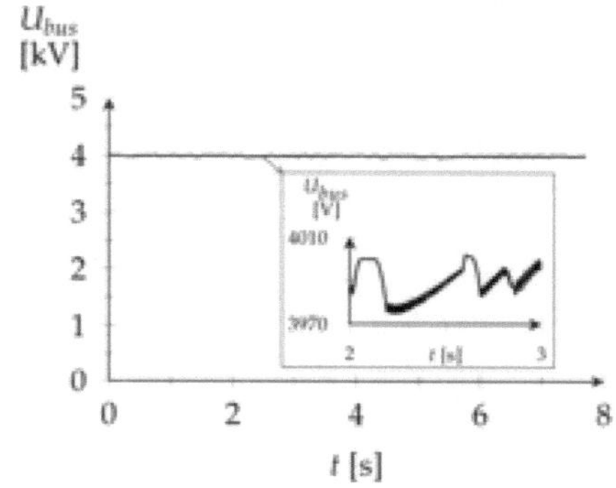

Figure 19.
TLSRG converter DC bus voltage.

6. Conclusion

In this chapter, a mathematical model to simulate the dynamic behavior of linear switched reluctance generators was presented, to apply in ocean wave direct drive converters. The model was developed according with the circuit configuration of the H-bridge asymmetric converter, adopted as power electronic converter to control the energy flow in the machine when operating as a generator. A hysteresis controller was applied to maintain the electric phase current close to a reference value during each conversion cycle, in order to allow the control of the generator damping force. Also, a DC/DC isolated converter was admitted to adjust the energy flow between the H-bridge asymmetric converter and the system electric load, in order to keep the DC bus voltage level near its nominal value. A PI controller was proposed to control the pulse width of the DC/DC conversion stage. A practical study case scenario was considered where the generator mathematical model was applied to simulate TLSRG operating as a power take-off system in an ocean wave point absorber device. The mathematical model of the TLSRG was integrated with the dynamic equations of the point absorber to evaluate the system behavior. In the simulation process, a regular ocean wave was assumed to drive the system. The system performance was evaluated for distinct values of I_{ref}, which implies distinct damping load profiles. The best performance was found for an I_{ref} of 35 A where an average electric power of 7.6 kW was generated with an efficiency of 43%. With the application of the DC/DC isolated converter, it was possible to maintain the DC bus voltage near its reference value with a maximum deviation of 1.05%.

Acknowledgements

The authors wish to thank Instituto de Telecomunicações for funding this work, with the investigation scholarship UID/EEA/50008/2019 and research grant BPD/N °19—29/01/2019—UID/EEA/50008/2019.

Author details

Rui Pedro Gouveia Mendes, Maria do Rosário Alves Calado and Sílvio José Pinto Simões Mariano*
University of Beira Interior and Instituto de Telecomunicações, Covilhã, Portugal

*Address all correspondence to: sm@ubi.pt

References

[1] Boldea I, Nasar S. Linear Electric Actuators and Generators. Cambridge, UK: Cambridge University Press; 2005

[2] Méndez S, Martínez A, Montaño C, Millan W, Oyarbide E, Pérez F. Design and characterization of a 1 kW AC self-excited switched reluctance generator. In: Proceedings of the IECON 2011—37th Annual Conference on IEEE Industrial Electronics Society, Melbourne, Victoria, Australia. Piscataway, NJ: IEEE; 2011. pp. 1789-1794

[3] Du J, Liang D, Xu L, Li Q. Modeling of a linear switched reluctance machine and drive for wave energy conversion using matrix and tensor approach. IEEE Transactions on Magnetics. 2010;**46**(6): 1334-1337

[4] Llibre FJ, Martinez N, Nogarède B, Leprince P. Linear tubular switched reluctance motor for heart assistance circulatory: Analytical and finite element modeling. In: 2011 10th International Workshop on Electronics, Control, Measurement and Signals (ECMS), Liberec, Czech Republic. Piscataway, NJ: IEEE; 2011. pp. 1-6

[5] Dio V, Miceli R, Trapanese M. The use of sea waves for generation of electrical energy: A linear tubular asynchronous electrical generator. In: OCEANS 2007, Vancouver, British Columbia. Piscataway, NJ: IEEE; 2007. pp. 1-4

[6] Moreau L, Zaïm M, Machmoum M. Electromagnetic design optimization of a low speed Slotted Switched Reluctance Machine using genetic algorithm. In: XXth International Conference on Electrical Machines (ICEM), Marseille, France. Piscataway, NJ: IEEE; 2012, 2012. pp. 233-237

[7] Torrey D. Switched reluctance generators and their control. IEEE Transactions on Industrial Electronics. 2002;**49**(1):3-14

[8] Lee D, Ahn J, Liang J. Classification and analysis of switched reluctance converters. Journal of Electrical Engineering & Technology. 2010;**5**(4): 571-579

[9] Krishnan R. Switched Reluctance Motor Drives: Modeling, Simulation, Analysis, Design, and Applications. Boca Raton, Florida, USA: CRC Press, Taylor & Francis Group; 2001

[10] Barnes M, Pollock C. Power electronic converters for switched reluctance drives. IEEE Transactions on Power Electronics. 1998;**13**(6): 1100-1111

[11] Boldea I. Variable Speed Generators. Boca Raton, Florida, USA: CRC Press, Taylor & Francis Group; 2015

[12] Shao B, Emadi A. A digital control for switched reluctance generators. In: 2011 IEEE International Conference on Mechatronics (ICM), Istanbul, Turkey. Piscataway, NJ: IEEE; 2011. pp. 182-187

[13] Ding W. Comparative study on dual-channel switched reluctance generator performances under single- and dual-channel operation modes. IEEE Transactions on Energy Conversion. 2012;**27**(3):680-688

[14] Llibre J, Martinez N, Leprince P, Nogarede B. Analysis and modeling of linear switched reluctance for medical application. Actuators. 2013;**2**(2):27-44

[15] Ding W, Liang D, Tang R. A fast nonlinear variable structure equivalent magnetic circuit modeling for dual-channel switched reluctance machine. Energy Conversion and Management. 2011;**52**(1):308-320

[16] Radun A. Design considerations for the switched reluctance motor. IEEE

Transactions on Industry Applications. September 1995;**31**(5):1079-1087

[17] Xue X, Cheng K, Ho S, Sutanto D. Precise analytical modelling magnetic characteristics of switched reluctance motor drives using two-dimensional least squares. In: 2003 IEEE 34th Annual Power Electronics Specialist Conference, PESC '03, Acapulco, Mexico. Vol. 1. Piscataway, NJ: IEEE; 2003. pp. 416-421

[18] Khalil A, Husain I. A Fourier series generalized geometry-based analytical model of switched reluctance machines. IEEE Transactions on Industry Applications. 2007;**43**(3):673-684

[19] Azongha S, Balathandayuthapani S, Edrington C, Leonard J. Grid integration studies of a switched reluctance generator for future hardware-in-the-loop experiments. In: IECON 2010—36th Annual Conference on IEEE Industrial Electronics Society, Glendale, AZ, USA. Piscataway, NJ: IEEE; 2010. pp. 3079-3084

[20] Miller T, McGilp M. Nonlinear theory of the switched reluctance motor for rapid computer-aided design. IEE Proceedings B—Electric Power Applications. 1990;**137**(6):337-347

[21] Ding W, Liang D. Modeling of a 6/4 switched reluctance motor using adaptive neural fuzzy inference system. IEEE Transactions on Magnetics. 2008; **44**(7):1796-1804

[22] Lee C. Flux linkage estimation in a switched reluctance motor using a simple reluctance circuit. Journal of Magnetics. 2013;**18**(1):57-64

[23] Filho P, Oliveira L, dos Santos Barros T, Villalva M, Filho E. Modeling and digital control of a high-power full-bridge isolated DC-DC buck converter designed for a two-stage grid-tie PV inverter. In: 2014 IEEE Energy Conversion Congress and Exposition (ECCE), Pittsburgh, PA, USA. Piscataway, NJ: IEEE; 2014. pp. 1874-1879

[24] Miller T. Electronic Control of Switched Reluctance Machines. ser. Newnes Power Engineering Series. Oxford, UK: Elsevier Science; 2001

[25] MacMinn S, Sember J. Control of a switched-reluctance aircraft engine starter generator over a very wide speed range. In: Proceedings of the 24th Intersociety Energy Conversion Engineering Conference, 1989, Washington, DC, USA. IECEC-89. Vol. 1. Piscataway, NJ: IEEE; 1989. pp. 631-638

[26] Cardenas R, Pena R, Perez M, Clare J, Asher G, Wheeler P. Control of a switched reluctance generator for variable-speed wind energy applications. IEEE Transactions on Energy Conversion. December 2005; **20**(4):781-791

[27] Rain X, Hilairet M, Bethoux O. Comparative study of various current controllers for the switched reluctance machine. In: 2010 IEEE Vehicle Power and Propulsion Conference, Lille, France. Piscataway, NJ: IEEE; 2010. pp. 1-6

[28] Choi D, Byun S, Cho Y. A study on the maximum power control method of switched reluctance generator for wind turbine. IEEE Transactions on Magnetics. 2014;**50**(1):1-4

[29] Mademlis C, Kioskeridis I. Optimal efficiency control of switched reluctance generators. IEEE Transactions on Power Electronics. 2006;**21**(4):1062-1072

[30] Pan J, Yu Z, Cheung N, Guang-Zhong C. On the voltage ripple reduction control of the linear switched reluctance generator for wave energy utilization. IEEE Transactions on Power Electronics. 2014;**29**(10):5298-5307

[31] Hasanien H, Muyeen S. Speed control of grid-connected switched reluctance generator driven by variable speed wind turbine using adaptive neural network controller. Electric Power Systems Research. 2012;**84**(1): 206-213

[32] Babarit A, Hals J, Muliawan MJ, Kurniawan A, Moan T, Krokstad J. Numerical benchmarking study of a selection of wave energy converters. Renewable Energy. 2012;**41**(0):44-63